Hodson's Horse

MAJOR HODSON.

From a photograph of the Statuette by Minton, Stoke-upon-Trent.

HODSON'S HORSE

1857-1922

MAJOR F. G. CARDEW

The Naval & Military Press Ltd

Published by
The Naval & Military Press Ltd
Unit 10, Ridgewood Industrial Park,
Uckfield, East Sussex,
TN22 5QE England
Tel: +44 (0) 1825 749494
Fax: +44 (0) 1825 765701
www.naval-military-press.com

PREFACE.

This book was begun many years ago, and to this fact it owes the great advantage of including personal notes and anecdotes which were furnished to the writer by the late Sir Charles and Sir Hugh Gough and Mr George Ricketts. The record has now been completed down to 1922. In that year the two regiments which were formed from Hodson's original corps were brought together again. Their reunion marked the end of an epoch and seemed a suitable point at which to close the narrative.

The work has been undertaken at the request of Lieutenant-Colonel A. J. Reynolds and the officers of the present regiment, but they have been good enough to give the writer so free a hand that he alone is responsible for the form and contents of the volume. No attempt has been made to chronicle all the doings of the regiments from year to year, but rather to concentrate on the most important events in their history, and thus if possible to present the records in a readable form.

The greater part of the biographical notes on British officers in Appendix III. has been prepared by Major V. C. P. Hodson. The writer wishes to express his thanks to him for this help and to Captain Stevens for the drawing of Hodson's grave. He also takes this opportunity of thanking General Sir Hubert Gough for his kindness in lending for reproduction the portrait of Sir Charles Gough, and Mrs Style for allowing him to use the original drawing by "Spy" of Sir Hugh Gough. Acknowledgment is also due to Miss Palliser for the loan of a photograph, and to the General Staff, War Office, for their courtesy in supplying two excellent maps.

A small glossary is included at the end of the book for the convenience of any readers who may not be familiar with India.

F. G. C.

CONTENTS.

PART I.

THE MUTINY CAMPAIGN.

PART II.

THE NINTH.

PART III.

The Tenth.

APPENDICES.

ILLUSTRATIONS.

PLANS AND MAPS.

The coloured maps in this reprint are placed at the end.

PART I.

THE MUTINY CAMPAIGN

CHAPTER I.

THE RAISING OF THE REGIMENT.

MORE than seventy years have passed since the English world was startled by the tidings of the sepoy mutiny. The lapse of time has dimmed the feelings of horror and consternation which were then aroused by the news from Meerut, Delhi, Jhansi, and Cawnpore, and none can now recall the sickening suspense and anxiety with which those at a distance watched the struggle of the little army of English, set in the midst of so many and great dangers. But the shadow of years cannot eclipse the glory that was won in that crisis, not only by the British forces, but scarcely less by many new regiments of India. The agonies of 1857 were the birth-throes of some of the finest units of the Indian army, and it will be a bad day for us and for India when we cease to remember the steadfast loyalty with which the soldiers of the Punjab followed their British officers through all the trials of the mutiny campaign.

The first to be raised and foremost in distinction among these new corps which, in the hour of our need, were recruited from our old adversaries of the Khalsa was the regiment of horse which was organised by Lieutenant W. S. R. Hodson of the 1st Bengal European Fusiliers in the camp before Delhi, which was led by him through some of the most stirring scenes of the next eight months, and which ever since his death at Lucknow in March 1858 has been proud to bear his name.

When the news of the outbreak at Meerut suddenly called the Commander-in-Chief from Simla in the opening days of the hot weather of 1857, Lieutenant Hodson, who was then with his regiment at Ambala, was immediately appointed to be Assistant Quartermaster-General in charge of the Intelligence Department with the field army.

This is not the place wherein to describe the brilliant services of the previous years which led to the nomination of so junior an officer for a post of such importance. Lieutenant Hodson's name stood high amongst those who had distinguished themselves by courage, determination, and resource in both military leadership and in civil administration since the annexation of the Punjab; and though at the moment of the sepoy outbreak he was serving as a simple subaltern with his regiment, it was a matter of course that in such an emergency he would be early selected for special employment.

Immediately after this appointment to the staff Hodson was ordered to raise one hundred horse and fifty foot for intelligence work and for duty as personal escorts when required. But the serious deficiency of cavalry soon made it necessary to do more than this to supply the need for mounted troops with the army. Hodson's special qualities of energy and daring as well as his close acquaintance with many of the leading families in the Punjab and his influence with them obviously fitted him to obtain recruits for this service, and without delay action was taken accordingly. On 19th May the following order (afterwards published in General Orders to the Army, dated Calcutta, 24th October 1857) was issued by the Commander-in-Chief: "Lieutenant W. S. R. Hodson, 1st European Fusiliers, and officiating Deputy Assistant Quartermaster-General, is appointed Commandant of a corps of Irregular Horse, which he is directed to raise at Kurnaul."

"My commission," wrote Hodson on the 25th

May, "is to raise a body of Irregular Horse on the usual rates of pay and the regular complement of native officers, but the number of troops to be unlimited—*i.e.*, I am to raise as many men as I please: 2000 if I can get them. The worst of it is, the being in a part of the country I do not know, and the necessity of finding men who can be trusted." [1]

In this perplexity Hodson wrote for assistance to Mr Robert Montgomery, then Judicial Commissioner of the Punjab,[2] and the ready response with which this appeal was met is described in the following extract from a letter, dated the 25th September 1860, from the Military Secretary of the Punjab Government to the Officer Commanding the 1st Regiment of Hodson's Horse:—

"The emergency was great; the crisis was imminent; Sir John Lawrence was absent; His Honour (Mr Montgomery) summoned Rajah Tej Singh and Nawab Imam Oodeen, and asked them each to raise a Rissalah. His Honour set about and raised one himself.

"On the 23rd June 1857 the three were ready, and sent off to Delhi under command of Sirdar Man Singh.

"One Rissalah (now 1st Troop, 1st Regiment, Hodson's Horse) commanded by Sirdar Man Singh, raised by His Honour.

"One ditto (now 3rd Troop, 2nd Regiment, Hodson's Horse) commanded by Mirza Attaoolla Khan, raised by Nawab Imam Oodeen.

"One ditto (now 1st Troop, 2nd Regiment, Hodson's Horse) commanded by Sirdar Bal Singh, raised by Rajah Tej Singh.

"Subsequently, as the siege went on, and reliable cavalry was required, His Honour had another urgent call, and sent, on the 9th July, two more Rissalahs.

[1] 'Hodson of Hodson's Horse,' by the Rev. George Hodson, p. 151.

[2] Afterwards Sir R. Montgomery, Lieutenant-Governor of the Punjab.

"One Rissalah (now 5th Troop, 1st Regiment, Hodson's Horse) commanded by Sirdar Tej Singh,[1] and raised by Sirdar Shum Shere Singh, Sindha-Wallah.

"Nawab Alli Ruzza Khan, a native gentleman of Lahore, and a friend of Captain Hodson, also raised a Rissalah (now 3rd Troop, 1st Regiment, Hodson's Horse) at his request, and sent it down to Delhi."

The five troops whose raising is thus described were the nucleus of Lieutenant Hodson's regiment; moreover the officers who commanded them included some of the most distinguished of all those who joined the corps, and these facts, coupled with the circumstance that they were the first officers, British or Indian, to be appointed to Hodson's Horse, is a sufficient reason for giving here some further details about the troops and the men who raised them.

The first, in regimental order, was raised by Mr Montgomery with the aid of Sardar Man Singh. This officer was the son of Dava Singh of Ruriala, in the district of Gujranwala, and youngest brother of Sardar Jodh Singh, a gentleman who had already done valuable service to the British Government before the outbreak of the mutiny. Man Singh had had considerable experience of war as an officer of Sikh cavalry, and had fought against the Afghans at Peshawar as well as against the English in the principal battles of the First Sikh War. In 1852 he entered the mounted police under Colonel Richard Lawrence, but eagerly embraced the opportunity of fresh military service when asked by Mr Montgomery to assist in raising sowars for Lieutenant Hodson. He served with distinction during the mutiny and was afterwards the first risaldar-major

[1] This seems to be a mistake; this troop appears to have been commanded by Sardar Jai Singh Chinah, as shown in an autograph note by Sir R. Montgomery, dated Murree, 19th September 1860, in possession of the regiment.

of the 9th. Associated with him in the command of the troop were Mirza Jiwan Beg (also from the police) as naib-risaldar and Nur Din as jemadar. The troop marched to Delhi with a total strength of eighty-four of all ranks.

The second troop was that raised in the Gujranwala district by Rajah Tej Singh, some time Commander-in-Chief of the Sikh army, and son of the Maharajah Ranjit Singh's celebrated favourite, Jemadar Khushal Singh. The officer selected by Tej Singh to command the troop was Sardar Bal Singh of Chuhar Kana, in the Gujranwala district, who had been a colonel of artillery in the Khalsa army. "This," wrote Sir Henry Daly in after years, "was the best of the troops in Hodson's Horse," and the officer in command was worthy of the tried and hard-fighting men of his troop. Bal Singh was the son of Sardar Budh Singh, who had been a prominent Ghorcharah [1] chief in the army of Maharajah Ranjit Singh. His family was one which represented in a remarkable degree the military characteristics of the Sikh people, for besides himself and his father all his three brothers had held commands in the army of the Khalsa. One of these brothers, Fateh Singh, with two sons and another nephew, joined Bal Singh's troop, and fought in Hodson's Horse through the Mutiny campaign. Fateh Singh was killed at Lucknow; one of his sons, Sardar Singh, lost a leg at the fight of Raniganj; the other, Sarjan Singh, died from the effects of the campaign a few months after its conclusion. Sardar Bal Singh himself, after most distinguished service, retired in 1864, and died a few years afterwards, but his family was for the next forty years well represented in Hodson's Horse, and the 2nd Regiment (10th Bengal Lancers) had a succession of officers from this distinguished stock.

Bal Singh's troop consisted in all of sixty-six officers

[1] The Ghorcharas held the position of Horse Guards in the Sikh army.

and men, the naib-risaldar and jemadar being respectively Sardar Jodh Singh and Sardar Jwala Singh.

The third troop, as already related, was raised by Nawab Imam-ud-Din, formerly Governor of Kashmir and celebrated equally with Rajah Tej Singh in the struggles which convulsed the Punjab between the death of Ranjit Singh and the English conquest. To command this troop a young man of Rajput descent and noble birth was appointed, Mirza Ata-ullah Khan, a cadet of the family of the Kashmir Rajas of Rajauri, who joined the troop with his brother, Mirza Abdulla Khan, and twenty-five others of his clan. He led the troop to Delhi, and served with great distinction throughout the Mutiny campaign, and both he and his brother afterwards rose to be risaldar-major of the 10th Bengal Lancers. The naib-risaldar of this troop was Sardar Hukm Singh, the jemadar was Mirza Ahmad Beg, and it numbered eighty-five of all ranks.

Of the two other troops mentioned in the Punjab Government letter quoted above, one (afterwards the 5th Troop of the 9th Bengal Lancers) was raised by Sardar Shamsher Singh Sindhawala, well known as one of the most prominent members of that powerful clan, and remarkable as much for his energy and ability as for his loyalty to the British Government after the Sutlej campaign. The officers of this troop were three brothers-in-law of Shamsher Singh named Jai Singh (risaldar), Harditt Singh (naib-risaldar), and Amar Singh (jemadar), sons of Sardar Sodh Singh of the China family (the second of these afterwards became risaldar-major of the 9th). The troop marched to Delhi ninety-two horsemen strong, attached to John Nicholson's movable column, with which it took part in the action against the Sialkot mutineers by the way.

There only remains to be described the troop raised by Sardar Ali Raza Khan, which was afterwards the third troop of the 9th Bengal Lancers.

This sardar belonged to a Kazalbash family from the western shores of the Caspian Sea, whose grandfather marched into India with Nadir Shah, and attained to considerable power under Ahmad Shah Durani. Ali Raza Khan was among the most influential of the Kazalbash nobles resident at Kabul at the time of the British occupation in 1839-42, and throughout that period, so disastrous to our arms and prestige, he displayed the most unremitting, as well as disinterested, friendship to the English. The hostility which he thus raised against himself on the part of the Barakzai family was such that he was compelled to accompany the British army on their retirement to India, and there he and his family continued to render good and loyal service to the British Government. The troop which he volunteered to raise and equip for Lieutenant Hodson, entirely at his own expense,[1] was commanded by his brothers, Muhammad Raza Khan and Muhammad Taki Khan, as risaldar and naib-risaldar respectively. His nephew, Sher Muhammad Khan, was its jemadar, and four other nephews fought in it—Abdulla Khan, Muhammad Hasan Khan, Muhammad Zaman Khan, and Ghulam Hasan Khan.

Muhammad Taki Khan was killed at Gangiri on the 14th December 1857, and Hodson wrote of him: "I have lost a fine old Ressaldar, our dear old friend Mohammad Raza Khan's brother." Muhammad Raza Khan served throughout the Mutiny with the greatest distinction. "He was twice wounded, at Malu and Shamsabad, and had two horses shot under him; and in every place where blows were thickest there was the gallant Muhammad Raza Khan to be found. After the campaign he received the first-class Order of Merit, the title of Sardar Bahadur, and the grant of his pension of Rs. 200 in perpetuity. He died at Lucknow, whither he had gone on leave, shortly afterwards."[1]

[1] 'The Punjab Chiefs,' by Griffin, vol. i. pp. 26 and 27.

The first five troops of his Horse to join Lieutenant Hodson on the Delhi ridge have thus been described in detail; but before proceeding to narrate the story of the part played by the new-formed corps in the campaign it will be convenient here to anticipate events by giving such particulars as are available of the other units which were afterwards added to the regiment. As has been seen, the third troop was raised by Nawab Imam-ud-Din, and shortly after its completion a second troop was formed by the same gentleman, the command of which was given to a sardar named Mirza Ghulam Muhammad Din, its naib-risaldar being Sharaf Ali Shah. This troop, ninety strong, marched to Delhi with Brigadier-General John Nicholson's movable column, together with that under Sardar Jai Singh, already noticed. A seventh troop was actually raised in the camp before Delhi, although a nucleus of it was brought from Lahore by Kahan Singh Rosa, a fine old officer of Sikh cavalry, who joined the Guide Corps immediately on the outbreak of the Mutiny and served with them till the fall of Delhi. The men whom he brought were drafted into a separate troop, which was then completed by overflows from other troops and by additional recruits brought by Risaldar Man Singh. The officers of this troop (afterwards the 4th Troop of the 2nd Regiment) were Risaldar Muhammad Taki Khan (transferred from his brother's, Muhammad Raza Khan's, troop), Naib-Risaldar Mirza Ahmad Beg (transferred from the 3rd Troop) and Jemadar Deva Singh. The eighth troop to be formed (afterwards the 2nd Troop of the 3rd Regiment) was raised by Nawab Jahan Fazan Khan of Sardhana in August 1857. Like the troop last described, it was raised at or about Delhi; its commander was a Pathan named Fateh Ali Shah, the officers being Naib-Risaldar Mahmud Shah and Jemadar Muhammad Sayyid Khan. Meanwhile, having secured risalas of the best sort from Gujranwala and from the Manjha, Lieutenant Hodson applied for assistance to another friend amongst the civilians of the Punjab, Mr George Ricketts, then Deputy Commissioner of Ludhiana, and the hero of a brilliant affair with the Jullundur mutineers on the night of the 19th June 1857. The manner of this application, and of the response which it met, is described in the following graphic letter from Mr Ricketts himself:—

"He (Hodson) asked me to get him as many good men as I could (this was as well as I can recollect about the beginning of August 1857)—a squadron if possible, and if possible with their own horses under them, or with sufficient money in their pockets to buy them; but on this point, horse or money, he was not very particular, for, as he said, he could always pick up the horses. . . . It was a curious business: there were the old Sikh *Ghorchurrahs* everywhere, and old artillerymen too. They were looking every way, certain that sooner or later they would take a hand one side or the other, and were just biding their time, and it was hard to get a beginning. This I got at last through the help of the Chowdries of Sulton (?), not far from Loodhianah. I had three of them (picked up in 1857) as jemadars, the handiest old fellows I ever saw . . . and through them I made a start, with some of their young relatives as troopers, and the son of one of them, Hira Singh, as a duffadar. . . . After the first start the men began to come in, and I had a pretty good number to select from. The test of their riding capabilities was to ride my grey mare, a country-bred, from my house verandah to the compound gate and back. She was a *junglee*, 14-3, and used to stand like a sheep until she was mounted, bare-back, and then the fun used to begin. She used to fly right and left and bound in the air, and *lumbai* all down the road, and get almost all of them off sooner or later (those who brought their own horses were excused). We soon found out those who had ridden before, and no others were accepted. These I sent down to Hodson to Umballa after the fall of Delhi, and to the best of my recollection some were sent to him at Delhi itself. . . . I was as careful as I could be in selecting them on enlistment, making all enquiries about them, getting men to answer for those I didn't know, &c.; but in spite of all this one conspicuous fellow slipped through my fingers and into the regiment, possibly under a false name. Had I caught him I should certainly have hanged him. After the capture of Lucknow, when a force under Brigadier R. Walpole was sent to march up the left bank of the Ganges and to recover Rohilcund, a few troopers of Hodson's Horse were retained as orderlies by Major Sarel, who had commanded one of Hodson's regiments for a time, and was then Brigade-Major to Colonel Haggard, commanding the cavalry under Walpole. In the camp was an old unattached native infantry major (I forget his name) doing some general duty. This major had with him two or three of his old Pandies, who had not mutinied when practically the whole regiment had mutinied at Jhansi. It was a notoriously vicious mutiny, and the ringleader of all, who headed the mutiny when the regiment broke out on parade, was this very Sikh. He had stepped out of the ranks and shot at once the Havildar-Major, and then the regiment 'went.' He must have lost little time in finding his way home, and then, finding that the Punjab and the Sikh people were not on the side of rebellion or of the mutineers, he must have kept perfectly quiet until this opportunity occurred to hide himself away in our service, and in a regiment where it was most improbable that enquiries after him would be made; and I may here add, this was not an uncommon practice. It was not a time to ask too many questions; men who would enlist and fight for us were welcome. I knew myself that several Sikhs, who were prisoners in Ludhiana jail when the mutineers threw the jail gates open, cleared out of the district and went and enlisted, and there I was content to leave them. Well, this fellow was sent with a note to the old officer of his old regiment. He was in the uniform of Hodson's Horse, but at the Major's tent were the two old Hindustani orderlies, and they recognised him at once and told their master. He, too, knew the man, and had him seized at once and marched off to the General Commanding. Before him the man admitted everything. 'But,' he said, 'we thought the *Raj* was over, and I have done good service under Hodson, as his officers will testify'; and then he begged that I might be referred to! But all this did no good. Men were all hard and bloodthirsty in the camp, and the court that tried him sentenced him to be blown away from a gun, and the sentence was promptly executed. It was a curious case, but it was a hard sentence. They might have turned him out, given him his life, and let him go; our ranks were full of men who had fought against us. . . . The other was a very different story. I was, some years after, talking over those times and 'Hodson's Horse' with C. Gough (now General Sir Charles Gough), who had been one of Hodson's officers. 'Do you know you sent us a man with a wooden leg,' said he. 'Never!' said I. 'Impossible!' 'You did,' said Gough, 'and we never found it out till he had been with us ever so long and done all his duty, and it only came out in action.' They had had a working day and had ridden down a lot of scattered Pandies, and the men were collecting together, when Gough saw one of his men ride at a fellow on foot in a very awkward manner, with his right leg stuck out towards him, and the man, to all appearance, got well home with his *tulwar* on the sowar's leg; but he passed on, swung his horse round, and went at him again as before, and missed him again, receiving apparently another sword-cut. Gough looked on with amazement, especially as his men looked on unconcerned—it was not their affair—and partly amused. ‘Phir dhoka dya,’ said one man.[1] At the third attempt the sowar got home and cut his man down, and came back to them grinning, with two big cuts through his leather boot into his wooden leg!”

Such was the manner in which recruits were obtained for Hodson from among the Sikhs of the Malwa. The men so enlisted were formed into two troops, which eventually became the first (or A) squadron of the 10th Bengal Lancers. Its first officers were Risaldar Sangat Singh, Naib-Risaldar Pakhar Singh of the Ambala district, formerly an adjutant of cavalry in the Khalsa army, and Jemadar Nihal Singh.

Four more troops remain to be mentioned. In October 1857, on the return of the headquarters of Hodson's Horse to Delhi from service in the Rewari district, Risaldar Man Singh was sent to Lahore to raise if possible 500 recruits for the regiment. “This he effected in about four months, using the utmost exertions and borrowing a considerable amount of the necessary funds on his personal security” (Griffin's ‘Punjab Chiefs’). The men enlisted at this time were formed into four troops, two of which were raised by Man Singh personally. Of these, the first was commanded by Risaldar Hussain Ali Khan, with Amar Singh as naib-risaldar; the second, afterwards the 2nd Troop of the 1st Regiment of Hodson's Horse, was commanded by Harsa Singh, nephew of Man Singh, and son of Sardar Jodh Singh. The naib-risaldar was Sardar Hukm Singh. The other two troops were raised by Sardar Shamsher Singh Sindhawala at Amritsar and Lahore. One of them afterwards became the 6th Troop of Hodson's 2nd Regiment, and was commanded by Risaldar Basawa Singh. Man Singh's two troops, when partly completed, marched to

[1] “Caught him again.”

Ambala. There further recruits were awaited, and the troops were meanwhile exercised, drilled, and equipped, and there they were joined by the two troops raised by Shamsher Singh. The whole then marched as speedily as possible southwards and joined the headquarters at Lucknow on the 13th March 1858, the day after the intrepid leader, in whose ranks they came to serve, had passed away.

Annexed is a table giving the outline of the above details of the formation of Hodson's Horse with as much accuracy as possible, together with the names of the officers of each *risala* so far as they can be traced. In compiling this list much assistance was received many years ago from the late Colonel Ata-ullah Khan. It has since been checked and in many respects corrected with the help of original rolls of the earliest troops, which were found among the records of the Punjab Secretariat, and which are now in the possession of Hodson's Horse. Further details have been obtained from the Medal Roll of the 2nd Regiment, Hodson's Horse, for service in the Mutiny, which is also preserved by the regiment.

[TABLE

Order of raising.	By whom raised.	Place.	Risaldar.	Naib-Risaldar.	Jemadar.	
1	Mr Montgomery and Man Singh	Lahore	Man Singh *	Mirza Jiwan Beg	Nur Din	1st Troop of 1st Regiment (Ninth).
2	Rajah Tej Singh	Lahore and Gujranwala	Bal Singh	Jodh Singh	Jwala Singh	1st Troop of 2nd Regiment (Tenth) (afterwards the 6th Troop of Tenth).
3	Nawab Imam-ud-Din	Lahore	Mirza Ata-ullah Khan †	Hukm Singh ‡	Mirza Ahmad Beg	3rd Troop of Tenth.
4	Ditto	Ditto	Mirza Ghulam-ud-Din	Sharf Ali Shah	Amar Singh (Khatri)	
5	Sardar Shumsher Singh	Amritsar	Jai Singh	Harditt Singh §	Amar Singh (Chinah)	5th Troop of Ninth.
6	Sardar Ali Raza Khan	Delhi	Muhammad Raza Khan	Sher Muhammad Khan	Agha Abul	3rd Troop of Ninth (later the 7th Troop Pathans)
7	Sardar Kahan Singh Roza and completed by Man Singh	Lahore and Delhi	Muhammad Taki Khan ¶	Mirza Ahmad Beg	Deva Singh	4th Troop of Tenth.
8	Nawab Jahan Fazan of Sardhana	Delhi	Fateh Ali Shah	Mahmud Shah	Muhammad Sayyid Khan	Afterwards in 3rd Hodson's Horse.
9	Mr George Ricketts	Ludhiana	Sangat Singh (son of Hari Singh Randhawa of Dodai)	Pakhar Singh	Nihal Singh	2nd Troop of Tenth (afterwards increased to a squadron and became A Squadron, Tenth).
10	Ditto	Ditto	—	—	—	
11	Man Singh	Lahore and Amritsar	Husain Ali Khan	Amar Singh	Ajit Singh	5th Troop of Tenth (afterwards absorbed).
12	Ditto	Ditto	Harsa Singh (nephew of Man Singh)	Hukm Singh		2nd Troop of Ninth.
13 } 14 }	Sardar Shamsher Singh	Ditto	Basawa Singh			6th Troop of Tenth (afterwards absorbed).

Note.—Numbers 1, 2, 9, and 12 are now merged in A Squadron, Hodson's Horse; Nos. 3 and 7 are represented by B Squadron

* Afterwards Risaldar-Major of the Ninth.
† Afterwards Risaldar-Major of the Tenth.
§ Succeeded Man Singh as Risaldar-Major of the Ninth.
‡ Killed at Mianganj, 23rd Feb. 1850.
¶ Killed at Gangiri, 14th Dec. 1857.

CHAPTER II.

THE SIEGE AND STORMING OF DELHI.

THE progress of the campaign against the sepoy mutineers prior to the arrival of Hodson's new recruits at the British camp before Delhi may be described very briefly.

The rising began on the 10th of May 1857, when practically the whole of the native troops at Meerut mutinied, murdered most of the British officers with their families, as well as many other Europeans, and marched the same night to Delhi. There they arrived early the following morning, and were joined by three other battalions of native infantry. Further massacres of Europeans took place, the sepoys and the mob from the bazars being encouraged in these outrages by the family of the old Emperor Bahadur Shah, who was living as a pensioner in the palace of the Moghals. In a few hours the fort and the city were entirely in the hands of the mutineers.

News of these events found the Commander-in-Chief, General the Honble. George Anson, at Simla. He hastened to Ambala, where he formed the available British troops into a small field force, and started with it as quickly as possible for Delhi. But on the 27th of May he was struck down by cholera, and died at Karnal. He was succeeded by Major-General Sir Harry Barnard, who pressed on southwards, and was joined on 7th June by a detachment from Meerut under Brigadier-General

Archdale Wilson. On the following day the rebels from Delhi were attacked at Badli-ki-Sarai, and driven back on the city in confusion, and on that evening the British force established itself on the celebrated Delhi Ridge, where for more than three months it maintained its precarious position, nominally besieging the capital, but actually being rather itself besieged by the large and constantly increasing numbers of the mutineers.

Here during the opening weeks of the operations Lieutenant Hodson was fully employed with his staff duties as well as with the command of the Guides, of which Corps he had formerly been second-in-command, and to the temporary command of which he was now appointed, while the permanent commandant, Captain H. Daly, was incapacitated by serious wounds; and here on the 12th July he was joined by the first troops of his new regiment from the Punjab.

"Three hundred of my new regiment," he wrote from the camp on that date, "have just arrived. One hundred more left Lahore on the 7th, and one hundred will be here very soon from the Sutlej. Mr Montgomery has done me most essential service, as I could never by myself have got so many men together; and everything he does is so complete. He sends figured statements giving all details regarding men and horses, which will save me much time and labour hereafter. . . . For officers I hope to have permanently McDowell, Shebbeare (now acting as my second-in-command of the Guides, and a most excellent officer), and Hugh Gough of the 3rd Cavalry."

Of the officers here named, Lieutenant C. T. M. McDowell of the 2nd Bengal European Fusiliers and Lieutenant R. H. Shebbeare were appointed in General Orders in July, but the latter never joined the corps, as the Guides, who had already suffered severely, could not spare his services. McDowell therefore took up the duties of second-in-command, though appointed as adjutant, and was Hodson's first assistant in training his new men. "I hope," writes the latter on the 17th July, "to have another officer or two in a few days, as more now devolves on poor Mac than his fragile frame can well stand. I wish his bodily strength was equal to his will and courage."

Even before McDowell joined the corps, some of Hodson's Horse had already followed their commander into action on the 14th July, only two days after the three risalas under Sardar Man Singh arrived in camp. On that date a determined attack was made by the rebels against the right and right rear of the British position at the southern extremity of the ridge, where our piquets were under the command of Major Reid of the Sirmur Battalion (now 2nd Gurkhas). Our troops remained on the defensive till 3 P.M., when a column was formed under Brigadier-General Showers to drive the enemy from the suburbs of the Sabzi Mandi. The troops engaged were six guns from the 2nd and 3rd Troops, 3rd Brigade Horse Artillery under Major Turner and Captain Money, the 1st Bengal European Fusiliers under Major Jacob, and the 1st Punjab Infantry under Major John Coke, and a few horsemen of the Guides Cavalry, Major Coke's Kohat Risala, and Hodson's Horse. With the piquets were detachments of the 60th Rifles and 75th Foot, the Sirmur Battalion, the Guide Infantry, and (at the Sabzi Mandi piquet) 180 men of the 8th and 61st Foot, while the 4th Company, 6th Battalion, of Artillery and some sappers and pioneers were in the batteries at Hindu Rao's house, which were the main objective of the enemy's attack. Brigadier-General Neville Chamberlain, Adjutant-General of the Delhi force, accompanied Brigadier Showers' column, which advanced against the Sabzi Mandi at about 4 P.M., taking with them the troops from the piquet on that flank. At the same time, Reid's troops from the Hindu Rao piquet advanced towards the Lahore Gate of

Delhi, thus threatening to cut off the retreat of the mutineers in that direction. This column it was which Hodson joined. "I had just returned," he writes, "from a long day's work with the cavalry, miles away in the rear . . . when I saw the column pass down. I joined it, and sent for a few horsemen to accompany me, and when we got under fire I found the Guide Infantry under Shebbeare had been sent to join in the attack. I accompanied them, and while the Fusiliers and Coke's men were driving the mass of the enemy helter-skelter through the gardens to our right, I went, with the Guides, Goorkhas, and part of the Fusiliers, along the Grand Trunk Road, leading right into the gates of Delhi." The enemy fled precipitately, and were pursued close up to the walls, where our troops came under a very heavy fire, and were compelled to fall back, their retirement being pressed (as was usual at this period of the siege) by a fresh advance of the rebels. But under Hodson's leadership the Guides, who formed the rear-guard, retreated steadily, holding the enemy in check until Hodson rode back and brought up two guns, on which "we stopped all opposition and drove the last Pandy into Delhi." "My Guides stood firm," adds Hodson, "and, as well as my new men, behaved admirably." [1]

Four days later, on the 18th July, Hodson was again in action on the outskirts of the Sabzi Mandi. He was immediately in command of the Guide Cavalry on this occasion; and although Colonel Jones of the 60th Rifles, in reporting the operations (Despatch No. 1424, dated Camp before Delhi, 12th August 1857), speaks of "Captain Hodson's Punjab Cavalry," yet it seems, from the casualty return as well as from Hodson's own narrative, that the Guides only are referred to. It is probable, however, that a few picked men from Hodson's Horse accompanied their leader on this as on other occasions. This was the last real contest in the

[1] 'Hodson of Hodson's Horse,' pp. 179, 180.

GENERAL SIR HUGH GOUGH, V.C., G.C.B.

From a drawing by Leslie Ward.

Sabzi Mandi, and with the rest of the force Hodson's regiment enjoyed comparative immunity from attack during the last days of July and beginning of August. Ample occupation, however, was found in teaching men and horses the elements of drill, a work in which the indefatigable McDowell was now assisted by another British officer, for about this time it was that the following appointments "to the regiment of irregular cavalry, under command of Lieutenant W. S. R. Hodson," appeared in Delhi Field Force orders dated the 3rd August (confirmed in General Orders by the Commander-in-Chief, dated Calcutta, 17th September 1857):—

"Lieutenant C. T. M. McDowell, Adjutant, to officiate as second-in-command during the absence on detached duty of Lieutenant R. H. Shebbeare.

"Lieutenant H. Gough of the 3rd Regiment light cavalry to officiate as Adjutant."

Lieutenant Hugh Gough joined the new corps towards the end of July. "On arrival at the camp at Delhi," he writes, "I at once found my way to Hodson's quarters, and reported myself to him. I shall not readily forget my first interview with this famous man. I found him sitting, booted and spurred, talking to a native, who was one of his spies from Delhi (for he was the chief of the Intelligence Department as well as commandant of his own body of Irregular Horse). He looked up with a quick sharp glance, which seemed to go through one—as he told me afterwards he 'liked my looks'—and then said, 'You are just the man I want, Gough; are you fit for a ride?' I promptly said, 'Yes, sir,' though as a matter of fact I was rather beat; and he then said, 'Well, come along with me; I am going out for a reconnaissance.' I had some breakfast, and we started with a small body of his men, and had a really long ride and a good reconnaissance through the enemy's country. We had no adventures, but I was struck with Hodson's marvellous knowledge of the language, and the quick way he seemed to extract all the information he wanted and his great powers of endurance."[1]

Meanwhile Hodson's exacting duties as A.Q.M.G. for Intelligence, coupled with the command of the Guides, left little opportunity for him to superintend the training and organisation of his new corps. It soon became evident that one or other of his several tasks must be abandoned, and so we find him writing on 20th July: "I have determined on giving up the Assistant-Quartermaster-Generalship. It gives me more work than I really can manage in such weather, in addition to the command of two regiments." But his talents in this particular service were so great that the authorities were not inclined to accept his resignation, and an alternative course was eventually adopted. "After many discussions *pro* and *con*, it has been arranged that I retain the Intelligence Department and give up the Guides," he writes on 24th July. "My own men require great attention, as they are now in considerable numbers."[2] In the meantime, as has been seen, McDowell was making such progress as he could with the new corps, aided by Hugh Gough and by Hodson himself when he could spare the time. Writing many years afterwards about these early days in the regiment, Gough says: "Fortunately a great number were old Khalsa warriors, and it did not take long to make them useful. The men were armed simply with tulwar and matchlock and an occasional long spear, according to each man's taste, and they brought their own horses. . . . [They] soon got rubbed up into shape, and were uniformed and remounted, as casualties occurred, on Government stud breds till they were fit for anything. The uniform was khaki picked out with red—*i.e.*, red puggrees and kummerbunds, and sash which was worn obliquely over the shoulder, and was meant to distinguish them from the 'catch-'em-

[1] 'Old Memories,' by Sir Hugh Gough.
[2] 'Hodson of Hodson's Horse,' pp. 184 and 186.

alive-ohs,' which abounded both in our force and the enemy's. From this they earned the name of Flamingoes." [1]

Notwithstanding the need for training, the necessities of the time obliged Hodson's Horse to take its share with other cavalry in the constant duties of long reconnaissances, patrolling, and the like.

" My own regiment," writes Hodson on 8th August, " is also in the Cavalry Brigade, and is very hard-worked. It is bad for a young and unformed corps, but there is such a scarcity of cavalry here that I cannot remonstrate, and I get no small amount of *kudos* for having so large a number of men fit to be put on duty within two months of receiving the order to raise a regiment. I shall have two more troops in with the 52nd, and Nicholson has given me 50 Afghans, just joined him from Peshawar, which added to thirty coming with Alee Reza Khan from Lahore, will complete an Afghan troop as a counterpoise to my Punjabees."

As has been described in Chapter I., the three troops with Hodson at this time were those which arrived from Lahore under Sardar Man Singh on the 12th July—namely, Man Singh's (afterwards 1st Troop, Ninth), Sardar Bal Singh's (afterwards the 1st and later the 6th Troop, Tenth), Sardar Mirza Ata-ullah Khan's (afterwards 3rd Troop, Tenth). The troops, which were expected to arrive a few days later with the 52nd (as Hodson says)—that is, with the Punjab Movable Column—were the second troop raised by Nawab Imam-ud-Din and that under Sardar Jai Singh Sindhawala.[2] Ali Raza Khan's troop, which was completed a little later, and of which the composition is described above by Hodson, was afterwards the 3rd Troop, Ninth. Finally the 7th Troop (afterwards 4th Troop, Tenth)

[1] Letter from Sir Hugh Gough, dated 7th October [1897].

[2] As a matter of fact these troops were delayed, and did not reach Delhi until the 26th August.

had already been partly completed at Delhi by the exertions of Sardar Khan Singh Roza. Hodson had good reason to be satisfied with the success of his efforts in raising in such difficult circumstances so valuable a body of men. "It is composed, more than anything we have hitherto had, of the old Sardars and soldiers of Ranjit Singh's time," wrote an officer from the camp, "in consequence of which and the skill of their commander, they are already an extremely efficient corps." Writing of the officers, Hodson says: "I trust there is every chance of our having a nice body of officers with 'Hodson's Horse,'" and a fortnight later, "We are getting on very comfortably, and are going to start a mess on our own account, so as to be ready to march without difficulty when required." Of McDowell himself he says: "He is doing admirably, and promises to be a first-rate officer of light horse. He rides well, which is a good thing, and is brave as a lion's whelp, which is another. I only fear whether he has physical strength for such work in such weather." It was about this time, or a little earlier, that Hodson startled the authorities in the Field Force by proposing to march southwards with a small column and to open communications with Cawnpore and Agra, and so to clear the way for reinforcements advancing from Calcutta. Writing on the subject, he says: "I proposed to take 600 of my Horse, 250 infantry of the Guides, and 4 guns; could I not have made my way with these?" And he adds the characteristic and admirable comment: "In Indian warfare I have always found 'toujours l'audace' not a bad motto."

This bold scheme, which by the way would quite probably have been successful under Hodson's leadership, was not approved, but it was not long before he found in another direction an opportunity for individual action and distinction for himself and his new regiment. As this was the first occasion on which Hodson's Horse was employed as a complete unit, it will not be out of place to give in full the official despatches dealing with the operations:—

"*From* Lieutenant W. S. R. Hodson,
Commanding Irregular Horse.

To the Assistant Adjutant-General,
Delhi Field Force.

Camp, Delhi,
24th August 1857.

Sir,—I have the honour to report the proceedings of the Cavalry Detachment which left camp under my command on the night of the 14th-15th inst., under verbal instructions from Major-General Wilson, commanding the field force.

European officers	6[1]
Guide Cavalry	103 sabres
Hodson's Irregular Horse . . .	233 "
Jhind Horse	25
Total . . .	361 "

"2. My instructions were to watch a party of the enemy who had moved out from Delhi by the Najafgarh road, with the avowed purpose of threatening our communications with Soneput and the Grand Trunk Road, or of marching to attack Hansi and the Rajah of Jhind; to ascertain their precise object and direction, and to afford support to either Soneput or the Jhind Rajah, as might be necessary. I was also to examine the state of the roads and country, with a view to the probable necessity of a larger force taking the field.

"3. On reaching Boanuh, by way of Azadpore and the canal bank, I ascertained that the enemy

[1] These were Lieuts. Hodson, McDowell, and H. Gough, Captain Ward, Lieut. D. W. Wise, 4th Cavalry, and Lieut. C. J. S. Gough, in command of Guides Cavalry.

had passed the 14th at Samplah, and were said to be moving towards Rhotuck. I therefore pushed on to Khurkowdeh on the road from Boanuh to that town, reaching it about noon on the 15th.

"4. Having been informed that a number of irregular cavalry men, whose horses were in the village, had arrived the day before from Delhi at Khurkowdeh, I took measures for securing the several entrances to it and attempting their capture, sending a small party of the Guide Corps to surprise and arrest the leading man, named Basharat Ali, a Resaldar of the 1st Regiment, Irregular Cavalry. Both objects were accomplished, only two sowars having had time to effect their escape before the village was surrounded. I then entered the village with a party of dismounted sowars. From information received from the villagers, I was able to seize several of the mutineer sowars before they had time to arm. A large party, however, took refuge in the upper story of a house belonging to one of the lambardars of the village, and defended themselves desperately. They were eventually overpowered and destroyed, but not without considerable difficulty and several casualties on our side, Lieutenant H. Gough and seven men being wounded. I subsequently caused those of the captured who were proved, on enquiry, to have been in the service of Government, and to have joined the rebels, to be executed.

"5. During the afternoon of the 15th the enemy broke up from Samplah and marched to Rhotuck, where they gave out that they were going to remain for two or three days. I marched after them on the morning of the 16th towards Rhotuck, by Sussaineh, Hamaioonpoor, and Balout. On reaching Bohur, five miles short of Rhotuck, I ascertained that the rebels had suddenly marched early in the morning towards Medinha on hearing of our movements. I therefore halted for the day, the rain being very heavy.

"6. On the morning of the 17th we moved on to Rhotuck. On approaching the town and riding on to reconnoitre with a small party, I found a large body of armed men drawn up at the old fort, in front of the place, accompanied by a few sowars. They immediately opened fire on us, and as we withdrew to bring up the detachment, followed us up the road firing and yelling in derision. The instant the head of the column arrived they were charged, dispersed, and driven into the town, leaving thirteen of their number dead. They subsequently turned out to be Rangurs, Kusais, and other turbulent inhabitants of the town, headed by Babur Khan, the chief of the Rangurs.

"7. After riding round Rhotuck, and reconnoitring the surrounding country and the approaches to it, I encamped in the open space in rear of the *kutcherry* buildings at the junction of the road by which we had marched from Bohur with the main road to Delhi. Some of the Zemindars and Hindus of Rhotuck came out to me immediately afterwards, and through their instrumentality the detachment was amply provided with all necessary supplies. No further attempt was made to annoy us.

"8. At about seven o'clock the next morning I received information that Babur Khan had gone during the night to the camp of the rebels on the Hansi road, and brought back 300 Rangur horsemen belonging to different irregular cavalry regiments, to assist him in an attack on us. Three or four minutes afterwards a large body of horsemen dashed up the roads from the town at speed, followed by a mass of footmen—armed with swords and matchlocks—certainly not less than 900 or 1000 in number. At the moment of the attack a party of twenty-five Jhind horsemen, who had come from Gohana on hearing from me of our approach, were crossing the road towards our camp, and found themselves suddenly charged by and intermixed with the enemy's horse. They defended themselves with their carbines, and thus checked the attacking party, two of their number being wounded. The whole of the horses of the detachment having been kept saddled, no time was lost in turning out, and the instant the twenty leading men were on their horses the enemy were charged and driven back in confusion towards the town, their flight being covered by the matchlock men, who had occupied some buildings and compounds between the *kutcherry* and the town. Directly the whole of the detachment was ready and formed up, I sent what little baggage and followers we had to the rear under a sufficient escort, and prepared for a further attack. I formed the main body on the road in three lines, the Guides in front, sending out a troop to the right front under Lieutenant Wise, and one to the left under Lieutenant McDowell, ready to take the enemy in flank should they again charge up the roads (of which there were three leading from the town to our position). These movements were covered by skirmishers and by the excellent fire of the Jhind horsemen armed with matchlocks, whom I desired to dismount and drive back by their fire any party of the enemy who might come from under shelter of the buildings. This service they performed exceedingly well and most cheerfully.

"9. Finding that our ammunition was nearly exhausted after some time had elapsed, and that there appeared little chance of the enemy coming from their cover to attack us again, I determined to draw them out into the open country behind our position and endeavour to bring on a fight there. Everything turned out as I had anticipated. My men withdrew slowly and deliberately by alternate troops (the troop nearest the enemy by alternate ranks) along the line of the Bohur road, by which we had reached Rhotuck, our left extending towards the main road to Delhi. The Jhind horsemen protected our right, and a troop of my own regiment the left. The enemy moved out the instant we withdrew, following us in great numbers, yelling and shouting and keeping up a heavy fire of matchlocks. Their horsemen were principally on their right, and a party galloping up the main road threatened our left flank. I continued to retire until we got into open and comparatively dry ground, and then turned and charged the mass, who had come to within one hundred and fifty to two hundred yards of us. The Guides, who were nearest to them, were upon them in an instant, closely followed by and soon intermixed with my own men. The enemy stood for a few seconds, turned, and then were driven back in utter confusion to the very walls of the town, it being with some difficulty the officers could prevent their men entering the town with the fugitives. Fifty of the enemy, all horsemen, were killed on the ground, and many must have been wounded.

"10. Nothing could be better than the conduct of all concerned. The Guide Cavalry behaved with their usual dashing gallantry, and their example was well emulated by the men of my new regiment, now for the first time engaged with an enemy. They not only remained under fire unflinchingly, but retired before the enemy steadily and deliberately, and, when ordered, turned and charged home boldly. It would have been hopeless to expect this but for the magnificent leading and admirable management of the officers in command of the several troops — Captain Ward and Lieutenants McDowell, Wise, C. J. Gough, and H. Gough. The difficulty of their task will be appreciated when it is remembered that, with the exception of the Guides, none of the party had been drilled or formed or knew anything of field movements."

The casualties suffered by Hodson's Horse in these operations were Lieutenant Hugh Gough, Naib-Risaldar Hukm Singh, Jemadar Ahmad Beg, and three sowars wounded.

In a private letter describing the operations, Hodson concludes his account of the successful affair at Rohtak with the words: "McDowell did admirably, as indeed did all. My new men, utterly untrained as they are, many unable to ride or even load their carbines properly, yet behaved beyond my most sanguine expectations for a first field; and this success, without loss, will encourage them greatly."[1] Two extracts are also worth giving from Sir Hugh Gough's description of the same operations. Having given a detailed narrative of the sharp encounter with the mutineers at Kharkhaodah, he relates how his own life was saved in the final mêlée there by the gallantry of his brother Charles. "When the enemy made their desperate rush I was rather in the forefront of the party awaiting them, and in the mêlée which took place I was forced backwards, and, suddenly making a false step from the roof on to a lower roof about a foot down, fell or was forced on my knees. While thus half falling one man made a cut at me with his heavy sword, which cut right down my riding-boot. Another was aiming a better directed blow, when my brother, seeing my danger, rushed forward and attacked the two, killing both, and thus undoubtedly saved my life. As it was, the hilt of my sword was forced into my wrist by a sword-cut, inflicting a slight wound."[2]

And again, in relating the details of the Rohtak fight, Sir Hugh writes: "I was in command of a troop of the Guides, and had with me my friend Kānon Khan, the native officer who was with me at Khurkowda. As we were retiring I saw him take out his small Koran, already alluded to, and begin to mutter his prayers. In my youthful arrogance and ignorance I rather chaffed him, asking him if he was afraid. He answered, 'No, sahib,

[1] 'Hodson of Hodson's Horse,' p. 203.
[2] 'Old Memories,' Sir Hugh Gough.

but a man should always be prepared,' a quiet rebuke which I felt I deserved."

Hodson and his officers might well congratulate themselves on the success of this brilliant expedition. "In three days," writes the former, "we have frightened away and demoralised a force of artillery, cavalry, and infantry some 2000 strong, beat those who stood or returned to fight us twice, in spite of numbers, and got fed and furnished forth by the rascally town itself."[1] He writes a day or two later that General Wilson gave him immense credit for what he had done, but did not "appreciate a tenth part of the effects which our bold stroke at Rohtuck, forty-five miles from camp, has produced." However this might be, Wilson was certainly not slow to praise his brilliant lieutenant for his success. He wrote as follows in an official letter: "The Major-General commanding the force having received from Lieutenant Hodson a report of his proceedings and operations from the 14th, when he left camp, till his return on the 24th, has much pleasure in expressing to that officer his thanks for the able manner in which he carried out the instructions given to him. The Major-General's thanks are also due to the European and native officers and men composing the detachment, for their steady and gallant behaviour throughout the operations, particularly on the 17th and 18th inst. at Rohtuck, when they charged and dispersed large numbers of horse and foot."

The despatch from Lieutenant Hodson, which has been already quoted, was published together with a number of other Delhi despatches in Governor-General's orders No. 1529 of 1857, dated Fort William, 4th December 1857, and republished in General Orders to the Army, dated the 10th December. The following is the covering letter which accompanied Lieutenant Hodson's report:—

---

[1] 'Hodson of Hodson's Horse,' pp. 203-4.

"No. 1489. *From* Maj.-General WILSON, Commanding Field Force.

To Lieut. H. W. NORMAN, Asst. Adj.-Genl. of the Army, Camp Field Force Staff Office.

"CAMP BEFORE DELHI,
27th Aug. 1857.

"SIR,—I have the honour to forward, for transmission to Major-General Gowan, C.B., Commanding in the Upper Provinces, and through him to Government, the accompanying report of the operations of a detachment of Irregular Cavalry I sent out under the command of Lieutenant Hodson on the 14th instant, to watch a party of the enemy, who had moved out from Delhi on the Rohtuck road, and to afford support if necessary either to Sonput or to our ally the Jheend Rajah.

"Lieutenant Hodson most fully carried out my instructions to my entire satisfaction, and his report will show that the whole of his detachment, both officers and men, behaved throughout in the most gallant and effectual manner.

"It must have been most gratifying to Lieutenant Hodson to find his new regiment so steady and staunch in their first engagement with the enemy."

In the despatches published under the order of the Governor-General in Council mentioned above, Lieutenant Hodson is four times singled out for special commendation, and his name is amongst those selected for the praise and acknowledgments of the Government of India in the order itself.

As has been seen, Hodson got back to the camp before Delhi on the 24th August, and during the remaining three weeks of the siege his regiment was not again actively engaged. Advantage was taken of the opportunity to continue the instruction of the corps, now fast increasing in numbers. The following describes one of Hodson's first cares:

"Camp, Delhi, 24th August.—I am to have a surgeon attached to my regiment at once, as I represented how cruel it was to send us out on an expedition without a doctor or a grain of medicine. We had eight wounded men, and two officers had fever on the road, and nothing but the most primitive means of relieving them."[1] The report of the Rohtak expedition has already shown that two more officers had been added to the staff of the regiment, and on the 1st September (Field Force Orders of that date) a further appointment was made: "I have to-day got a new subaltern," writes Hodson, "a Mr Baker, late of the 60th Native Infantry, and a doctor, so we are seven in all. . . ."[1]

Speaking of the training of his men, he says, on 31st August:[1] "The very day the active portion of the work is over, I shall ask to go to some good station, and organise and discipline my regiment, and get it properly equipped and fit for service. At present it is merely an aggregation of untutored horsemen, ill-equipped, half clothed, badly provided with everything, quite unfit for service in the usual sense of the term, and only forced into the field because I have willed that it shall be so; but it would take six months' constant work to fit it properly for service. . . . My idea of being able to raise a regiment when in the field, and on actual and very active service, was ridiculed and pooh-poohed, but I stuck to it that it could be done, and General Anson was only too willing that I should try, hitherto with success, and with the considerable gain, to an army deficient in cavalry, of having a good body of horsemen brought at once on duty in the field." Again: "I am sending for Heratees, and Candaharees, the farther from Hindoostan the better. Mr Ricketts, too, is collecting men from his district. I have at present 200 spare horses, but as I am to raise 1200 or 1400 men, I fear mounting them will be a difficulty; it is very difficult to work in a camp

[1] 'Hodson of Hodson's Horse,' pp. 205-209.

on service, where so little can be got or bought. . . ."

"2nd September: Hodson's Horse made a very respectable show indeed last evening, when paraded altogether for the first time."

Meanwhile attacks and counter-attacks between the British on the ridge and the mutineers in the city had been of almost daily occurrence, and during the first fortnight of September the malarial climate, so fatal at Delhi in the season following the annual rains, began to tell seriously on the British force, exhausted as were all ranks by the months of incessant exposure and hard work. Hodson's Horse suffered like all the rest. McDowell, whose delicacy was, according to Hodson, his only fault, was so ill that his commander feared he would have to take leave. Hodson himself had warnings that his constitution, notwithstanding its natural strength, would not stand much more of such a strain. On the 11th September, Captain Ward was very ill, and Hodson adds: "The natives too are very sick, and a large number are in hospital; in short, we want to be in Delhi."

They had not much longer to wait. On the 3rd September a train of heavy siege guns, long and anxiously expected, reached the camp in safety. At dawn on the 8th the first battery opened fire on the defences of the city, and on the night of the 13th the breaches were declared practicable for assault on the following day. At last the toil and suspense of the past three months was to end. Every man in the sorely tried force on the Delhi ridge was stirred with enthusiasm at the prospect, and the prevailing excitement was shared to the full by Hodson's Horse. Sir Hugh Gough thus recalls the emotions of the moment: "On the evening of 13th September orders were issued for the assault to take place the next day. The various columns were formed, and great was our excitement. We had been so long sitting before this doomed city, in the most trying heat and with apparently fruitless labour, that the immediate hopes of an end gave us all a most pleasurable feeling. Knowing, as all did, that a desperate struggle was at hand, few probably felt anything but intense excitement and delight. I happened that evening to have a talk with one of our senior native officers, Ressaldar Man Singh—a grand old Sikh, who himself had fought against us in the Sutlej and Punjab campaign, and we discussed the question of to-morrow's big fight. As the old man was fond of telling the story even to his dying day, to my own boys amongst others, it runs in his words as follows: 'Gough sahib came to me on the day before the assault and said, "Man Sing, there is going to be a great battle to-morrow, and we are going to take Delhi. Hodson says he will ride to Jehannum after the Pandies. I wonder how it will end." I said to Gough sahib, "Well, sahib, wherever Hodson goes we'll all go," whereupon Gough sahib said, "Well, Man Sing, salaam; then we'll all go to Jehannum together."' I cannot vouch for the strict accuracy of this story, but one and all of us were prepared to follow Hodson to the very death—and I am sure there was not a desponding heart in the whole force." [1]

The story of the storming of Delhi by the scanty British forces on the 14th September 1857 is one which has been told and retold with never-failing interest, but it is necessary to confine this narrative to the part played by the Cavalry Brigade, which, though comparatively subordinate, was nevertheless not the least striking among the events of the day.

The brigade was under the command of Brigadier Hope Grant of the 9th Lancers, and consisted of 200 lances of that regiment, and 410 sabres made up from the Guides Cavalry, 1st, 2nd, and 5th Punjab Cavalry and Hodson's Horse, together with the guns of the 1st Troop, Horse Artillery, and four

[1] 'Old Memories.'

guns of the 2nd Troop. Their orders were at first to take up a position in rear of No. 1 Field Battery, where they would be at hand to meet and check any attempt on the part of the mutineers to make a counter-attack against the British camp or to cut off the columns assaulting the breaches. There the cavalry remained until between 6 and 7 A.M. By this time the 4th column, under Major Charles Reid of the Sirmur Gurkhas, to which was allotted the difficult task of advancing through the walled gardens and close-packed houses of the Kishenganj and Teliwara suburbs, had been checked by the enormous numerical superiority of the enemy, and had fallen back to the line of the Sabzi Mandi suburb, the extreme right of the British position during the siege. As they retired, the rebels swarmed out of the city and crowded the enclosed ground about Kishenganj and Teliwara, threatening to overwhelm the 4th column and endangering the safety of the batteries and of the whole British camp. This was the contingency to meet which Hope Grant's position had been selected. The brigade immediately received orders to advance and "went there at a gallop," writes McDowell, "bang through our own batteries, the gunners cheering us as we leapt over the sand-bags." Forming up under the Mori bastion, which had already been taken by the British stormers, the cavalry advanced towards the Kabul Gate, the 9th Lancers being in advance and the Irregular Cavalry in reserve. Wheeling to the left under the Shah bastion and arriving within musket shot of the Kabul Gate, the brigade became exposed to a heavy fire from the city as well as from the walled gardens and houses of the suburbs, where the enemy had succeeded in establishing two guns, while the guns at the Lahore Gate also opened against the column with grape. An effort was made to maintain the position so occupied. The Horse Artillery, under Major Tombs, was sent to the front, and, coming into action, drove the enemy out of the gardens at Teliwara and spiked the two guns there. But the hostile fire was so galling that the Brigadier-General attempted to withdraw his force to a point where they would be less within reach of the musketry and grape from the city. As they fell back, however, the enemy again swarmed into the suburbs, and it was evident that any further retirement would endanger not only the British batteries near this point, but also the communications between the assaulting columns and the camp. It was necessary therefore to stand fast, however great the losses incurred. No more trying position can be conceived, especially for cavalry, who, sitting on their horses, were unable to take advantage of any cover offered by the inequalities of the ground, and were equally incapable of replying to the fire from guns and musketry with which they were assailed. The Horse Artillery indeed continued in action and kept the enemy in check, but for more than two hours the cavalry were compelled to remain inactive. It was indeed a crucial test of discipline and endurance, and the steadiness of the whole brigade well deserved the praise which it earned from Hope Grant, whose usually matter-of-fact despatch becomes quite enthusiastic on the subject. "I beg leave to state," he wrote, "I have never in the whole course of my life seen so much bravery, and so much noble conduct displayed by men, as was the case in the brigade I had the honour to command. . . . Nothing could be finer than the conduct of the 9th Lancers. . . . Not a man flinched from his post though under this galling fire for two hours. . . . The behaviour of the native cavalry was also admirable. Nothing could be steadier, nothing could be more soldier-like than their bearing. . . . Lieutenant Hodson commanded a corps raised by himself, and he is a first-rate officer, brave, determined, and clear-headed." The position of the 9th Lancers at the head of the brigade, as well as the fact of their being British soldiers, naturally made them especially the object of the enemy's fire, and the losses amongst the native cavalry were therefore small compared with those suffered by the Lancers. But the situation was nevertheless trying in the extreme to all, far more so than can be estimated from the actual number of the casualties, which but for the inaccuracy of the enemies' fire would certainly have been much heavier.

At length came relief. A small party of Guides Infantry assumed the offensive in the gardens of Teliwara, and were afterwards reinforced by a detachment of the Baluch Battalion. The enemy meanwhile finding their rear threatened by the capture of the Kabul Gate, retreated into the city, their retirement bringing to an end the share of the Cavalry Brigade in the operations of the day. The total losses of the brigade in killed and wounded were 10 officers and 116 men and 56 horses. Hodson's Horse got off lightly, losing only 3 wounded, and 2 horses killed and 6 wounded.

CHAPTER III.

DELHI AFTER THE SIEGE.

THE fighting on 14th September, notwithstanding that it involved very heavy losses, only gained for the British a small portion of the walls of Delhi, and four anxious days were to pass before the city was finally won. In those days, however, and in the severe street fighting with which they were occupied, there was no opportunity for the use of cavalry. It was not until the forces of the mutineers had been driven into the open country outside Delhi that a period succeeded in which Hodson was able to exercise all his ability as an intelligence officer and his daring as a leader, with results so startling that in a very brief space of time he and his regiment became famous not only in India but even throughout the world.

On the morning of 19th September, the day on which the city was finally evacuated by the rebels, Hodson rode with Brigadier-General Hope Grant from the British camp to the Idgah Hill, due west of the Lahore Gate, whence could be surveyed the camp of the Nasirabad and Bareilly mutineers and all the country south of the city. A bustle in the camp quickly caught his eye, and within a few minutes his suspicions were confirmed by information that the rebels were retreating. Galloping back to Delhi to inform General Wilson, Hodson immediately obtained permission to reconnoitre the enemy's camp more closely. The rest of the day's proceedings may be told in his own words :—

"McDowell and I started with seventy-five men, and rode at a gallop right round the city to the Delhi gate, clearing the roads of plunderers and suspicious-looking objects as we went. We found the camp, as I had been told, empty, and the Delhi gate open ; we were there at 11 A.M. . . . I brought in the mess plate of the 60th Native Infantry, their standards, drums, and other things, and McDowell and I had been for five hours inside the Delhi gate, hunting about, before a guard was sent to take charge of it."

But this daring ride into the enemy's camp, almost ere it had been evacuated, was forgotten in the interest aroused by the sensational achievements of the two following days. In all the turmoil, excitement, and anxiety of the Mutiny campaign, when every day brought fresh stories of heroism, fortitude, and gallantry, no deeds aroused so much attention, both by reason of the extraordinary audacity displayed and on account of the importance of their results, as did Hodson's rides to Humayun's tomb on the 20th and 21st September and the capture by him there of the king and princes of Delhi. These exploits cannot be better described than in the following letters written from Delhi at the time: the first to Hodson's brother, the Rev. George Hodson, and published by him in the 'Times,' the second written by Lieutenant McDowell. Both letters are printed in Mr Hodson's book, 'Hodson of Hodson's Horse.'

"The very day after we took possession of the palace, Captain [1] Hodson received information that the King and his family had gone with a large force out of the Ajmeer gate to the Kootub. He immediately reported this to the General commanding, and asked whether he did not intend to send a detach-

[1] So written throughout this letter. Hodson was not, however, gazetted Captain till some months later.

HODSON WITH SOME OF HIS OFFICERS.

From a photograph taken during the Mutiny campaign.

The English officer on Hodson's left is probably Lieut. McDowell ; the Sikh seated in the foreground is Risaldar Man Singh.

ment in pursuit, as with the King at liberty and heading so large a force our victory was next to useless, and we might be besieged instead of besiegers. General Wilson replied that he could not spare a single European. He then volunteered to lead a party of the Irregulars, but this offer was also refused, though backed by Neville Chamberlain. . . . General Wilson at last gave orders to Captain Hodson to promise the King's life and freedom from personal indignity, and make what other terms he could. Captain Hodson then started with only fifty of his own men from Humayun's Tomb, three [1] miles from the Kootub, where the King had come during the day. The risk was such as no one can judge of who has not seen the road, amid the old ruins scattered about, of what was once the real city of Delhi.

"He concealed himself and men in some old buildings close by the gateway of the tomb, and sent in his two emissaries to Zeenat Mahal (the favourite Begum) with the ultimatum—the King's life and that of her son and father. After two hours, passed by Captain Hodson in most trying suspense, such as (he says) he never spent before, while waiting the decision, his emissaries came out with the last offer—that the King would deliver himself up to Captain Hodson only, and on condition that he repeated with his own lips the promise of the Government for his safety.

"Captain Hodson then went out into the middle of the road in front of the gateway, and said that he was ready to receive his captives and renew the promise.

"You may picture to yourself the scene before that magnificent gateway, with the milk-white domes of the tomb towering up from within, one white man among a host of natives, yet determined to secure his prisoner or perish in the attempt.

[1] This is inaccurate. The tomb is 3½ miles from the Delhi Gate of the modern city, and about 5½ miles from the Kutb.

"Soon a procession began to move slowly out, first Zeenat Mahal, in one of the close native conveyances used for women. Her name was announced as she passed by the Maulvie.[1] Then came the King in a palkee, on which Hodson rode forward and demanded his arms. Before giving them up, the King asked whether he was 'Hodson Bahadur,' and if he would repeat the promise made by the herald? Captain Hodson answered that he would, and repeated that the Government had been graciously pleased to promise him his life, and that of Zeenat Mahal's son, on condition of his yielding himself prisoner quietly, adding very emphatically that if any attempt was made at a rescue he would shoot the King down on the spot like a dog. The old man then gave up his arms, which Captain Hodson handed to his orderly, still keeping his own sword drawn in his hand. The same ceremony was then gone through with the boy (Jumma Bukht); and the march towards the city began, the longest five miles,[2] as Hodson said, that he ever rode in his life, for, of course, the palkees only went at a foot pace, with his handful of men around them, followed by thousands, any one of whom could have shot him down in a moment. His orderly told me that it was wonderful to see the influence which his calm and undaunted look had on the crowd. They seemed perfectly paralysed at the fact of one white man carrying off their King alone. Gradually as they approached the city the crowd slunk away, and very few followed up to the Lahore gate. Then Captain Hodson rode on a few paces and ordered the gate to be opened. The officer on duty asked simply as he passed what he had got in his palkees. 'Only the King of Delhi' was the answer, on which the officer's enthusiastic exclamation was more

[1] Maulvi Rajab Ali, who was one of Hodson's most useful assistants in collecting information about the rebels.

[2] It is 5 miles from Humayun's Tomb to the Lahore Gate, 3½ to the Delhi Gate.

emphatic than becomes ears polite. The guard were for turning out to greet him with a cheer, and could only be repressed on being told that the King would take the honour to himself. They passed up that magnificent deserted street to the palace gate, where Captain Hodson met the civil officer (Mr Saunders), and formally delivered over his royal prisoners to him. His remark was amusing: 'By Jove! Hodson, they ought to make you Commander-in-Chief for this.'

"On proceeding to the General's quarters to report his successful return and hand over the royal arms, he was received with the characteristic speech, 'Well, I am glad you have got him, but I never expected to see either him or you again!'"[1]

The second letter, from Lieutenant C. T. McDowell, Hodson's second-in-command, and the only Englishman who accompanied him on this daring adventure, runs as follows:—

"On the 21st a note from Hodson, 'Come sharp; bring one hundred men.' Off I went: time, six o'clock A.M. He told me he had heard that the three princes (the heads of the rebellion, and sons of the King) were in a tomb six miles off, and he intended going to bring them in and offered me the chance of accompanying him. Wasn't it handsome on his part? Of course, I went. We started at about eight o'clock, and proceeded slowly towards the tomb. It is called Humayoon's Tomb and is an immense building. In it were the Princes and about three thousand Mussulman followers, in the suburb close by about three thousand more, all armed; so it was rather a ticklish bit of work. We halted half a mile from the place and sent in to say the Princes must give themselves up unconditionally or take the consequences. A long half-hour elapsed, when a messenger came out to say the Princes wished to know if their lives would be

[1] 'Hodson of Hodson's Horse,' pp. 226-228.

promised them if they came out. 'Unconditional surrender' was the answer. Again we waited. It was a most anxious time. We dared not take them by force, or all would have been lost, and we doubted their coming. We heard the shouts of the fanatics (as we found out afterwards) begging the Princes to lead them on against us. . . . At length, I suppose imagining that sooner or later they must be taken, they resolved to give themselves up unconditionally, fancying, I suppose, as we had spared the King we should spare them. So the messenger was sent to say they were coming. We sent ten men to meet them, and by Hodson's order I drew the troop up across the road ready to receive them and shoot them at once if there was any attempt at rescue. Soon they appeared in a small 'ruth,' or Hindoostanee cart drawn by bullocks, five troopers on each side. Behind them thronged about two thousand or three thousand (I am not exaggerating) Mussulmans. We met them and at once Hodson and I rode up, leaving the men a little in rear. They bowed as we came up, and Hodson, bowing, ordered the driver to move on. This was the minute. The crowd behind made a movement. Hodson waved them back; I beckoned to the troop, which came up, and in an instant formed them up between the crowd and the cart. By Hodson's order I advanced at a walk on the people, who fell back sullenly and slowly at our approach. It was touch-and-go. Meanwhile Hodson galloped back and told the sowars (ten) to hurry the Princes along the road, while we showed a front and kept back the mob. They retired on Humayoon's Tomb, and step by step we followed them. Inside they went, up the steps, and formed up in the immense garden inside. The entrance to this was through an arch, up steps. Leaving the men outside, Hodson and myself (I stuck to him throughout), with four men, rode up the steps into the arch, where he called out to them to lay down their arms. There was a murmur. He reiterated the command, and (God knows why, I never can understand it) they commenced doing so. Now, you see, we didn't want their arms, and under ordinary circumstances we would not have risked our lives in so rash a way, but what we wanted was to gain time to get the Princes away, for we could have done nothing had they attacked us but cut our way back, and very little chance of doing even this successfully. Well, there we stayed for two hours collecting their arms, and I assure you I thought every moment they would rush upon us. I said nothing, but smoked all the time to show I was unconcerned; but at last when it was all done, and all the arms collected, put in a cart and started, Hodson turned to me and said, 'We'll go now.' Very slowly we mounted, formed up the troop, and cautiously departed, followed by the crowd. . . . Well, I must finish my story. We came up to the princes, . . . close to Delhi. The increasing crowd pressed close on the horses of the sowars, and assumed every moment a more hostile appearance. 'What shall we do with them?' said Hodson to me. 'I think we had better shoot them here; we shall never get them in.' We had identified them by means of a nephew of the King, whom we had with us; besides, they acknowledged themselves to be the men. Their names were Mirza Mogul, the King's nephew, and head of the whole business; Mirza Kishere Sultamed (Khizr Sultan), who was also one of the principal rebels, and had made himself notorious by murdering women and children; and Abu Bukr, the Commander-in-Chief nominally and heir-apparent to the throne. . . . There was no time to be lost. We halted the troop, put five troopers across the road, behind and in front. Hodson ordered the Princes to strip, and get again into the cart. He then shot them with his own hand. So ended the career of the chiefs of the revolt, and of the greatest villains that ever shamed humanity." [1]

The act described above occasioned so much criticism at the time and even for many years after Hodson's death brought upon him and his memory such a load of obloquy that some brief consideration of the surrounding circumstances cannot be avoided. In the first place, it may be said that a careful examination of the evidence leads to the conclusion that Hodson genuinely believed that the success of his enterprise, if not the lives of himself and his men, depended on his instant action. Secondly, it cannot be too strongly urged that the whole atmosphere of the time was abnormal. Horrible stories were current of the murderous outrages committed by the mutineers at Delhi, and encouraged by the very persons who were now Hodson's captives. These reports, supported by the dreadful news of what had been done at Cawnpore, Jhansi, and elsewhere, were everywhere believed. If they were exaggerated, it took prolonged and careful inquiry to elicit evidence to that effect. Meanwhile the minds of all in the English forces were filled with indignant horror at their narration. On a man of Hodson's fiery temperament, such stories could not but have an inflammatory effect. Having succeeded in capturing those who were believed to have been the prime instigators of the excesses, he was determined that by no chance should they escape, and rather than run any risk of that accident he was prepared to take the law into his own hands. While we recognise the justice of the retribution, we may regret in Hodson's own interests that he should have taken upon himself the responsibility of the deed. But in such an emergency he was not a man to shirk what he conceived to be his duty for fear of what might be thought of his action

[1] 'Hodson of Hodson's Horse,' Introduction, pp. xiii-xvi.

Note.—The senior native officer with Hodson on this occasion was Risaldar Man Singh.

by others. It was notably characteristic of him that he was ever prepared to trust to his own judgment and to abide the consequences, a quality which is essential to great achievement but which is also apt to incur the hostility of cool and careful observers.

For readers unacquainted with the locality it will be useful to pause here and to give some further description of Humayun's tomb and its surroundings, which may be explanatory of the foregoing narratives. The road to the tomb runs due south from the Delhi gate of the modern city, and passes directly through the midst of the ruins of ancient Delhi. About half a mile from the gate the jail is passed on the right, and on the left stand the ruins of Firoz Shah's Kotla. After another half mile the lofty walls of the old fort of Delhi are reached, which skirt the left of the road for some six hundred yards. Thence to the gateway of Humayun's tomb is a distance of about two miles, several ancient buildings being passed on the way. Just outside the gateway of the tomb enclosure stands a considerable suburb of lofty buildings known as the Arab Sarai, while the tomb of Nizam-ud-Din and a number of other mosques and tombs are scattered about in the immediate neighbourhood. The present road from the city is made on an embankment, and runs at a distance of two or three hundred yards from most of the buildings named; but that which existed at the time of the Mutiny winds immediately below the walls of the ancient fort, and is completely commanded there as well as at several other points. It will be seen that the account given of the capture of Bahadur Shah states that Hodson, with his prisoners, entered Delhi by the Lahore gate, and passed up the Chandi Chauk to the palace. This would entail a circuit by a cross road from Humayun's tomb, passing Nizam-ud-Din's tomb and cutting into the road from Delhi to the Kutb at about two and a half miles from the Ajmir and three and a half from the Lahore gate. But though longer, Hodson may have preferred it to the danger of passing so close to the walls of the old fort, where he would be liable to be attacked at any moment and at a great disadvantage. On the occasion of his second visit Hodson had either become more reckless by reason of his success of the previous day, or he relied for safety on the fact that he held the greater number of the hostile crowds at bay at Humayun's tomb, while his prisoners were being hurried by their escort towards Delhi. Whatever may have been the reason, he attempted to enter the city by the Delhi gate, with the result that a mob of malcontents collected from the ruined forts through which the road lay, until, when Hodson and McDowell overtook the cortège at a point where the road passes through a slight hollow between the jail and Firoz Shah's Kotla, the chance of the gate being reached in safety seemed so doubtful that he considered it necessary to forestall any attempt at rescue by instant action.

The tomb of Humayun itself is a magnificent building of red sandstone surmounted by a white marble dome some 150 feet in height. It stands on a double-terraced plinth, the lower terrace being 400 feet square and about 4 feet above the ground level; the upper terrace is some 25 feet higher, and is 330 feet square. This terrace is cloistered in each of the four faces; a great number of tombs of the family of the Emperor Humayun are in these surrounding cloisters, while in the centre of the south face a passage leads with many windings into a lofty vault below the very heart of the building. Here is the actual sarcophagus of the Moghal Emperor, and here in the fancied security of the gloomy crypt his wretched descendant took refuge from the just revenge of his conquerors.

The whole building stands in an extensive quadrangle 380 yards square, formerly laid out as a garden, surrounded on three sides by a strong

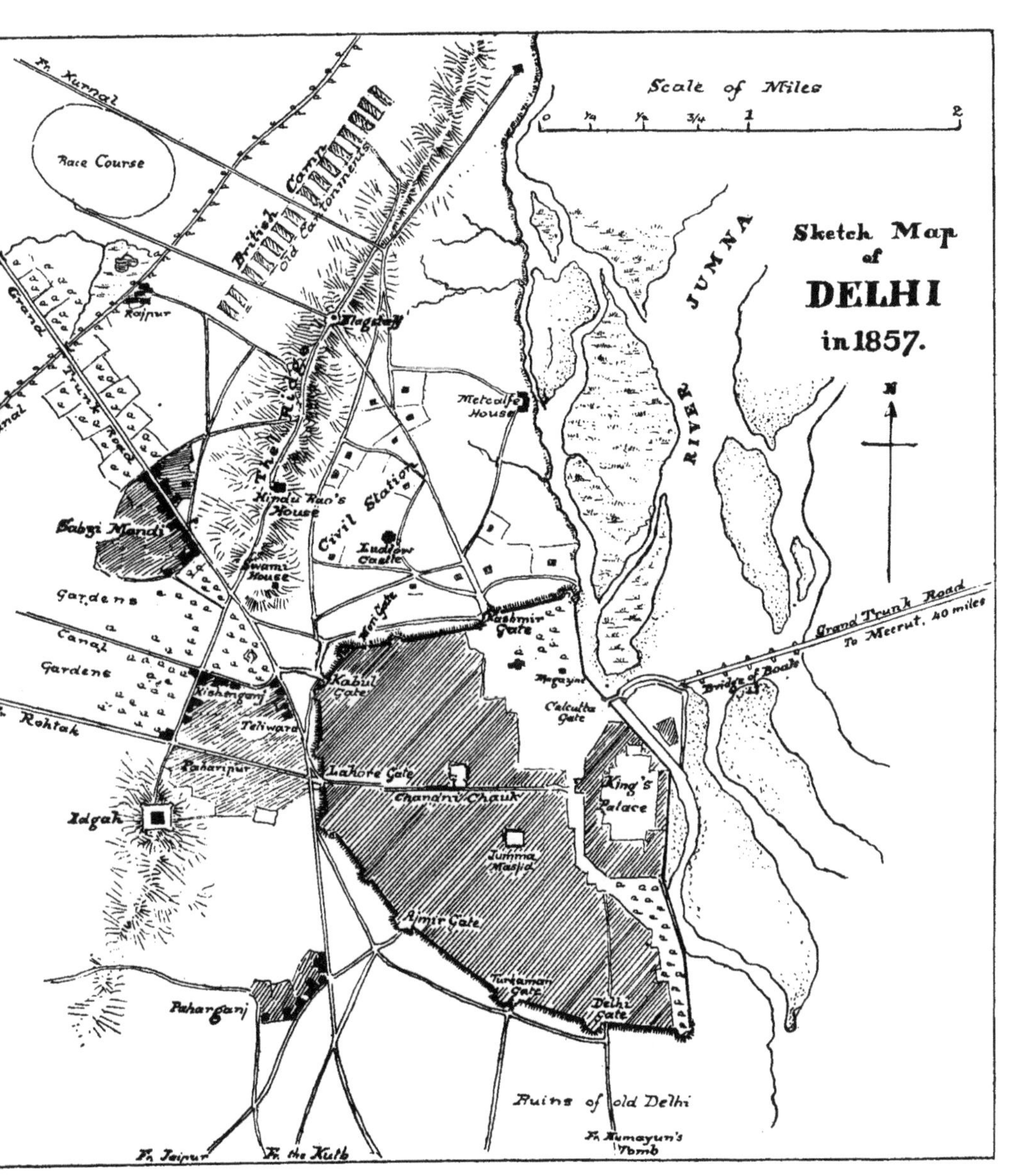

Scale of Miles
0 1/4 1/2 3/4 1 2
Sketch Map of DELHI in 1857.
N
Fr. Kurnal
Race Course
British Camp
Old Cantonments
Rajpur
Grand Trunk Road
Canal
The Ridge
Flagstaff
Metcalfe House
Hindu Rao's House
Civil Station
Ludlow Castle
Sabzi Mandi
Swami House
Gardens
Canal
Gardens
Kishenganj
Teliwara
Rohtak
Paharipur
Idgah
Kashmir Gate
Kabul Gate
Magazine
Calcutta Gate
Lahore Gate
Chandni Chauk
King's Palace
Jumma Masjid
Ajmir Gate
Turkaman Gate
Delhi Gate
Paharganj
JUMNA RIVER
Grand Trunk Road
To Meerut, 40 miles
Bridge of Boats
Ruins of old Delhi
Fr. Jaipur
Fr. the Kutb
Fr. Humayun's Tomb

cloistered wall, but open on the east to the fertile fields which border the Jumna a mile away. In the centre of the west face of the quadrangle wall is a massive and handsome gatehouse of red sandstone, with arched doorway, approached by a low flight of steps, the scene of the eventful incidents which have been described.

On the day following the events just narrated General Archdale Wilson forwarded to the Army Headquarters his final despatch dealing with the capture of Delhi. In it he writes as follows :—

"I beg also to bring very favourably to notice the officers of the Quartermaster-General's Department . . . Captain W. S. R. Hodson, who has performed such good and gallant service with his newly raised Regiment of Irregular Horse, and at the same time conducted the duties of the Intelligence Department, under the orders of the Quartermaster-General, with rare ability and success."

These words were repeated by the Governor-General in Council in the Order which published General Wilson's despatch (G.G.O. 1383, dated 5th November 1857).

On the 27th September, Hodson, with a detachment of his Horse and some other irregular cavalry, left Delhi with a small column under Brigadier Showers, with orders to scour the country south and west of the city and to restore order in the disaffected Gujar villages.

After four days of uneventful marching the force returned on 30th September to Delhi, and the headquarters of Hodson's Horse, under their commanding officer, were immediately attached to another and larger column, as shown below, which under the same Brigadier marched from Delhi on 2nd October to operate farther afield in the districts west of Delhi, and especially against the rebel chiefs of Rewari, Jajhar, and Dadri. Each of these places were occupied without resistance, as well as the strong fort of Kanaund; the Nawab of Jajhar was taken prisoner, a number of rebel villages were destroyed, and seventy guns with much ammunition and treasure were taken. But the expedition was unmarked by any fighting, and the column returned to Delhi on the 28th October. A day or two afterwards Hodson obtained leave of absence, and went to Ambala, the regiment remaining at Delhi under Lieutenant McDowell until ordered to march down country with General Seaton's column a month later.

No. 14 Lt. Field Battery.
2 18-pr. Guns.
2 5½-pr. Mortars.
2 Companies of Sappers.
6th Dragoon Guards.
Hodson's Horse.
2nd Eur. Bengal Fusiliers.
1st Punjab Infantry.
Kumaon Battalion.

CHAPTER IV.

OPERATIONS FROM SEPTEMBER TO DECEMBER 1857.

On the same day (27th September) that Hodson left Delhi with Brigadier-General Showers' column, a portion of his corps, consisting of three troops, was detailed to accompany the force detached from the Delhi army to pursue such of the rebels as had crossed the Jumna and were making their way by Bulandshahr and Aligarh towards Rokilkhand and Oudh. Lieutenant H. H. Gough was selected to command this detachment, with Lieutenant G. A. A. Baker as his subaltern. The movable column, which was commanded by Brigadier Edward Greathed, was composed as follows:—

2 Troops of Bengal Horse Artillery . .	120 men
1 Bengal Light Field Battery	60 „
1 Company of Sappers	200 „
9th Lancers	300 „
Detachments each of 1st, 2nd, and 5th Punjab Cavalry and Hodson's Horse	400 „
8th Foot (King's) . . . . . 75th Foot	} 450 „
2nd and 4th Punjab Infantry	1200 „

The first object of the column was the attack of Malagarh fort, which was held by a Nawab named Walidad Khan, and in which a number of the rebels from Delhi, reinforced by some of the Jhansi mutineers, were expected to make a stand. The fort was some four miles from the town of Bulandshahr, and it was in front of the latter place that the hostile force took up a position on 28th September with six guns entrenched at the junction of two roads leading into the town. From this they were driven by the infantry of the British force, which advanced under cover of the field battery on the right and a troop of Horse Artillery on the left, and the rebels fell back through Bulandshahr towards Malagarh fort. But a timely movement round the left of the town by the cavalry under Major Ouvry, 9th Lancers, intercepted their retreat, and effectually scattered them with great loss. Immediately after the action Lieutenant Gough with his detachment was sent forward to Malagarh, which place was found deserted, and was occupied without further resistance.

The conduct of the irregular cavalry in the successful action of Bulandshahr was described by Major Ouvry as meriting the highest praise, and the name of Lieutenant Gough was amongst those mentioned by the same officer as having performed most gallant service. The losses of Hodson's Horse were only two men wounded, two horses killed and two wounded.

From Bulandshahr the column marched to Aligarh, where a skirmish took place on the 5th October with a body of rebels, who, however, offered but slight resistance. Here one troop of Hodson's Horse was left under the command of Lieutenant Baker, while the remainder continued the advance with the rest of the column by forced marches to Agra. That city was reached on the morning of the 10th October, and on the same morning was fought what is known as the action of the Agra parade ground. The British force, just pitching camp after an exhausting march of forty-four miles in twenty-eight hours, was completely surprised by a sudden attack from the Nimach and Gwalior mutineers. But the latter, whose intelligence and precautions were no better than those of their opponents, were equally taken by surprise to find themselves confronted by an unlooked-for force from Delhi instead of by the Agra troops only. Greathed's regiments, accustomed to the constant alarms of the Delhi ridge, were only a few minutes in forming up, and, supported by the 3rd Bengal European Regiment and a battery of 9-pounders from the fort, they not only checked the attack of the enemy but converted the repulse of the latter into a complete rout. This result was mainly due to the brilliant conduct of the irregular cavalry, who pursued them across the Kala Naddi with great slaughter. The conduct of Lieutenant Hugh Gough was again mentioned in the despatches describing the action, while he has himself written in the following terms of the manner in which the detachment of Hodson's Horse bore themselves in the fight:—

"I was well pleased with the behaviour of my men on this occasion. They had very readily pulled themselves together when the first alarm took place, and behaved with considerable dash. It was always necessary to bear in mind that these men were utterly undisciplined and untaught soldiers, according to our ideas, being either raw recruits or disbanded soldiers of the old Khalsa army who had fought against us in the Punjab some eight years previously. They were indifferent riders, as Sikhs usually are (till taught), and at least half of them used with one hand to clutch hold of the high knob in front of the Sikh saddle as they galloped along. They had no knowledge of drill or of our words of command; in fact all I attempted to teach them were 'Threes right,' or 'Threes left' (never 'Threes about'!), and 'Form line,' 'Charge.' However, with all their want of knowledge and training, they had plenty of pluck, and their success lay in that, combined with readiness and goodwill for any amount of work."[1]

The losses of Hodson's Horse in the action at Agra were one sowar killed, one Indian officer and

[1] 'Old Memories.'

two sowars wounded, and one sowar missing, as well as one horse wounded and one missing.

The movable column resumed its march towards Cawnpore on the 13th October, but Lieutenant Gough was left behind with his detachment for a day or two, with orders to follow the column with a convoy of supplies on camels as soon as the latter were collected. While at Agra, Lieutenant R. D. Craigie Halkett of the 72nd Native Infantry was appointed to do duty with the detachment, and as soon as the convoy was ready to start the two officers hastened to overtake the movable column with their valuable but troublesome charge.

Before the detachment rejoined the command of the column had been transferred from Brigadier Greathed to Brigadier Hope Grant, a circumstance which threatened incidentally to be for a time disastrous to the reputation of this portion of Hodson's Horse. The story may be given in Sir Hugh Gough's own words:—

"Just one march before we reached Cawnpore a very unpleasant incident occurred, which caused me great grief at the time, and which I feared would utterly ruin all my chances of distinction, but which, as so many things unexpectedly do in one's daily life, afterwards proved just the reverse. I have already alluded to the undisciplined state of my men, and that their idea of 'orders' was about as vague as could be well conceived. Our commander —Brigadier Hope Grant, C.B., a man who had been brought up in the strict routine of the 9th Lancers— could not appreciate the fact that such a 'rabble,' as he was pleased to term us 'Hodson's Horse,' could be worth anything as soldiers. As ill-luck would have it, on visiting the piquets one afternoon the General (Hope Grant) found the one supplied by Hodson's Horse sadly wanting in that alert smartness so dear to the heart of the energetic cavalry commander. He was very angry, and 'pitched into' the native officer roundly and justly, as he deserved, and then sent for me, when he gave me as rough a rubbing-up as his naturally kind old heart and tongue was capable of, and, visiting all the sins of my men, who never dreamt they were to blame, on my devoted head, passed the order that the detachment of Hodson's Horse under Lieutenant Hugh Gough was to be placed on perpetual rear-guard till further orders."

Cawnpore was reached without further opposition from the enemy on the 26th October, and here Brigadier Hope Grant halted for four days in order to communicate with the Commander-in-Chief, Sir Colin Campbell, who was on his way up-country. In pursuance of orders received, Grant left Cawnpore on the 30th October, his force being now augmented to an effective strength of some three thousand men. On the 1st November he crossed the Ganges by the Banni Bridge, and here a sharp skirmish took place with a body of mutineers, who endeavoured to oppose the passage. Hodson's Horse were still, as Sir Hugh Gough relates, occupying their place of punishment as perpetual rear-guard, and both leader and men were cursing their ill-luck at being thus deprived of all chance of distinguishing themselves, when an unexpected chance threw an opportunity in their way, of which they were not slow to take advantage. "Fortune," writes Gough, "sent an enterprising enemy round by the rear to see what they could do in the way of loot and damage. This party, numbering over two hundred horsemen, suddenly appeared on our left flank and made a dash towards the line of baggage. Captain Wheatcroft of the Carabineers—a gallant officer, who had just come out from England in hopes of joining his regiment at Meerut or Delhi and who, on reporting his arrival at Cawnpore, had been posted to the Military Train—was then commanding the rear-guard. He desired me to reconnoitre the enemy's cavalry, and see what they were up to. I went forward therefore with some fifteen troopers, and soon came in full view of the enemy—a body of our own mutinied Irregular Horse—who seeing the smallness of my party promptly came at us and saluted us with a volley from their carbines, which, as they fired from on horseback, was ill directed and harmless. In the meantime, I ordered up as many of my regiment as I could quickly gather together and as soon as I got about forty men charged them with a tremendous cheer and soon got into the thick of them. They could not stand the shock of the charge which we were able to deliver home, and broke and fled. We pursued them some way and cut up numbers of them. My men were mad to retrieve their disgrace and the rear-guard punishment and behaved most splendidly. Wheatcroft, in the meantime, seeing me disappear over the undulating ground with a cheer and a charge and knowing our small numbers, was in a desperate state of anxiety and alarm and was about to start to my assistance when he saw us returning in triumph and safety. To me this little affair gave the deepest joy, for I felt that my men had shown what they could do and that if they had been slack on piquet duty they were not slack in a charge. Wheatcroft was full of praise and congratulation, and made a very flattering report of my little achievement to General Hope Grant, who made amends for all we had suffered by saying we should have the post of advance-guard on the march of the force to the Alam Bagh."

North of the Banni Bridge the force halted and awaited the arrival of Sir Colin Campbell, who joined it on the evening of the 9th November. After a further delay of two days, to enable a few more reinforcements to reach the column, the march was resumed on the morning of 12th November. The composition of the force was as follows:

Naval Brigade. 8 Heavy Guns.
Bengal Horse Artillery. 10 Guns.
Bengal Field Artillery. 6 Guns.
Royal Artillery (Heavy). 1 Battery.
Bengal and Punjab Sappers. Detachments.
H.M. 9th Lancers.
1st, 2nd, and 5th Punjab Cavalry. } Detachments.
Hodson's Horse. }
H.M. 8th Foot.
53rd Light Infantry.
75th Foot.
93rd Highlanders.
2nd and 4th Punjab Infantry.

But the regiments were so weak that the total strength amounted only to about 700 cavalry and 2700 infantry. Only six miles intervened between the camp and the Alam Bagh, which was held by Havelock's troops, and it was not expected that any opposition would be encountered in this march. Hodson's Horse, according to Sir Hope Grant's promise, occupied the post of honour in advance, and had proceeded some distance when it suddenly encountered a force of the enemy on the right flank, numbering about 2000 infantry with two guns. "After a smart skirmish the guns were taken, Lieutenant Gough, commanding Hodson's Irregular Horse, having distinguished himself very much in a brilliant charge by which this object was effected." [1] Of this combat, so creditable to the regiment, Sir Hugh Gough gives a more detailed account as follows :—

"Suddenly, as our column was advancing up the road an attack developed itself on the right flank, where a body of the enemy calculated at about two thousand strong with two guns had taken up a position, having apparently come out of the fort of Jellalabad. As these guns were troublesome, and perhaps from the wish to give me a chance, the General—for it was Hope Grant who was commanding the advance—rode up to me and desired

[1] Sir Colin Campbell's despatch, dated 18th November 1857.

me to take my squadron and see if I could capture the guns. He further gave me an order to spike them if I found I could not get them away, and to carry out this order I was provided with a hammer and spikes or large nails. Of how I disposed of them I have not the slightest recollection, but I rather suspect I threw them away!

"With my small body of men, my only chance of success was by making a flank attack, and if possible a surprise. With this object I made a considerable *détour*, and managed, under cover of some fields of growing corn or sugar-cane, to arrive on the left flank of the enemy perfectly unseen. The guns were posted on a small mound and a considerable body of the enemy had an admirable position in rear of this mound, in front of and amidst some trees and shrub. Between us and them lay a marshy *jheel*, with long, reedy grass—an unpleasant obstacle, but which served admirably to cover our movements. I then advanced my men through this jheel and long grass at a trot, and so concealed our movements till we got clear, when I gave the word 'Form line' and 'Charge.' My men gave a ringing cheer, and we were into the masses. The surprise was complete, and owing to its suddenness they had no conception of our numbers, and so the shock to them and victory to us was as if it had been a whole brigade.

. . . .

"The men followed me splendidly, and in a very short time the affair was over,—the guns were captured, the enemy scattered, and the fight became a pursuit. Our loss was very trifling, as is often the case in a sudden surprise, but we cut up numbers of the enemy, and should have accounted for more but for the nature of the ground.

. . . .

"Sir Colin Campbell had just ridden up to the front as the affair took place, and witnessed the charge. I was very proud, both for my men and myself, when a little later he sent for me, and, complimenting me highly, said he should be glad to promote the man I would recommend for conspicuous gallantry. Sir Colin Campbell afterwards made particular mention of my name in his despatches, thereby gaining for me the honoured and most coveted distinction of the Victoria Cross."

The relieving force camped at the Alam Bagh the same evening. The operations of the next few days, which resulted in the final relief and rescue of the Lucknow garrison, need not be described here. In the house-to-house fighting of those operations there could be no opportunity for the use of cavalry, and the duties of Hodson's Horse, as well as of other mounted troops, were confined to protecting the flanks and communications of the British force. While employed in this manner the corps was never actually engaged, but it was constantly exposed to fire from buildings and enclosures, and in this manner suffered the loss of Lieutenant R. D. Craigie-Halkett and one sowar, who were killed, and two sowars wounded. It may also be recorded that on the night of the 15th November, Lieutenant Gough with his detachment formed part of the escort which returned with Lieutenant Fred Roberts from the Martinière to the Alam Bagh for ammunition, a dangerous and difficult undertaking which is graphically described by Lord Roberts in his Memoirs. On the 22nd November the Residency was evacuated, and on the 27th the return march to Cawnpore was begun, a force under General Outram being left at the Alam Bagh.

Although the operations at Lucknow afforded little opportunity for distinction to the detachment, yet the name of Lieutenant Gough was twice brought to notice by the Commander-in-Chief in the despatches recording those events. (G.G.O. 1546, dated Fort William, 10th December 1857).

Shortly after leaving camp on the morning of the 28th November news reached the Commander-in-Chief that General Windham, who had been left in command at Cawnpore, had suffered a reverse at the hands of the Gwalior rebels under Tantia Topi, and was hard pressed. The march was accordingly hastened as much as the long train of sick, wounded, women, and children would allow, but it was not until the evening of the 30th that the whole force was encamped at Cawnpore. Here it was found that the enemy had been able to occupy a strong position, and were in possession of the city, but it was impossible to resume the offensive against them until a convoy, with all encumbrances, could be despàtched towards Allahabad.

This was successfully effected on the 3rd December, while secure accommodation was provided on the 4th and 5th for those of the wounded to whom a further march would be dangerous. On the evening of the latter day preparations were made to attack the enemy the following morning.

The battle which ensued on the 6th December was fought mainly by the infantry of the force, and Sir Hugh Gough records that "to our disgust the enemy's cavalry, the well-known Gwalior troops from whom we had hoped for better deeds, never showed in action at all." But the opportunity for the cavalry came at length when the hostile infantry, completely beaten, began a headlong retreat southwards. "The pursuit," wrote Sir Colin Campbell, "was pressed with the greatest eagerness to the fourteenth milestone on the Kalpi road, and I have reason to believe that every gun and cart of ammunition which had been in that part of the enemy's position which had been attacked now fell into our possession."

Two days afterwards Lieutenant Gough, with 109 of his men, formed part of a column under Sir Hope Grant which was despatched in pursuit of such of the rebels as had retreated northwards from Cawnpore along the Bithur road. After a march of twenty-four miles a body of the enemy was encountered at Suraj Ghat, in the act of crossing the

Ganges into Oudh. They were immediately attacked by our artillery, whose concentrated fire obliged them to abandon the guns, and drove them in full retreat across the ford. Meanwhile a force of rebel cavalry attempted to effect a diversion by an attack on our guns, but they were promptly met and put to flight by the cavalry under Brigadier Little. Fifteen guns were captured, as well as the whole of the rebels' baggage and ammunition. Lieutenant Gough was mentioned by General Hope Grant for the able way in which he commanded his detachment, an opinion which was endorsed in Sir Colin Campbell's despatch (G.G.O. 1649, dated 24th December 1857). Hodson's Horse were fortunate in coming out of the actions of 6th and 8th December without a single casualty.

After the complete defeat of the rebels at Cawnpore the force halted at that place until the 24th December, awaiting the return of the transport which had conveyed the women, children, and sick to Allahabad. On that date active operations were recommenced, and the Commander-in-Chief started for Fategarh.

But at this point it is necessary to return to the headquarters of Hodson's Horse, which, as has been seen, remained with the Delhi force until the beginning of December, and was then attached to the column under Brigadier-General Thomas Seaton.

The latter was ordered to march from Delhi to Cawnpore in charge of a large convoy, and left Delhi on the 7th December. Aligarh was reached on the 11th of the same month, and here some reinforcements from Meerut joined, and with them Captain Hodson, who (as related) had been on leave at Ambala, and who had just obtained his promotion *vice* Major Jacob of the 1st Bengal European Fusiliers, killed at the storming of Delhi. The other officers with Hodson's Horse were the same who had been attached during the siege of Delhi, with the addition of Lieutenant C. J. S. Gough, who joined from the

Guides on the eve of the departure from Delhi. The total strength of the column was now 1959 men and 11 guns as follows :—

2 18-pounder guns	233 men
1 8-inch howitzer	
6 9-pounder Horse Artillery guns . . .	
2 6-pounder Horse Artillery guns . . .	
1st Troop, 6th Carabineers	140 ,,
Detachment, 9th Lancers	
Hodson's Horse	550 ,, [1]
1st European Bengal Fusiliers	376 ,,
7th Punjab Infantry	540 ,,
2 Companies of Sappers	120 ,,
	1959 men

The Aligarh district was at this time threatened by a considerable force of rebels, while its only protection was the Baluch Battalion, 300 strong, under Colonel Farquhar, with 200 Afghan Horse and two 6-pounder guns. On arrival at Aligarh, Brigadier Seaton at once placed his convoy in safety there, so as to be free to operate against the enemy. On the 15th December, having taken a reinforcement of 100 men of the 3rd European Bengal Regiment from the garrison of Aligarh fort, he joined Colonel Farquhar at Gangiri, and from him learnt that the rebels were believed to be advancing towards the neighbouring town of Kasganj. Whilst preparations were made for camping at Gangiri, Captain Hodson was sent out with part of his regiment to obtain reliable information and if possible to gain touch with the enemy. Only a short time elapsed ere

[1] This is the number given by Colonel Seaton in his despatch of the 28th December (G.G.O. 55 of 1858). It does not agree with that given by Hodson in a letter from Aligarh, dated the 11th December: "I have 596 sabres with me now, 50 more coming from Delhi, besides the 140 with Gough." It is probable, however, that Hodson's number, 596, includes the troop under Lieut. Baker, which had been left at Aligarh by Gough, and which does not appear to have gone on with Seaton's column.

Captain Light, the Brigadier's orderly officer, galloped back with the news that the rebels had been descried, apparently advancing to attack, and this information was presently confirmed by Captain Hodson in person. The Brigadier immediately fell in his force and sent out his artillery to the front. Meanwhile the enemy's columns appeared in sight, with three guns on their right and a body of horse on their extreme left, working round towards our flank. The guns on each side at once came into action, but those of the British were at a disadvantage by reason of the lie of the ground, and they were unable at once to crush the enemy's artillery. Seeing this, the squadron of British dragoons and lancers charged the latter in the most gallant manner, cutting up the gunners and capturing all three guns. At the same time, Hodson with his Horse charged the enemy's cavalry on the British right, put them to flight, and pursued them for some distance until they were entirely dispersed. The rebel infantry broke and fled without waiting to try further conclusions.

The fact was that the enemy were taken completely by surprise. They had no intelligence of the arrival of Seaton's column, and expected only to be opposed by Colonel Farquhar's small force. Their advance was consequently made with extreme confidence, and their overthrow all the more complete.

The losses in the cavalry of the British column were, however, serious, those of the Carabineers and 9th Lancers amounting to three officers and six men killed, one officer, two non-commissioned officers, and thirteen men wounded; Hodson's Horse lost one native officer (Muhammad Taki Khan) and three men killed, and two non-commissioned officers and seventeen men wounded. "The General will see by the list of casualties," runs Seaton's despatch to Major-General Penny (G.G.O. 340 of 1858), "that Captain Hodson's newly raised body of horse was not at all backward. It rendered excellent service;

less it could not do under its distinguished commander, whom I beg particularly to mention to the Major-General as having, on every possible occasion, rendered me the most effective service, whether in gaining information, reconnoitring the country, or leading his regiment."

The defeat of the rebels at Gangiri did not, however, free the Aligarh district of their presence. A large number of them had remained with almost the whole of their artillery and baggage in an entrenched camp at the town of Patiali. Against these Brigadier Seaton now moved, passing through Kasganj on the 16th December and encamping that night at Sahawar. At daylight on the 18th the march to Patiali was resumed. Hodson with part of his regiment was as usual sent far to the front to obtain accurate intelligence of the enemy's whereabouts, and in the course of the morning he reported that the rebel camp was in an entrenched position immediately in front of Patiali and approached by a road which defiled through a small village about half a mile from the front of the camp. Seaton at once made his dispositions to leave this village on his left and massing his cavalry on his right to turn the enemy's left flank. In order that the victory which he anticipated might be more complete, he determined to delay offensive action until the whole of his column could get up.

"Having sent orders to Captain Hodson not to engage the enemy further, I took the opportunity of serving out bread and grog to the men, thus allowing time for the heavy guns under Lieutenant Gillespie to arrive. By this time the enemy had opened fire from three guns on Captain Hodson's party, unreplied to by us until I had got my infantry near enough to act effectively should the enemy stand. I now directed Colonel Kinleside to bring his artillery into action, which was at once and most effectually done.

"Four Horse Artillery guns under Lieutenants

Deroin and Griffin dashed to our immediate front, whilst four more under Lieutenant Bishop, making a *détour* to the right and wheeling into line, took the enemy in flank. The enemy in the meantime had shown his strength and opened with eleven guns, some of heavy calibre, upon us. By this time the infantry and heavy guns topped the rise which had previously hidden them, and the sun being to the back of the foe shone directly on us, exposing a front so formidable as to deprive him of all further courage to resist. His fire slackened, and at last ceased altogether." [1]

This was the critical moment. The Horse Artillery limbered up and charged, with some of Hodson's Horse, led by the Brigadier in person, against the hesitating enemy. In a moment the rebels broke and fled, and the gallant Brigadier with his small party galloped right through their camp, cutting down numbers and pursuing the fugitives up to Patiali.

The Brigadier continues his narrative: "My object in keeping so large a body of cavalry on my right was now attained, as instead of going through the town and being checked by the numbers of enclosures, gardens, &c., which extend for more than a mile to its east, they went round, and thus came into the open plain beyond on more even terms with the enemy. Colonel Kinleside got quickly through the town, and joining the cavalry beyond, carried on the pursuit with them. This was kept up for seven miles. The enemy were cut down on all sides, whilst many were drowned in endeavouring to cross the *jheels* of the Boor Gungha to our left, or shot in the water. Several guns, tumbrils, &c., were taken in this pursuit. . . . In this engagement twelve guns and many tumbrils were captured, a list of which I have the honour to enclose, and I estimate the loss of the enemy to be over 700 killed. . . . After the action at Gun-

[1] Brigadier Seaton's despatch, published in G.G.O. 340 of 1858.

geree, I specially mentioned Captain Hodson and his regiment. I can but repeat what I then said, and beg that the Major-General will be good enough to bring this officer and his great and important services to the special notice of His Excellency the Commander-in-Chief."

The loss of Hodson's Horse in this action was only one dafadar killed, and one sowar slightly wounded.

The movable column halted on the 18th, 19th, and 20th December at Patiali, but Hodson's Horse got little rest during the interval as they were daily employed on long reconnaissances through the surrounding country. It was soon evident that the result of the victories of Gangiri and Patiali had been to free completely the Aligarh district from the rebels, and accordingly on the 21st December Brigadier Seaton began to retrace his steps towards the Grand Trunk road. Passing through Kasganj on the 23rd, Etah on the high road was reached on 24th December, and thence the column moved by Malaun and Kurauli to Mainpuri, while a detachment went back to Aligarh to escort the captured guns and to bring on the convoy which had been left there.

On approaching Mainpuri the Brigadier was informed that the rajah of that place with a small force of rebels was prepared to dispute his progress and had occupied a position barring the road. Seaton promptly made dispositions so as to turn the enemy's left as at Patiali and advanced to the attack; but the rebels did not wait to join battle. "They only waited," writes Hodson, "until the Horse Artillery guns had opened on them and then fled precipitately, so we had to ride hard to overtake them." The pursuit was continued for sixteen miles, and Hodson concludes his brief account with the words: "No one hurt but two of my sowars. We have got all their guns (six in number), and the Doab is clear now to Fatehgarh."

Seaton halted at Mainpuri on the 28th, 29th, and 30th December, and from this place Hodson, accompanied by McDowell, rode forward to open communications with the Commander-in-Chief (Sir Colin Campbell), a daring adventure so characteristic of the men as well as of the work which Hodson's Horse were called on to do in the guerilla warfare of the Mutiny Campaign that it deserves to be recorded here in full in the words of Lieutenant McDowell himself.[1]

"On the night of the 29th," he writes, "Hodson came into my tent about nine o'clock and told me a report had come in that the Commander-in-Chief had arrived with his forces at Goorsahaigunge, about thirty-eight miles from Mynpooree, and that he had volunteered to ride over to him with despatches, asking me at the same time if I would accompany him. Of course, I consented at once, and was very much gratified by his selecting me as his companion. At 6 A.M. the next morning we started with seventy-five sowars of our own regiment. I do not wish to enhance the danger of the undertaking, but shall merely tell you that since Brigadier Grant's column moved down this road towards Lucknow it had been closed against all Europeans, that we were not certain if the Commander-in-Chief's camp was at Goorsahaigunge (which uncertainty was verified as you will see), and that to say the least of it there was a chance of our falling in with roving bands of the enemy.

"We started at 6 A.M., and reached Bewar all safe, fourteen miles from our camp. Here we halted and ate sandwiches, and then leaving fifty men to stay till our return, pushed on to Chibberamow, fourteen miles farther on. Here we made another halt, and then, leaving the remaining twenty-five men behind, we pushed on by ourselves unaccompanied for Goorsahaigunge, where we hoped to find

[1] 'Hodson of Hodson's Horse,' pp. 261 *et seq.*

the Commander-in-Chief. On arriving there (a fourteen miles' stage) we found the Commander-in-Chief was at Meerun-ke-Serai, fifteen miles farther on. This was very annoying, but there was no help for it, so we struck out for it as fast as we could, the more so as we heard that the enemy, 700 strong with four guns, was within two miles of us. We arrived at Meerun-ke-Serai at 4 P.M., and found the camp there all right. We were received most cordially by all, and not a little surprised were they to hear where we had come from. Hodson was most warmly received by Sir Colin Campbell, and was closeted with him till dinner-time. . . . We had a very pleasant dinner, and at 8 P.M. started on our long ride (fifty-four miles) back. We arrived at Goorsahaigunge all safe, and pushed on at once for the next stage, Chibberamow. When we had got half-way we were stopped by a native who had been waiting in expectation of our return. . . . He told us that a party of the enemy had attacked our twenty-five sowars at Chibberamow, cut up some, and beaten back the rest, and that there was a great probability some of them (the enemy) were lurking about the road to our front. . . . We deliberated what we should do, and Hodson decided we should ride on at all risks. 'At the worst,' he said, 'we can gallop back; but we'll try and push through.' The native came with us, and we started. . . . Taking our horses off the hard road on to the side where it was soft, so that the noise of their footfalls could be less distinctly heard, we went silently on our way, anxiously listening for every sound that fell upon our ears and straining our sight to see if, behind the dark trees dotted along the road, we could discern the forms of the enemy waiting in ambush to seize us. It was indeed an anxious time. We proceeded till close to Chibberamow. 'They are there,' said our guide in a whisper, pointing to a garden in a clump of trees to our right front. Distinctly we heard a faint hum in the distance—whether it was the enemy or whether our imagination conjured up the sound I know not. We slowly and silently passed through the village, in the main street of which we saw the dead body of one of our men lying stiff and stark and ghastly in the moonlight; and on emerging from the other side dismissed our faithful guide with directions to come to our camp; and then, putting spurs to our horses, we galloped for our dear life to Bewar, breathing more freely as every stride bore us away from the danger now past. We reached Bewar at about two o'clock A.M., and found a party of our men sent out to look for us. Our troopers had ridden in to say they had been attacked and driven back, and that we had gone on alone, and all concluded we must fall into the hands of the enemy. We flung ourselves down on charpoys, and slept till daylight, when our columns marched in. . . .

"It appears from the reports afterwards received that the party that cut up our men were fugitives from Etawah, where a column of ours under General Walpole had arrived. They consisted of about 1500 men with seven guns, and were proceeding to Futteypore. We rode in at one end of Chibberamow in the morning—they rode in at the other. They saw us, but we did not see them, as we were on unfavourable ground. Thinking that we were the advanced guard of our column they retired hastily to a village two miles off. Meanwhile Hodson and I, unconscious of their vicinity, rode on. They sent out scouts and ascertained that only twenty-five of our sowars were in the village, upon which they resumed their march, sending a party to cut up our men and, I suppose, to wait for our return. . . . We rode ninety-four miles. Hodson rode seventy-two on one horse, the little dun, and I rode Alma seventy-two also."

On the 4th January, Brigadier Seaton, who had

been joined on the 3rd by another column under Brigadier Walpole, brought his convoy safely to the Commander-in-Chief's camp at Fatehgarh, and here Lieutenant H. Gough rejoined the headquarters of the regiment and Hodson's Horse was once more united.

CHAPTER V.

THE OPERATIONS IN OUDH AND THE DEATH OF HODSON.

THE next three weeks were marked by no conflict with the enemy, though detachments of the British force were constantly on the move. On the 6th January, Hodson, with part of his regiment, accompanied a brigade under Colonel Adrian Hope, which marched through Shamshabad and Kaimganj, returning to Fatehgarh on the 12th. On the 13th a detachment under Lieutenant H. Gough formed part of a cavalry force sent to Miran-ki-Sarai to watch the roads and fords of the Ganges which approached Oudh from the north-west, an uneventful duty from which the detachment did not return till the 28th January. Again on the 15th the headquarters of the regiment under Hodson marched with Brigadier Walpole's column to Allahganj on the Ramganga to make a demonstration as if against Bareilly. Here the column remained inactive for some days and while here an incident occurred which might have had serious consequences, and which is thus described by Sir Charles Gough, then a lieutenant in Hodson's Horse :—

"Hodson was ordered to pitch his camp on a spot which was about 800 yards from a position occupied by the enemy on the other bank of the Ramganga. Hodson pointed this out to the D.A.Q.M.G., but he took no heed of it, and so the camp was pitched. The enemy left us alone the first day, but on the next Hodson ordered the regiment to parade for inspection in the centre street of the camp. The opportunity was too tempting for the enemy's gunners, and they opened fire. The very first shot struck immediately behind the line and passed through it, carrying off the leg of one of the men. Their other shots were not so good, but they dropped all about the camp. Of course, the camp was then moved."

On the 25th, Hodson was suddenly recalled to Fatehgarh in order to join another column under Adrian Hope, which was about to march against a force of some five thousand rebels at the village of Shamshabad. The column was composed as follows :—

9th Lancers. Two squadrons.
Hodson's Horse. 200 men (with four officers—namely, Captain Hodson, Lieut. McDowell, Lieut. C. J. S. Gough, and Lieut. Wise).
Bengal Horse Artillery. One troop.
Bengal Field Battery. 4 guns.
42nd Royal Highlanders.
53rd Light Infantry.
4th Punjab Rifles.

"I marched from this camp," runs Brigadier Hope's despatch (G.G.O. 226 of 1858), "leaving the tents standing, on the 26th instant at 11 P.M. . . . and halting for three hours, short of the village of Kurshinabad, proceeded at daylight on the 27th towards Shumshabad. A thick fog compelled us to move cautiously, and it was 9 o'clock before the column closed up, under cover of the village of Shumshabad. The rebels in considerable force had taken post about three-quarters of a mile beyond that place. They occupied a commanding knoll on the edge of the plateau overlooking the plain, which stretches towards the river some six miles distant. On the knoll was a brick building, the shrine of a Mussulman saint, and the place was surrounded with the remains of an old entrenchment, upon which they had raised a sandbag battery. Their front was defended by a ravine, impassable for cavalry or guns, which runs at right angles across the road to Mhow, along which we moved, to the right of which their position was."

The action which followed may be described in the words of Sir Charles Gough :—

"Brigadier Hope moved off the road (to the right), and with his staff and accompanied by Hodson and McDowell rode up to a solitary tree close by, from whence he began to examine the enemy's position. Immediately their guns opened fire from the high ground, and the very first shot plumped into the middle of the group, striking poor McDowell just under the right knee, smashing his leg and passing through his horse. We saw his horse rear up and fall back, McDowell crying out, 'Doctor, doctor.' The battery of Horse Artillery, which had been moved up, immediately galloped forward (along the Mhow road), crossing the ravine in our front by a stone bridge, and took up a position on the other side of the ravine, coming into action at close quarters—about 700 yards—with the enemy's battery on the mound. Hodson immediately led his regiment over the bridge, and formed up on the left flank of the battery. The ground here was undulating; the guns were placed on the crest of a ridge, whilst we were slightly sheltered by the rising ground. The roar of the round shot and shell flying just over our heads was terrific. Almost immediately a body of the enemy's cavalry was seen coming over some high ground on our left. Hodson ordered the left squadron to wheel to the left, calling out to me to keep in support and remain with the guns, but immediately afterwards he shouted out, 'Come on, Gough.' I ordered the right squadron (I think the ressaldar was a stout old Sikh named Bal Singh) also to wheel to the left, and led on upon the enemy on our proper left. The officers with the regiment were Hodson, Wise, and myself. As we advanced at a gallop I saw another body when we got on the rising ground to our then right (proper front), who came down upon us; their leader carrying a carbine and challenging us. We turned our horses to the right to meet this. For my own part, being on the right I now got the lead, and rode straight at the enemy's leader, thinking to run him through the body before he could swing his carbine round on me. The point of my sword struck him full in the breast, but to my surprise the next moment I found myself unarmed, and looking round saw my sword sticking fast in the man's body, he still on his horse; but in the midst of the enemy there was no time to look after him. I drew my revolver and rode down to where I saw Hodson and Wise and some of our men hotly engaged with the enemy. I shot down one man attacking Wise, and the next moment felt a spear strike me on the left rear. Throwing my body round to defend myself, I saw Hodson's sword descend upon my assailant's head, and down he went, falling half over me. My throwing myself round like this had, it so happened, exposed me to some fellows I had had my eye on just before—three of them side by side, shouting 'Allah! Allah!' One of them seeing me turn rode up quite close and fired. I felt a tremendous bang on my right shoulder, and for a moment was not quite sure what had happened. For a few minutes it was a very sharp scrimmage, but then the survivors of the enemy turned. Hodson received a severe sabre cut on the fleshy part of the forearm, but made very light of it. The enemy was pursued and dispersed with considerable loss, and their guns captured."

Writing of the same mêlée, Hodson himself says:—

"They were very superior in numbers and individually so as horsemen and swordsmen, but we managed to 'whop' them all the same and drive them clear off the field; not, however, until they

GENERAL SIR CHARLES GOUGH, V.C., G.C.B.

As a young man, in the uniform of the 8th Bengal Light Cavalry.

had made two very pretty dashes at us, which cost us some trouble and very hard fighting. It was the hardest thing of the kind in which I was ever engaged. . . . I got a cut which laid open my thumb from a fellow after my sword was through him, and about half an hour later this caused me to get a second severe cut which divided the muscles of the right arm, and put me *hors de combat*, for my grip on my sword handle was weakened, and a demon on foot succeeded in striking down my guard or rather his *tulwar* glanced off my guard on to my arm." [1]

Of the progress of events in another part of the field the Brigadier wrote:—

"Captain Blunt brought his guns into action in an excellent position on the right bank of the ravine, and the line of infantry was formed on the right, consisting of the 4th Punjab Rifles under Captain Wylde, the 42nd Royal Highlanders under Lieut.-Colonel Cameron being on the extreme right, the 53rd Regiment under Major English being in second line. The enemy were unable to withstand the admirable fire of our guns, and I ordered the infantry, who had been screened in a hollow of the ground, to advance, and soon after seeing the camp nearly abandoned the 4th Punjab Rifles were directed to secure it, which they did in a very spirited manner, shooting down many of the retreating enemy."

Turning again to Sir Charles Gough's account of the affair, we read:—

"When we returned we found poor McDowell lying desperately wounded; he was so weak from the frightful shock, the shot having completely shattered his leg, that the doctors found he could not stand the necessary amputation, and he died in the course of the evening."

Well might Hodson write, "My usual fortune deserted me on the 27th at Shumshabad." The first fight of 1858 had opened the year with grievous

[1] 'Fraser's Magazine,' February 1859.

loss and ill-omen for Hodson's Horse. Short of losing their leader himself, no blow could have fallen so heavily on men and officers as the death of their second-in-command. Not only had he been with the corps ever since the first troops under Man Singh had reached the Delhi ridge from the Punjab, but he had also gained in no common degree the love and admiration of all who had served with him. "He was," says Sir Charles Gough, "a most delightful companion, always bright and gay, full of life, and with very considerable ability, a good writer and a brave soldier." And Hodson wrote: "What grieves me most is the loss of poor Mac: he was invaluable to me as a brilliant soldier, true friend, and thorough gentleman—I mourn him as for a brother." Lieutenant McDowell was twenty-eight years old at the time of his death.

The total losses of the regiment on the 27th January were Lieutenant McDowell and four men killed; Captain Hodson and one man severely wounded; Lieutenant C. J. S. Gough, Jemadar Jawala Singh, and nine men slightly wounded; and three men missing. Captain Hodson was specially mentioned by Brigadier Adrian Hope for having boldly met the enemy's attempt to outflank the British, and his local knowledge and information were stated to have been most valuable.

The next day, 28th January, Brigadier Hope's column returned to the headquarters camp at Fatehgarh. Hence after three more days of inactivity the Commander-in-Chief moved to Cawnpore, preparatory to an advance against Lucknow. Hodson was, however, incapacitated by his recent wound, and was left till 4th February at Fatehgarh, when he accompanied Brigadier Walpole's column to Cawnpore. Meanwhile the regiment was employed in detachments on various duties, escorting convoys, attached to flying columns and the like, but without being engaged with the enemy. In this manner the greater part of February passed. Sir Colin Campbell still delayed the final advance against the capital of Oudh, and the headquarters of Hodson's Horse remained part of the time at Cawnpore and afterwards at Onao.

During the latter part of this month reinforcements were gradually massed at the Alam Bagh, within striking distance of Lucknow, and at length the inaction of Hodson's corps was ended by an order for him to advance to the same rendezvous. He was by this time cured of his wound, or sufficiently so to take command of his men, and he marched from Onao on the morning of 24th February with the headquarters of his regiment. The march was a forced one of thirty-six miles and, as Hodson wrote next day, terribly dusty; but the toil was well repaid, for it enabled the regiment to be present on 25th February at the last and most determined effort by the rebels to overwhelm Outram's garrison at the Alam Bagh, which for more than three months had held at bay the whole of the main army of mutineers, varying in strength from 30,000 in November to 96,000 in February.

"The despairing attack of the rebels," writes Malleson, "was made with all the pomp and circumstance of war." The Begum of Oudh herself, attended by the Wazir, accompanied the army into the field, and after a violent cannonade against the Alam Bagh, directed an extended turning movement against the British right with a force of some 20,000 to 30,000 men, while a strong detachment was simultaneously sent to threaten Outram's left. The British commander was not slow in perceiving which was the main attack, and advancing from his right at about 10 A.M. with detachments of mounted troops and infantry numbering 771 sabres and 874 bayonets, he first succeeded in cutting off the advanced force of the enemy from their supports, and then drove the isolated detachments back one after the other in considerable confusion towards Lucknow. An effort was made later in the day to press the attack on Outram's left, but without success. In this action 374 of Hodson's Horse accompanied their leader into action, forming part of a mixed force of cavalry under Colonel Campbell of the Queen's Bays, which by a wide *détour* cut off the most advanced portion of the rebel army in the neighbourhood of Jalalabad fort. The share which Hodson's Horse had in the fortunes of the day is thus described by Sir Hugh Gough :—

"This was my first day in action with Hodson's Horse as a complete regiment, for when at the siege of Delhi the corps was in its infancy, and when I left Delhi with my wing it was certainly not weaned ; but now we were a full-blown regiment, men better equipped, clothed, and drilled, and the horses of a better stamp, and with decent saddlery and accoutrements. . . .

"No time was to be lost, as the enemy had already heard of the reinforcements which had come during the night, and were in full retreat to Lucknow. Our camp was not far from the Alum Bagh, and our route to meet the enemy lay by the village of Jellalabad, passing close by the scene of my previous encounter with them at the relief of Lucknow ; in fact we passed so close that I was able, *en passant*, to give Hodson a hurried description of the fight. When we now came in view of the enemy, they were passing in rather a disorganised mass right across our front as we advanced. We could see they had a couple of field-guns, one gun being about 600 yards ahead of the other. The main body was almost entirely infantry, and all were mutineers, arrayed in uniform. Our rapid approach had a great effect upon them ; they seemed to make no effort to rally and stand, and, as we advanced and charged, we got well into them, and the whole affair seemed over. The rearmost gun was in our possession, and the enemy, as far as we had encountered them, in full flight ; but somehow, owing to the ardour of the charge and pursuit, our regiment got quite out of hand, lost all formation, and scattered ; and they, seeing our condition, and probably having a leader with a good cool head, rallied round their remaining gun, regained their formation as we lost ours, and, pouring in volleys of musketry with discharges of grape from their gun, rendered our confusion worse confounded. Our men, gallant and forward in pursuit or a charge, could not stand being hammered at a disadvantage. There was a din of shouting and noise, officers doing their best to bring the men up, but all to no effect ; and it looked sadly probable that Hodson's Horse would in their turn retreat. Hodson at this crisis managed to get a few brave spirits together—not more than a dozen. Well I remember him, with his arm in a sling from his wound at Shumshabad, shouting to the men to follow him as he made an attempt to charge. He and I were riding close together, and, as we advanced with our small following, I saw his horse come down with him, and the next instant my own charger, my beloved 'Tearaway,' reared straight up and fell dead. The fire was most deadly. The range was short, and just suited to the point-blank fire from the smooth-bore musket under which we were exposed, so that nearly every one of our small party was killed or wounded. Fortunately I fell clear of my horse, and catching a sowar's whose rider had just been killed I speedily mounted and, as good luck would have it, was able to rally our men to a certain extent, who seeing our supports coming up (7th Hussars and Military Train), now came on with a will, and charging the remaining gun, scattered the enemy in all directions. My temporary charger—a small, grey, country-bred mare—carried me well, and we followed the enemy in pursuit, the British cavalry also cutting in. It was no easy matter, as they (the enemy) had got amongst trees and low jungle and were guarded by a village, where cavalry were not of much use. In the ardour of pursuit I had got ahead of my men, when I came upon a couple of sepoys on their way to the village. They had their bayonets fixed, and seeing me unsupported, stood—one in my direct front and the other on my right. I made for the former, but the one on the right took aim at me as I passed, and shot me clean through the thigh, the bullet going through my saddle and my horse, killing her dead. Fortunately I fell clear, though helpless. My opponent was just coming up to finish me off, when he was sabred by a trooper of the Military Train.

"The affair was now over. The enemy suffered severely and were driven back into Lucknow, and we returned to camp, and I was much pleased to think that our men had retrieved their previous discomfiture. Their temporary 'funk' was really due to their having got out of hand after their first charge, and not having time to rally before they had again to face the enemy's heavy musketry fire. The steadiest cavalry in the world might have found it difficult, and to an absolutely newly raised regiment the position was a very trying one."

The casualties of the regiment on the 25th February were three men and five horses killed; one officer, six men, and twenty-three horses wounded.

From that date until the 4th March, Hodson's Horse remained in camp at the Alam Bagh. But meanwhile a detachment of the corps under Captain C. J. S. Gough had taken part in an affair of some importance with a column commanded by Brigadier-General Sir J. Hope Grant. The column in question was sent out from Cawnpore during the latter half of February to clear the country on the left bank of the Ganges and to protect the road to Lucknow. Its strength was as follows :—

Artillery. E Troop Royal Horse Artillery and detachments of Bengal Artillery . . .	326 men
Cavalry. Detachments of 7th Hussars, 9th Lancers, 1st Punjab Cavalry, and Hodson's Horse	636 ,,

Infantry. 34th, 38th, and 53rd Foot, and a detachment of Sappers . . . . 2284 men

Captain Gough was attached to the force, with one squadron of Hodson's Horse. On the 23rd February the column arrived before an old fortified town named Mianganj, which was held in strength by the rebels. Sir Hope Grant disposed his guns to breach the wall, and prepared to storm the place with the infantry as soon as the breach was practicable; meanwhile he placed the cavalry detachments at different points to watch the various exits. The enemy did not wait to meet the stormers, but, as had been expected, endeavoured to escape into the open country. Here they were met by the cavalry, and a smart melée ensued. A large body of rebels streamed out of the gate, which was watched by Gough. The latter, having been joined by Captain Anson's squadron of the 9th Lancers, advanced to the attack. "The enemy opened a heavy but desultory fire on us as we came down, and when we got to close quarters we charged home and got in among a lot of fellows, who all fought resolutely for their lives. We cut up a good number, with some casualties on our side. Hope Grant brought the rest of the cavalry to follow up, and the enemy dispersed among the *dhal* fields and woods, where they could find shelter. One large *dhal* field was pretty full of them armed with muskets and bayonets. Hope Grant called on me to clear it out. I confess I thought it a not very pleasant task, seeing that the *dhal* was so high and thick that we could not possibly charge through it. However, it had to be done, so I formed the men up in line, and, placing myself in front of the centre, we pushed our way at a walk into the field. I had one of the native officers riding near me, when suddenly bang! went a musket right under my horse's head without hitting me, and up sprang a Pandy with his bayonet fixed. I succeeded in reaching out and giving him a cut with my sword, when he began to load again. I then drew my revolver, whilst he brought his bayonet to the charge and came at me, but I managed to drop him with a shot in his chest. There was a deal of peppering in the field, but eventually we cleared it all out. . . . We lost old Naib-Ressaldar Hukm Singh, who was shot through the body, and two or three other men." (From an account by General Sir Charles Gough.) The loss of the detachment was Naib-Risaldar Hukm Singh of Sardar Ata-ullah Khan's troop killed, and three men wounded.

From Mianganj, Sir Hope Grant's column rejoined the army at Alam Bagh, and Captain Charles Gough and his squadron returned to regimental headquarters.

For their gallant conduct at Shumshabad and Mianganj respectively (as well as on several previous occasions) the brothers Hugh and Charles Gough were subsequently awarded the Victoria Cross.

On the 2nd March, Sir Colin Campbell's operations against Lucknow began in earnest, but, as has been already mentioned, Hodson's Horse were not at first actively employed. On the 4th the regiment was moved up to the Commander-in-Chief's camp at Dilkusha, only to return to the Alam Bagh the next day. On the 6th, Hodson was placed in charge of the communications between the headquarters camp and Alam Bagh and Jalalabad, and the regiment accordingly again moved to a point midway between the two first-named camps. Meanwhile Hodson himself was still suffering from the effects of the wounds received at Shamshabad, and his adjutant, Hugh Gough, was *hors de combat* by reason of the severe wound received on the 25th February.

"On the 11th March," writes Sir Charles Gough, "orders came for one wing of the regiment to proceed to Alambagh, and the other wing to remain to watch the line. Hodson arranged to take the wing to Alambagh, and told me to remain in command of the wing where we were, and he drove away in a light dogcart to army headquarters, telling me he would be back in time to march the wing away. However, time went on, and it was getting late and he had not returned. I accordingly ordered the movement to take place." The events of that fatal afternoon can be followed in detail from the various accounts of eye-witnesses. Sir Hugh Gough writes in his 'Old Memories':—

"Hodson, who was also unable to ride, was one day driving into Lucknow, and asked me to accompany him. I tried, but was unable to get up into the dogcart, and so he started by himself. He nodded a cheerful 'Good-bye' to me as he drove off with his orderly Nihal Singh. Little did I think at the moment I should never see him again! Had I been able to accompany him it is possible events might have turned out otherwise; he would probably have remained to look after me, and thus avoided his fate." Hodson therefore drove off to the headquarters camp alone. General George Hutchinson, at that time Brigade-Major of Engineers with Sir Colin Campbell's force, has related how "Hodson came to my tent in Lord Clyde's camp about 12.30 or 1 P.M. and asked, 'Where is Napier?' I replied, 'He is in the city, and will probably be taking the Begum Kotee this afternoon,' and added, 'Will you stay and lunch with me and go into the city afterwards . . . ?' He had only one sowar with him, and was dressed in his usual military costume. He did stay to lunch, and I well remember the deeply interesting account I led him to give of his hand-to-hand conflicts. . . . He left me after luncheon to see Napier." Brigadier-General Robert Napier (afterwards Lord Napier of Magdala) then takes up the story. "I was reconnoitring the breach," he wrote in a letter to Mrs Hodson, "whilst the guns were making it practicable, and waiting for the moment when I could send word for the troops to advance, when your husband suddenly stood beside me, and said laughingly: 'I am come to take care of you.' The signal was given for the troops to advance, and we watched their progress and entry into the building. All serious opposition ceased, and we followed through the breach into the Palace. None of the enemy remained except a few parties shut up in houses, whom our troops were despatching. Your poor husband, Captain Taylor, and I were together then. I got separated from them in the crowd, and proceeded to push on our advantage. When I returned General Lugard told me that both your husband and Taylor were wounded, and that he was earnestly asking for me. I went immediately to Banks's House, and found him in a dooly, not suffering much pain. His wound had been dressed, and he had all necessary attendance. I was obliged to leave him to go to the Commander-in-Chief, but then returned and remained with him until the approach of morning obliged me to return to my duties."

The end has been told by Dr T. Anderson, the surgeon with the regiment. After describing how an urgent message from Hodson reached him at the Alam Bagh, whither he had proceeded with one wing of the regiment, and how after some delay he at length got to headquarters and found that Hodson was lying wounded at Banks's House, he goes on: "I found him in a dooly and Doctor Sutherland with him, whom I at once relieved, and learnt the following particulars from him and from the orderly who remained with Hodson, and who had been by his side when hit. He had arrived at Banks's House just as the party going to attack the Begum's Palace was starting, and fell in with them. The place had been taken before he was wounded. When the soldiers were searching for concealed sepoys in the courtyard and buildings adjoining, he said to his orderly, 'I wonder if any of the rascals are in there?' He turned the angle of the passage, looked into a dark room which was full of sepoys; a shot was fired from inside. He staggered back some paces, and then fell. A party of Highlanders, hearing who had been hit, rushed into the room and bayoneted every man there. The orderly, Nihal Singh, a large powerful Sikh, carried him in his arms out of danger, and got a dooly and brought him back to Banks's House, where his wound was looked to and dressed.

"He was shot through the right side of the chest in the region of the liver, the ball entering in front and going out behind. There had been profuse bleeding, and I saw that the wound was most likely mortal.

"He was very glad to see me, and began talking of his wound, which he thought himself was mortal. I lay beside him all night, holding his hand on account of the great pain he suffered. . . . About 9 A.M. I had the dooly lifted into a room which I had had cleared out, where he was much quieter. At 10 A.M., however, bleeding came on again profusely, and he rapidly became worse. I told him recovery was impossible. He then sent for General Napier, to whom he gave directions about his property and messages to his wife. After this he rapidly sank, though he remained sensible, and was able to speak till a quarter past one, when he became too weak, and in ten minutes more the sad scene was over. He died most quietly without a struggle. He merely ceased to breathe."

He was buried on the evening of the 12th of March in the grounds of La Martinière School. The Commander-in-Chief and the whole of his staff were present at the ceremony, the service being read by the Rev. Dr Smith. On the monument that marks his grave are inscribed the fitting words: "Here lies all that could die of William Stephen Raikes Hodson."

Hodson's death was something more than one of the daily casualties of war, regrettable but inevitable.

To his own regiment it was a calamity immeasurable and irreparable, the loss of an inspiration. His men adored him, and none could take his place. To his British officers (as Sir Charles Gough wrote in after years) he was endeared by "his gallantry in leading, his astounding energy, activity and resource in difficulties, his coolness in danger, and his genial, cheerful, and kind disposition." Nor were such feelings confined to those immediately in contact with him. To every soldier in the army he was as a hero of romance, the very personification of daring and bravery. How great a loss his death meant to the public service is perhaps best indicated by the three following quotations, each a striking testimony to his brilliant qualities. Of these the first is the letter of condolence which Sir Colin Campbell sent to his widow:—

"MARTINIÈRE, *March* 13, 1858.

"MADAM,—It is with a sentiment of profound regret that I am compelled to address you for the purpose of communicating the sad news that your gallant and distinguished husband, Major Hodson,[1] received a mortal wound from a bullet on the 11th instant. He unfortunately accompanied his friend, Brigadier Napier, commanding Engineers, in the successful attack on the Begum's Palace. The whole army, which admired his talents, his bravery, and his military skill, deplores his loss, and sympathises with you in your irreparable bereavement. I attended your husband's funeral yesterday evening in order to show what respect I could to the memory of one of the most brilliant officers under my command.

C. CAMPBELL,
Com.-in-Chief in the East Indies."

The next quotation is in the words of Sir Robert Montgomery, Judicial Commissioner and afterwards

[1] Two days before Hodson's death the mail from England brought the announcement of his promotion to a brevet majority.

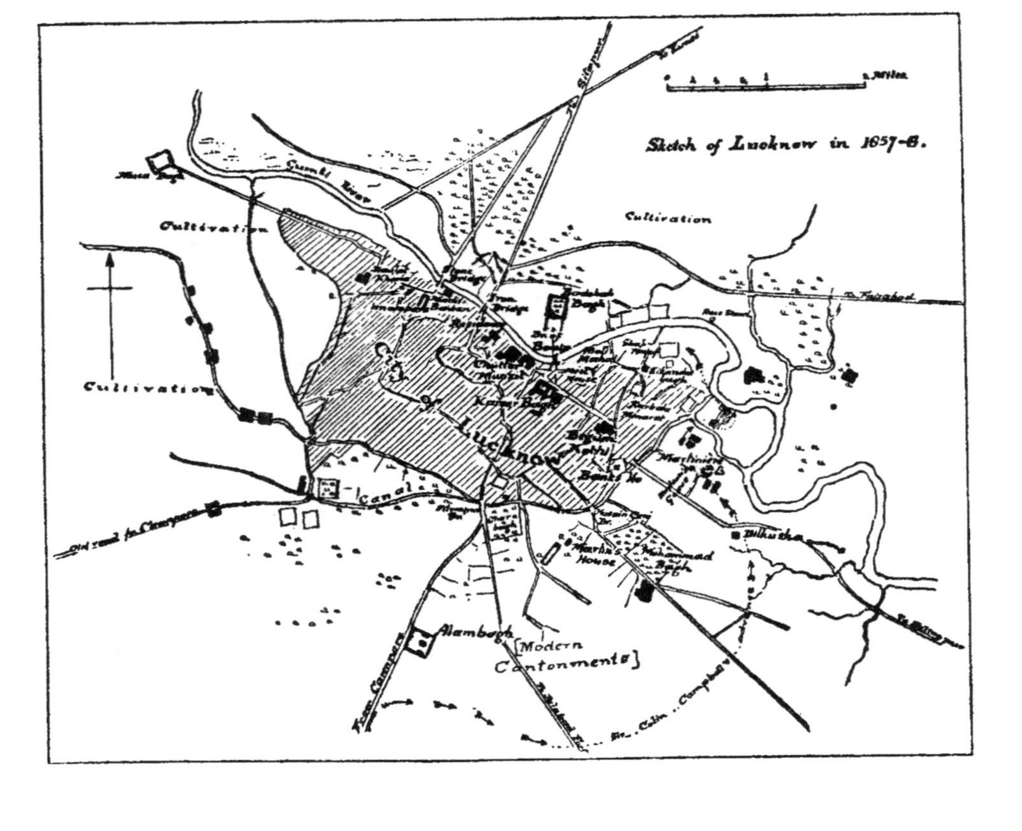
Sketch of Lucknow in 1857-8.
Cultivation
Cultivation
Cultivation
Gumti River
To Fyzabad
City of Lucknow
Canal
Kaisar Bagh
Begum Kothi
Banks Ho
Martinière
Dilkusha
Martin's House
Muhammad Bagh
Alambagh
[Modern Cantonments]
From Cawnpore
Old road to Cawnpore
Sir Colin Campbell

Lieutenant-Governor of the Punjab, who had been intimately acquainted with Hodson and his work for some years, and who (as described in the first chapter of this book) had been largely instrumental in raising his corps of horse :—

"I look round and can find no one like him. Many men are as brave, many possess as much talent, many are as cool and accurate in judgment, but not one combines all these qualifications as he did."

Last of all comes the tribute paid by Lord Stanley in the House of Commons on the occasion of the vote of thanks to the army in India on the 14th April 1859 :—

"There are two names which are specially distinguished. The first is that of Major Hodson of the Guides, who in his short but brilliant military career displayed every quality which an officer should possess. Nothing is more remarkable in glancing over the biography of Major Hodson that has just appeared than the variety of services in which he was engaged. At one time he displayed his great personal courage and skill as a swordsman in conflict with Sikh fanatics; he was then transferred to the civil service, in which he performed his duties as though he had passed his whole life at the desk; afterwards recruiting and commanding the Corps of Guides; and lastly taking part in the operations before Delhi, volunteering for every enterprise in which life could be hazarded or glory be won. He crowded into the brief space of eleven eventful years the services and adventures of a long life. He died when his reward was assured, obtaining only that reward which he most coveted—the consciousness of duty done and the assurance of enduring military renown."[1]

[1] See Appendix I.

HODSON'S GRAVE AT LUCKNOW.

From a drawing by Captain C. F. L. Stevens, M.C., Hodson's Horse.

CHAPTER VI.

THE END OF THE MUTINY CAMPAIGN.

On 13th March, the day after Major Hodson's death, Major Henry Daly, C.B., formerly of the Guides, was appointed to command the regiment. The post was one of considerable difficulty. In the first place it was no easy task to succeed a leader possessed of such remarkable qualities as Hodson, nor was the task lightened by the circumstances surrounding the formation and previous service of the regiment. "Never before," wrote Daly, "was a corps raised as this has been raised. Troop by troop, detachment by detachment, the Punjab supplied them, and down they marched to Delhi." At the date of Hodson's death the regiment numbered about 750 sabres at Lucknow, while some 400 more, mostly without horses, were at Meerut, and four troops under Man Singh arrived from the Punjab on 13th March. Of administrative machinery for this considerable force there was almost nothing.

"No English muster-roll[1] or pay abstract had been framed. No English paper of any sort—record, receipt, account, or statement—was forthcoming. The only English paper known was the registry of a few names in Hugh Gough's handwriting, which is now, as then, in the muster-roll book.

"The enrolment of men for the corps had begun in June 1857. Rs. 119,557 . 10 . 4 had been taken up,

[1] Memorandum left by Daly for the information of his successor in command.

and all save the 16,000 [in the treasure chest] had been disbursed. It was alarming to know that for the account of this one was entirely in the hands of the munshis. The times were pressing; work and movement the lot of all, and in this vocation no man had been more active or more zealous than poor Hodson. He possessed a rarely clear head and keen memory, and on these and the munshis was his sole reliance. No European officer knew anything of the affairs of the regiment. Everything was done by himself."

It is not surprising that Daly regarded with some misgiving the responsibility involved in assuming a command in such conditions. His difficulties were increased by the loss of the two officers, Captain Charles Gough and Lieutenant Hugh Gough, who had been longest with the corps. The latter was sent away on sick leave owing to his serious wound and the former was appointed to another regiment. Major Daly thereupon appointed Lieutenant C. H. Mecham second-in-command and gave him charge of the treasure chest, with instructions to obtain from the two regimental munshis without delay a statement of the expenditure up to that date. At the same time he made an urgent application for the appointment of an officer for the special purpose of setting the accounts in order. At the moment no officer was available, but subsequently Lieutenant R. B. Anderson, 1st Bombay Fusiliers, was appointed for this duty. The work proved difficult and prolonged, but it was completed early in 1859 to the entire satisfaction of the authorities, thanks to the ability and energy of Lieutenant Anderson and, it should be added, to the zeal and honesty of the principal munshi, Dumichand, to which Major Daly in his memorandum quoted above bears grateful testimony.

But the task of improving the organisation and administration of Hodson's Horse was not confined to the accounts of the corps. At the close of the

Lucknow operations there were, as has been mentioned, 400 men on the rolls of the regiment at Meerut without horses, a fact which is not surprising in view of the numbers of recruits sent from the Punjab to the Meerut depot. Lieutenant L. F. Wells, who was the only British officer with the depot from the end of 1857 until August 1858, reported on 3rd August that he had received nearly 700 recruits during that period, of whom he had sent on some 400 to Lucknow after giving them such initial training as was within his power. Meanwhile with the headquarters of the regiment when Daly assumed command no less than 100 more were dismounted, and Daly writes that upwards of 1000 of those with horses had no saddles! He adds the following curious description of the personnel attached to the corps: "Plunder and the tales of golden floods had enticed many of the relatives and friends of the sowars from the Punjab; the lines of the regiment were full of these amateurs; they wore the uniform, and have sometimes, in the absence of the sowars, actually attended parades and taken duty. In a skirmish I was at first surprised to see the great array at the commencement. Their occupation, however, quickly thinned the gathering. I had much difficulty in breaking through this combination."

All these irregularities and the unwieldy size to which the corps had been swelled by the arrival of reinforcements from the Punjab led Major Daly very soon after assuming command to recommend that it should be organised in two regiments, a proposal which was officially approved in July. Finding, however, that, even after weeding out all unsuitable men, there still remained on the rolls a considerable number in excess of the proper strength of two regiments, as well as sufficient troop officers for no less than sixteen troops, Daly submitted a further recommendation that three regiments instead of two should be formed, their strength being completed by the enrolment of a Pathan squadron for each regiment. This proposal was also approved and Daly sent an officer to the Punjab frontier to superintend the enlistment of Pathans, at the same time applying to Major H. B. Lumsden for assistance in the project. But on the circumstances being made known to Sir John Lawrence he protested that the number of men drawn from the Punjab and its borders for service in the British forces was already too large, and he urged that no further enlistments should be permitted. This view was accepted by the Government of India, with the result that only about a hundred Pathans were enlisted, and these were mostly drafted to the 3rd Regiment, Hodson's Horse.

The division of the corps into three regiments was formally approved in the following General Order by His Excellency the Commander-in-Chief, dated Allahabad, the 26th August 1858 :—

"With the sanction of the Right Honourable the Governor-General, the Commander-in-Chief directs the formation of the Corps of Irregular Cavalry (Hodson's), commanded by Major H. D. Daly, into 3 Regiments of 6 Troops each, of the usual strength.

"The Governor-General has further been pleased to decide that such of the Ressaldars of the aforesaid regiment as obtained that rank for enrolling Ressalahs, shall retain their rank, although they may be in excess of the established proportion of that grade allowed to a regiment of irregular cavalry.

"2. The Commander-in-Chief is pleased to make the following appointments to these regiments :—

"*1st Regiment of Hodson's Horse.*

Brevet-Major C. H. Barchard of the 25th Regiment Native Infantry to be second-in-command.

Lieutenant S. G. Warde of the 11th Regiment Native Infantry to be Adjutant.

2nd Regiment of Hodson's Horse.

Brevet-Major H. A. Sarel, Her Majesty's 17th Light Dragoons, to be second-in-command."

A fortnight later appeared the following order (dated 9th September 1858) :—

" *3rd Regiment of Hodson's Horse.*

Brevet-Major Sir H. M. Havelock, Bart., of Her Majesty's 18th Foot to be second-in-command."

We must now return to the active services of Hodson's Horse subsequent to the lamented death of their founder. In the week which passed between that event and the final occupation of Lucknow the regiment was not employed in the actual attack on the city, but it formed part of a cavalry brigade under Brigadier-General Campbell which made an attempt to cut off the retreating rebel forces. The operations, however, were muddled and abortive, and the brigade returned to Lucknow without having achieved much success.

Immediately after the final capture of Lucknow the following force was detailed to garrison that place, with Major-General Sir Hope Grant in command :—

Horse Artillery. Two troops.
Field Artillery. Two battalions.
Garrison Artillery. Four batteries.
Engineers. One company.
Pioneers. Three companies.
Cavalry. 2nd Dragoon Guards, 1st Sikh Cavalry, Hodson's Horse, Lahore Light Horse.
Infantry. 20th and 23rd Fusiliers, 38th and 53rd Foot, 90th Light Infantry, 97th Foot, 1st Madras Fusiliers, 27th Madras Native Infantry.

On the 11th April a portion of this force, including a squadron of Hodson's Horse under Lieutenant

Lawford, and strengthened by the 7th Hussars and the 2nd Battalion, Rifle Brigade, marched under General Hope Grant's personal command to attack a body of rebels under the notorious Maulvi of Faizabad at the village of Bari. The column met with little or no opposition, and having passed through Bari to Muhammadabad and so to Ramnagar and Bitauli, all of which places were evacuated by the rebels as the British approached, Sir Hope Grant led his force back on to the Cawnpore and Lucknow road, where he had one or two unimportant skirmishes in the early part of May. On the 24th April, Grant submitted a despatch describing these operations, in which Lieutenant Lawford was favourably mentioned. After a fortnight or so of inaction the column again moved against the bands of rebel troops, who continued to hinder the pacification of Oudh and to threaten the safety of the Grand Trunk road. At length, on Sunday, the 13th June, a considerable force of the enemy, reckoned at 15,000 in number, was brought to bay in a very strong position at Nawabganj, eighteen miles from Lucknow on the Faizabad road. Hope Grant had left his baggage at Chinhat under a separate column and reached Nawabganj by a night march with the following force :—

Horse Artillery. One troop.
Field Artillery. Two light batteries.
Cavalry. 2nd Dragoon Guards (two squadrons), 7th Hussars, Hodson's Horse, 1st Sikh Cavalry (one squadron), Mounted Police (one troop).
Infantry. 2nd and 3rd Battalions Rifle Brigade, 5th Punjab Rifles, Engineers and Sappers (detachments).

The rebel position lay along a nullah which crossed the Lucknow road at right angles about four miles from Nawabganj. The road traversed the nullah by an old stone bridge, but some two miles higher up was a ford which led on to the right of the enemy's line, and it was against this point that Sir Hope

Grant directed his attack. The difficult march across country for twelve miles from Chinhat was accomplished successfully, and the advanced guard arrived within a quarter of a mile of the nullah about half an hour before daybreak on the morning of the 13th June. After a short rest the passage of the nullah [1] was forced as soon as it was light, the enemy being completely surprised. They did not, however, yield without a gallant struggle. Whilst Sir Hope Grant was directing the attack against the main position, a large body of the enemy moved round the British right, expecting to find a baggage convoy to be attacked. They were met by Hodson's Horse, a squadron of Police Horse and the 2nd Rifle Brigade, afterwards supported by Major Carleton's battery of artillery, and were held in check despite the most desperate efforts to break the resistance of the British force. Their attack having been successfully repulsed, Lieutenant Mecham and Lieutenant the Honourable J. Fraser were detached with a hundred of Hodson's Horse to attack the enemy's left, while Major Daly charged them in front. In this manœuvre Lieutenant Mecham was severely wounded while gallantly leading his men, but the enemy were forced to retire, their retreat being hastened by the fire of Major Carleton's guns.

The attack against the enemy's main position had meanwhile been successfully pushed home under the Major-General's personal direction. But a final effort was made by the rebels before abandoning the field. A very large body of Ghazis with two guns advanced against Hodson's Horse and a section of Major Carleton's battery on the British right. The onslaught of the fanatics, led by standard-bearers, was not to be checked by the small body of troops before them; but Sir Hope Grant, seeing

[1] Malleson, whose account is full of inaccuracies, says that Hope Grant's attack was across the bridge, which would have been a frontal attack and quite opposed to the General's scheme. The account in the text is taken from the despatch.

the danger in time, brought up the rest of the battery and the 7th Hussars, and at length the enemy finally retreated, having lost nine guns and six hundred men killed.

Sir Hope Grant, in his despatch describing the action, mentioned the following officers: "Major Daly, to whom I am greatly indebted for his excellent conduct in the field, and for the good information he brought me." . . . "Lieutenant and Adjutant Baker, Hodson's Horse, is particularly mentioned by Major Daly for his gallantry." . . . "Lieutenant Mecham, Hodson's Horse, who was severely wounded in leading his squadron to the charge." . . . "I would now report the good and gallant conduct of Ressaldar Man Singh and Jemadar Hussain Ali, both of Hodson's Horse; the former came to the assistance of Lieutenant Baker, and was severely wounded; the latter dismounted and, sword in hand, cut up some gunners who remained with their guns."

The losses of the regiment in this action amounted to three men killed, two British and two Indian officers and nineteen non-commissioned officers and men wounded. There were also eleven horses killed and twenty wounded.

Hodson's Horse continued to form part of Sir Hope Grant's force during the rest of the Oudh campaign, but for three months or so after the battle of Nawabganj it had no serious engagement with the enemy. A detachment accompanied Brigadier Horsford's column against Sultanpur (afterwards reinforced by Hope Grant in person), and was present at the passage of the Gumti and the occupation of Sultanpur on the 25th August; but the rebel force dispersed without offering any considerable resistance, and the only casualty in the detachment was the loss of one man and four horses drowned in crossing the river. Major Daly was mentioned in Sir Hope Grant's despatch (G.O.C.C., dated 9th November 1858), and the General also brought "particularly to notice the great assistance rendered by the Punjaub Rifles and Major Daly's corps in swimming across the artillery and 7th Hussars' horses." [1]

The date of the last-mentioned affair brings us to the time when, as already described, Hodson's Horse was, on Major Daly's advice, divided into three separate regiments, with Daly himself in command of the whole. At this point therefore the narrative common to both the 1st and the 2nd regiments, afterwards the Ninth and Tenth Bengal Cavalry, properly speaking comes to an end. But during the guerilla warfare which continued in Oudh for some months longer detachments from the three several units of the corps were so intermingled that it will be best to complete the story of these operations before proceeding to deal with the separate records of the 1st and 2nd Regiments.

To return therefore to the movements of Sir Hope Grant's force in the autumn of 1858, we find a detachment of seventy sabres from the 2nd Hodson's Horse, under Lieutenant C. M. MacGregor, taking part in an action near Dariabad on 18th September, when one sowar was killed and eighteen of all ranks were wounded. Major Hume, commanding the column, mentioned in terms of high praise the gallantry of Lieutenant MacGregor, who was himself severely wounded, as well as the conduct of Ressaidar Mirza Ahmad Beg, 4th Troop, who "behaved most gallantly, and led his men well after Lieutenant MacGregor was wounded" (G.O.C.C., 19th December 1858). The native officer thus referred to was promoted to risaldar for his fine behaviour on this occasion.

[1] From a present state of Hodson's Horse in Sir Hope Grant's Field Force on 30th August 1858 we find that there were with the main force 4 British and 12 Indian officers, 288 non-commissioned officers and men, and 299 horses, as well as 12 men and 9 horses on the sick list. With the Faizabad brigade were one British and 7 Indian officers, 98 non-commissioned officers and men, and 111 horses, and 9 men and 13 horses on the sick list.

On 20th October seventy-six sabres of the 1st Regiment, under Lieutenant C. H. Palliser, formed part of a column under Brigadier-General Horsford which attacked a considerable body of rebel sepoys known as the Nasirabad Brigade, upwards of 4000 strong with six guns, at Daudpur on the Sultanpur-Lucknow road. The rebels did not await the attack of the British, but evacuated their position in confusion as soon as they found their flanks threatened by General Horsford's cavalry. They were vigorously pursued by the 7th Hussars, Hodson's Horse, and two Royal Horse Artillery guns, and lost very heavily, while the casualties in the British force were only seven wounded, among whom were a native officer and one man of Hodson's Horse. Lieutenant Palliser was favourably mentioned by Brigadier Horsford in the latter's report on the operations (G.G.O., 1858, No. 513, p. 1671).

Again on 27th October the headquarters of the corps (which had now been rejoined by Lieutenant Palliser and his detachment) were present with Sir Hope Grant's column in an action at Dohlpur. Here the rebels had occupied a strong position on broken ground near a small tributary of the Gumti, named the Khandu. The enemy made but small show of resistance, but during the difficult work of clearing the jungle-clad ravines and nullahs of lurking fugitives, Lieutenant Palliser was shot by a sepoy concealed in the thicket, and would probably have been killed but for the timely intervention of a Pathan sowar who accompanied him, and who, springing from his horse, cut the sepoy down. Jemadar Man Singh, an officer of Gurkha birth, was also among the wounded on this day.

During October and November detachments of Hodson's Horse took part in various other skirmishes and small engagements, as for instance near Nawabganj on 10th October, at Jabrauli, south of Lucknow (25th October), where Lieutenant Mitford was severely wounded and earned a special mention in despatches by his gallantry, at the fort of Kuali, near Mahora (23rd November), at Jagdispur (30th November) and elsewhere.

Meanwhile the combined movements of the Commander-in-Chief and Sir Hope Grant to drive the rebel forces entirely out of Oudh were steadily progressing. At the end of November Sir Hope Grant crossed the Ghaghra, having in his column the headquarters and the 1st and 2nd Regiments of Hodson's Horse, and on the same day (25th November) engaged and routed a large force of the enemy under the rajah of Gonda and Mehndi Husain, taking four of their guns and inflicting on them considerable loss. Gonda was then occupied, and the combined movement northward and westward was continued until the scattered forces of the rebels were hemmed in by the British columns at Muhamdi, Shahjahanpur, Pilibhit, &c.

The pacification of Oudh was now almost complete. In the remaining operations a detachment of the regiment was included in a small column under Colonel Christie of the 80th Foot, which had a smart and successful skirmish with a rebel force at Basantpur on the 23rd December. The detachment suffered in this affair the loss of one risaldar (Ghulam Muhammad Khan) killed, one naib-risaldar and three men wounded, four horses killed and two wounded. On the 4th January 1859 the headquarters took part in a small skirmish near Kamdakot in the Bahraich district; and after an interval of three months the headquarters and a wing of the 1st Regiment were present on the 31st March in an action under Brigadier Horsford near the Jarwa Pass on the Nipal frontier. On this occasion a mixed detachment, including thirty sabres of the 1st Hodson's Horse, was pushed forward from Tulsipur to meet a reported inroad by the rebels. News was afterwards received that the enemy were in greater strength than had been anticipated. Brigadier Horsford therefore hastened on with the remainder of Hodson's Horse, and, arriving at a critical moment, turned a threatened reverse into a success. The enemy were compelled to retreat precipitately, and were driven back with considerable loss over the Jarwa Pass. The casualties in the corps were one man and one horse killed, nine men and seven horses wounded. Lieut.-Colonel Daly, C.B., was mentioned in Brigadier Horsford's despatch, and Dafadar Changan Singh was recommended for the 1st Class Order of Merit by Lieut.-Colonel Gordon, 1st Sikh Infantry, commanding the advanced party. This honour was awarded in G.G.O. 577 of 1859.

Finally, in the month of May 1859, the 1st Hodson's Horse took part in two small affairs on the Oudh frontiers, the first being on the edge of the Nanpara jungle on the 2nd May, and the other, in which a detachment of fifty sowars under Lieutenant S. G. Warde was engaged, at the village of Lalpur, which had been raided by one of the few bands of rebels still in arms. The enemy were driven out of the village and dispersed with some loss, the only casualty in the detachment being one sowar wounded.

Meanwhile detachments of the 3rd Regiment had been present at two sharp skirmishes in April—namely, at Gonda on the 13th under Lieutenant Mecham, and at the fort of Bungaon on the 27th of the same month, where a troop was commanded by Risaldar Fateh Singh.

On the 20th January orders had been issued for the 1st Regiment to proceed to Faizabad, and the headquarters now marched to that place, and in June 1859 they there took up quarters in cantonments for the first time since the formation of the regiment.

Some three months later they were again ordered to take the field and joined a column under Colonel Brett, 54th Foot, for operations in the trans-Ghagra districts. But the regiment was never actively engaged, and it finally returned to its quarters at Faizabad on the 4th January 1860.

Similarly the 2nd Regiment received orders to take up quarters at Gonda, where it arrived on 25th May 1859, and there it remained for the next twelve months, with an interval of some weeks in the cold weather, during which it was employed with a column of observation under Brigadier-General Holdich, C.B., on the Nepal frontier.

So ended the active services of Hodson's Horse during the Mutiny, which bitter struggle had been the occasion of its formation, and in the progress of which the regiment, no less than its daring leader, had earned well-merited distinction. In the course of the campaign, in addition to the repeated recognition accorded in various despatches to Captain Hodson, and to the brevet-majority bestowed on him just before his death, two officers, the brothers Charles and Hugh Gough, won the Victoria Cross, and Lieutenant R. C. W. Mitford was recommended for the same decoration; Captain Charles Gough received a brevet-majority, and it was notified that Lieutenant C. H. Palliser would receive a brevet-majority on his substantive promotion to the rank of captain; fourteen Indian officers and twenty-five non-commissioned officers and men won the Order of Merit.

The casualties among British officers amounted to three killed and nine wounded; among Indian officers three were killed and seven wounded.

Both the 1st and 2nd Regiments were subsequently permitted to bear on their appointments the names "Delhi" and "Lucknow."

Some few further details about the interior economy and administration of Hodson's Horse during the first year of its history may here be added.

A full description has been given in the first chapter of the manner in which the corps was raised, and in Chapter II. mention has been made of the uniform which the men wore during the siege of Delhi. Towards the end of the year clothing more suitable for camp work in northern India during the cold season was made up at Ambala and was served out to the men. This consisted of dark-blue quilted coats, worn with a red kummerbund, a red pugri and *kulla,* khaki pyjamas, and Punjabi shoes. The saddlery was of brown leather with brass mountings. The next year, when the heat during the campaign in Oudh made the wadded coats too warm to be serviceable, khaki coats were again substituted, with red facings covering the whole breast. New saddlery for some of the corps was procured at the same time from Cawnpore, the saddles being of rough leather without flaps or panels.

The first army list in which Hodson's Horse is mentioned is the Bengal List, dated the 20th October 1857, wherein the regiment is named as part of the Delhi Field Force. The first list in which a detail of the officers is given is that of the 10th July 1858, page 194, headed as follows :—

"Hodson's Horse—2 Regiments. At the disposal of the Chief Commissioner in Oude."

This was published a month before the formation of the corps into a brigade of three regiments, and the next army list, dated 1st October 1858, contains a detail of the new formation.

The composition of each regiment is shown in the Bengal Army List of the 24th January 1859 as follows : "1 Trumpet Major, 1 Woordie Major, 3 Ressaldars, 3 Ressaidars, 6 Naib Ressaldars, 6 Jemadars, 6 Kote Duffadars, 48 Duffadars, 6 Nishanburdars, 6 Trumpeters, 500 Sowars." A contemporary regimental record shows 3 Trumpeters and 3 Nagarchis, 420 Sowars and 2 native doctors ; also 1 Nakeeb, 1 Munshi, 1 Chowdry, 1 Mutsaddi, 3 Flag or weighmen, 2 Lascars, 6 Bhisties, all the latter forming the bazar and menial establishment.

It remains to be recorded that the organisation of the three regiments as a separate brigade did not long survive the end of the Mutiny Campaign. In April 1859, Major Daly was obliged to take

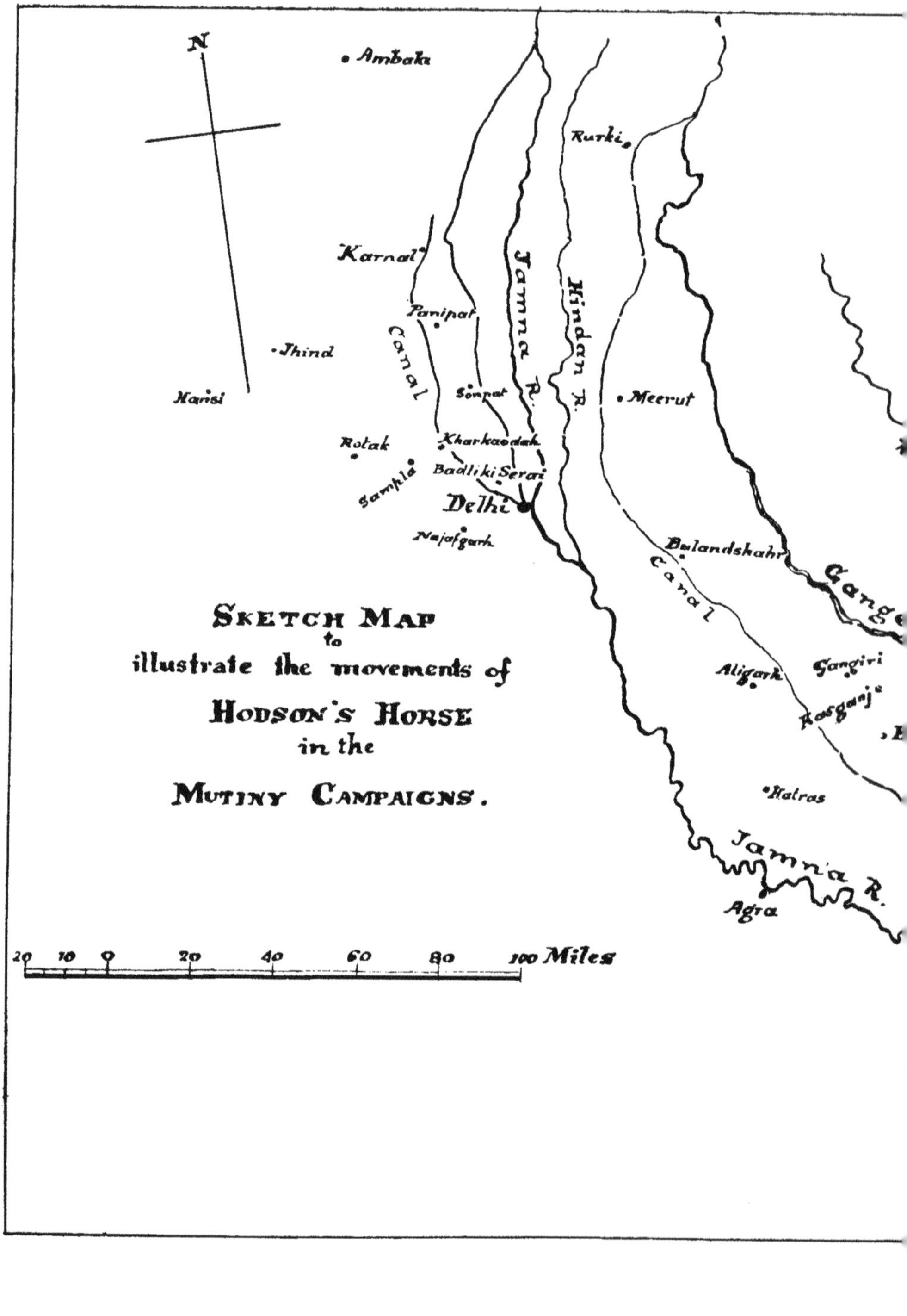

N

Ambala

Rurki

Karnal

Panipat

Jamna R.

Hindan R.

Canal

Jhind

Hansi

Sonpat

Meerut

Rotak

Kharkaodah

Sampla

Badli ki Serai

Delhi

Najafgarh

Bulandshahr

Canal

Gange

Aligarh

Gangiri

Kasganj

Hatras

Jamna R.

Agra

20 10 0 20 40 60 80 100 Miles

SKETCH MAP
to
illustrate the movements of
HODSON'S HORSE
in the
MUTINY CAMPAIGNS.

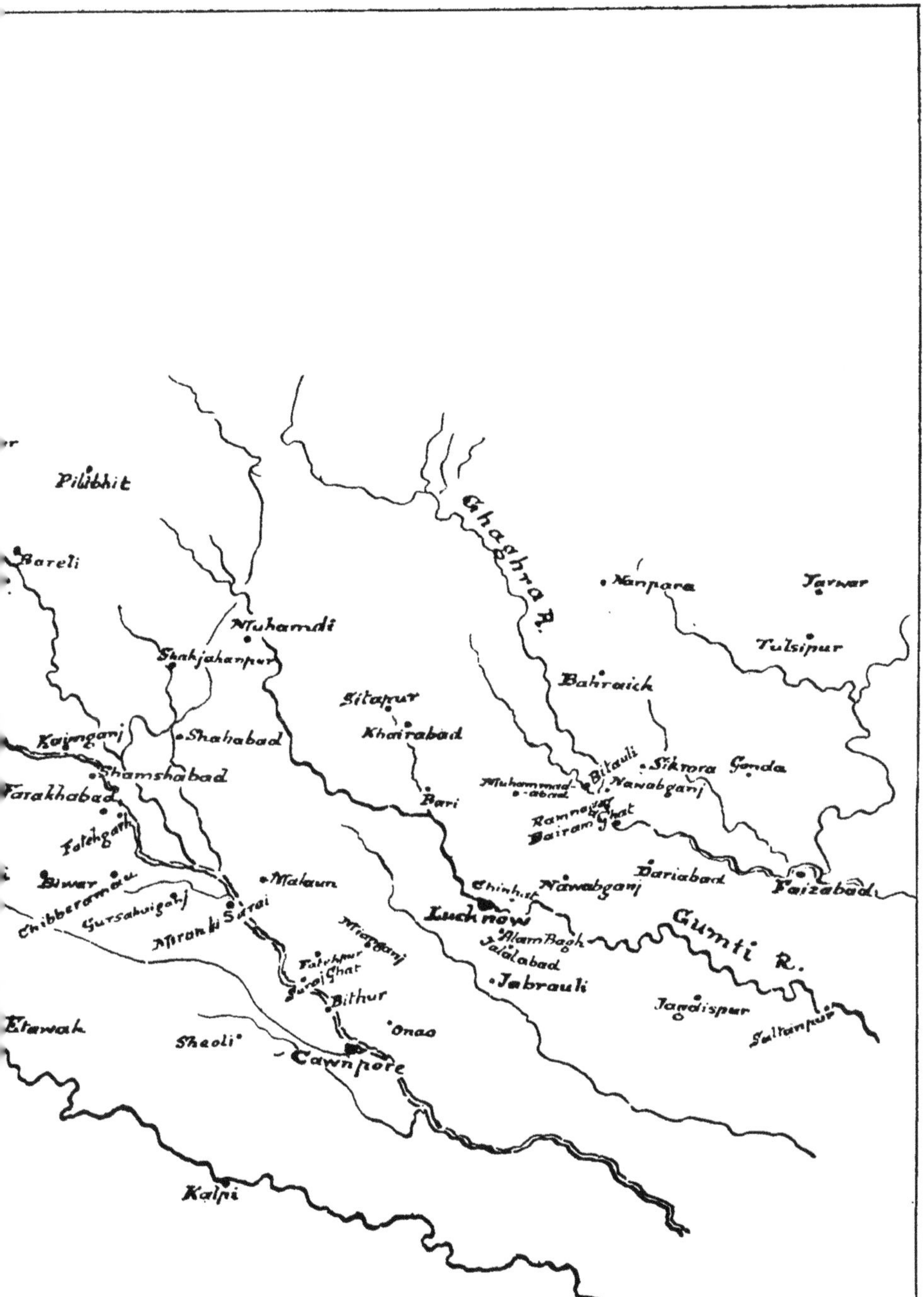

Pilibhit
Bareli
Muhamdi
Shahjahanpur
Shahabad
Kaimganj
Shamshabad
Farakhabad
Fatehgarh
Biwar
Chibberamau
Gursahaiganj
Miran ki Sarai
Malaun
Miaganj
Fatehpur
Sarai Ghat
Bithur
Etawah
Sheoli
Cawnpore
Onao
Kalpi
Ghaghra R.
Sitapur
Khairabad
Bari
Nanpara
Tulsipur
Bahraich
Bitauli
Sikrora
Gonda
Nawabganj
Muhammad-abad
Ramnagar Ghat
Bairam Ghat
Dariabad
Faizabad
Chinhat
Nawabganj
Lucknow
Alam Bagh
Jalalabad
Jabrauli
Gumti R.
Jagdispur
Sultanpur

leave to England on medical certificate, Major W. T. Hughes of the 1st Punjab Cavalry being appointed to officiate for him in command of Hodson's Horse. But on the 12th December of the same year the post of commandant was abolished and the three regiments were placed on the same footing as other units of irregular cavalry.

Only a month later, on the 5th January 1860, orders were issued for the 3rd Regiment of Hodson's Horse to be disbanded in pursuance of a policy of military economies in India. Neither the orders nor the policy in question, however, seriously prejudiced the future prospects of the Indian officers and men who at the time were serving in the disbanded regiment. The continuance of hostilities in China throughout the year 1859 necessitated at this moment the despatch thither of considerable reinforcements, and on the same day on which the disbandment of the 3rd Hodson's Horse was decreed a further Government order called for volunteers to form a new cavalry regiment for service in China. A large proportion of the officers and men from the former regiment volunteered *en masse* for the new service, and the regiment so formed, known at first as "Fane's Horse" and afterwards as the 19th Bengal Lancers, earned distinction in many fields, and to-day survives in the 19th King George's Own Lancers.

We have now followed the story of Hodson's Horse from its formation in the summer of 1857 down to the time when the two units which sprang from the parent stock began their period of entirely separate existence. They retained, it is true, the strong connecting link of their common origin and the name of their common founder, but for upward of sixty years they had no other connection, and the records of their services are independent one of the other. They will thus be dealt with in the following pages.

PART II.

THE NINTH HODSON'S HORSE

CHAPTER I.

1859-1885.

As has been related, the First Regiment of Hodson's Horse arrived at Faizabad at the end of May 1859, and after having again taken the field in the trans-Ghagra district for the last three months of the year, the regiment finally took up its quarters at that station on 4th January 1860, and remained there for two years.

This period was very fully occupied in completing the organisation, equipment, and training of the corps, for which work there had been little opportunity during the previous crowded months. The whole of the native army of Northern India was in the melting-pot, and a "new model" was emerging of which the cavalry was dealt with in a General Order of Government, dated 31st May 1861. In accordance with that order the First Regiment of Hodson's Horse became the 9th Bengal Cavalry, and was organised in six troops, with a strength of 13 native officers and 480 of other ranks.

On 8th February 1862, in consequence of the threat of trouble in the state of Bhutan, the Ninth, under Major J. P. Caulfield, left Faizabad at very short notice and moved towards the frontier, one squadron with the headquarters halting at Jalpaigori, the second at Barhampur, and the third at Raniganj. At these places the regiment was kept until April 1863. The trouble in Bhutan did not at the time come to a head (it broke out more seriously in the following year), and nothing occurred to break the monotony of a very uncongenial service. The unfortunate squadron at Barhampur meanwhile suffered severely from sickness and there was great mortality among the men there.

At length a welcome order was received for the regiment to move *viâ* Benares to Cawnpore, but ill-fortune still seemed to dog its march, and while at Benares the death of the commanding officer, Major Caulfield, plunged all ranks in deep and genuine mourning.

Arriving at Cawnpore on 29th April, the Ninth remained there for only seven months and then marched under the command of Captain C. M. Mecham to Peshawar, which place was reached in January 1864. Here the regiment was quartered till November 1866, when it moved to Mian Mir, Captain Mecham having meanwhile been replaced by Major H. L. Campbell as commandant.

For the next four years the permanent station of the Ninth was at Mian Mir, but this period was broken by an interlude in the autumn of 1868, when a turbulent outbreak on the part of the tribes of the Black Mountain in Hazara compelled the Government of India to send a punitive expedition into those regions. The regiment, in response to urgent orders, started from Mian Mir on 4th September, and moving at first by forced marches reached Deoband on 23rd September, whence detachments were sent out to Abbottabad, Mansurah, and Kaki. But the hope of taking part in the subsequent operations was not realised. The columns which moved into the Hazara country were composed only of artillery and infantry, nor indeed did they meet with any but the most feeble opposition. When the expedition was at an end the headquarters of the Ninth were moved to Abbottabad and one squadron was sent to Ughi in the Agror Valley (Hazara). A little later the headquarters and two squadrons returned to Mian Mir, arriving there on 15th January 1869, but the detached squadron remained in Agror until August of that year.

The regiment moved in relief from Mian Mir to Rawalpindi in November 1870, and thence to Deoli in the cold weather of 1873-74, a march which lasted no less than three whole months, in the course of which the second squadron was detached for duty at Jhansi. After four years at Deoli it marched to Meerut, where it arrived on 8th March 1878.

In the course of these years no incident occurred more notable than at rare intervals an occasional camp of exercise, but certain changes in uniform, armament, and constitution require mention.

In January 1863 an order was issued directing that the regiment should in future be composed as follows :—

1 squadron of Sikhs.
1 troop of Pathans and border tribesmen.
1 troop of Dogras.
1 squadron of Punjabi Muhammadans.

Owing, however, to the circumstances in which Hodson's Horse had been raised the proportion of Sikhs in the ranks was a good deal in excess of that authorised above, and some ten or twelve years elapsed after the Mutiny before this excess was absorbed.

Some notes about the makeshift uniform worn by Hodson's Horse during the Mutiny campaign have been given in Part I. Nothing very definite can be traced with regard to the dress of the Ninth in the years which immediately followed except that (according to Captain Sampson, the Adjutant, writing in 1868) the men in 1860-61 wore alkalaks of salmon colour and turbans of red saloo. Details of the uniform of the officers at this time are lacking. The facings were dark blue, and (until 1864) the lace worn was silver. In the matter of dress in those days each commanding officer in the Bengal Cavalry had a good deal of independent authority, with results which were sometimes rather startling. This avoidance of any sort of sealed pattern was encouraged by the authorities, so much so that in a volume of 'Standing Orders for the Bengal Cavalry,' published by order of the Commander-in-Chief in 1866, we find the following curious instructions: "Sowars should be allowed to use any kind of bit which gives them full control over their horses. In like manner uniformity is not required either in the handles or blades of swords. They should be allowed to wear any kind of sword they like, provided it is of good quality."

In 1868, however, detailed orders of dress and equipment for the Bengal Cavalry were published by the Adjutant-General. In accordance with these, the uniform of the Ninth was to be blue with scarlet facings and (for the officers) gold lace. The British officers now wore in full dress a tunic of dark-blue cloth with collar and cuffs of scarlet, the collar edged with gold lace and the cuffs heavily laced with gold. Across the breast of this tunic were "four quadruple rows of black cord, hanging loose as in a French staff jacket." Shoulder cords of curb chain completed the coat. Round the waist was worn a red kummerbund of Kashmir shawl pattern, the embroidered ends hanging loose on the right side, while over this was a gold-laced sword belt with slings and a square silver buckle. The breeches worn in the Ninth at this time were dark blue, and the dress was finished off with jackboots cut so low as to be level with the top of the knee when bent. The head-dress was a grey felt helmet with gilt chain, spike, and ornaments. Etceteras included a gold-laced pouch-belt with a pouch of black enamelled leather. Sabretaches do not seem to have been in use at this time, but they were introduced a little later, and were not discarded until 1883.

It appears from Captain Sampson's notes that the men also were dressed in similar blue tunics with

RISALDAR MAJOR MAN SINGH, C.I.E.,

Sardar Bahadur, Order of Merit.

From a portrait painted not long before his death, by
Lieut. E. M. Molyneux, 12th B.C.

red facings and red kummerbunds. But it is evident from old photographs that these gave place very soon to loose blue frocks, or blouses. For wear by the Indian officers these blouses were still adorned with the "four quadruple rows of black cord hanging loose," but those of the rank and file were plain, with red braid on the neck and cuffs. Their head-dress was, as before, a turban of red saloo. They wore pyjamas dyed "Multani mutti" colour and jackboots.

Up to 1863 a proportion of the men carried lances; after that date all ranks were armed with carbines and swords. As regards saddlery, the British officers used a hunting saddle with brass-bound cantle, Crimean wallets, and a red cloth shabrack of light dragoon pattern. For all other ranks the Nolan saddle was in use.

In 1875 revised orders of dress and equipment were published in the 'Bengal Cavalry Standing Orders.' These included the substitution of a plain loose blouse or kurta for the frogged coat hitherto worn by Indian officers, and a blue lunghi for the red turban. Moreover, the use of this Indian dress was now for the first time introduced for British officers when in "marching order B" (*i.e.*, when on duty with the men, and not in full dress). The breeches to be worn with this and with full dress were now white. The facings, kummerbund, &c., remained as before. At the same time considerable alterations were made in the full dress of British officers. The quadruple rows of hanging cord were now to be gold instead of black, and five instead of four in number; the helmet was white instead of French grey; the breeches were white melton instead of blue; and for all officers a dress pouch was introduced of red cloth embroidered with gold. Last but not least, new and careful instructions were issued about the boots of "Napoleon"[1]

[1] The difference between Napoleon and jackboots is that in the former the tops are rounded at the corners, whereas in the latter they are brought to a point.

pattern, which were in future to be worn by all ranks. The tops in front were to reach 2 to 2½ inches above the knee-cap; at the back they were to be "as high as possible." These clumsy articles, so unsuitable even fifty years ago for field conditions, remained in use in the Ninth until 1902, when progress in the ideas about the use of cavalry at last necessitated their disappearance. To any one who may have seen cavalry soldiers endeavouring to do dismounted work thus impeded, it will seem surprising that they were retained so long even for full-dress purposes.

The only other point to be mentioned here is that at the period under notice, 1874, the summer dress of the 9th Bengal Cavalry was a kurta or blouse and pyjamas of blue drill. Khaki was not introduced for this purpose until 1880.

To return now to the narrative. The regiment had been only just a month at Meerut in 1878 when all ranks were pleasurably excited by the receipt of a warning that it should hold itself in readiness for foreign service. No indication of its intended destination was given, but the political situation in Europe at the moment was such as to suggest that employment of the greatest importance was likely to fall to the lot of the troops. War had been raging for several months past between Russia and Turkey. After serious initial reverses the Russians had at length broken down the resistance of their opponents, and were advancing towards Constantinople, which lay apparently at their mercy. The determination of the Government then in power in England to preserve if possible the integrity of the Turkish capital was well known, and the announcement that an expeditionary force was to leave India under sealed orders was generally accepted as an indication that Great Britain was prepared to intervene by force of arms in the European war. These circumstances were sufficient to create the greatest enthusiasm with regard to the coming adventure, while the fact that Indian troops had never before been employed in Europe increased the interest of the event and the gratification of those regiments which were selected for the service. Definite orders were received by the Ninth on 18th April 1878, in accordance with which the regiment entrained for Bombay on the 23rd and 24th of that month.

In order to bring the regiment up to full field service strength after the elimination of all invalids, recruits, and young horses, a squadron from the 10th Bengal Lancers (2nd Hodson's Horse) under Captain H. C. Greenaway was attached to the Ninth and joined the regiment at Bombay. As a further preparation for the field the old Victoria carbines which had served as the firearms of the regiment since 1869 were withdrawn before the departure from Meerut, and Snider carbines were handed over to the Ninth by the 15th Hussars, who were about to be armed with the Martini-Henry.

The regiment was embarked and left Bombay in the vessels and on the dates shown below :—

Date.	Transport.	Commander.	Number of	
			Officers and Men.	Horses.
1st May	"Kilkennan," towed by s.s. "Nankin"	Capt. J. L. N. Willis	83	84
2nd "	"Narcissus," towed by s.s. "Maxima"	Capt. D. H. Robertson	100	104
3rd "	"Aros Bay," towed by s.s. "Macedonia"	Headquarters under Col. H. L. Campbell	114	124
3rd "	"Citadel," towed by s.s. "Macedonia"	Capt. H. C. Greenaway, 10th B.L.	87	88
3rd "	s.s. "Macedonia," towing the two ships above named	Lt.-Col. M. H. Heathcote	11 with followers and ponies	3
3rd "	"Seaforth," towed by s.s. "Trinacria"	Lieut. A. Burlton-Bennett, 10th B.L.	47	59

201 camp followers and 194 ponies were also embarked on the ships, 137 men and 188 ponies being on board the *Macedonia*.

The original instructions directed the transports to proceed to Aden, there to receive further orders. Thence they went on to Suez, and here final orders were received to make for Malta. The voyage up the Red Sea had been so much delayed by strong head winds that it became necessary to put both men and horses on a reduced allowance of water, a serious hardship at this the hottest season of the year; but after reaching the Suez Canal nothing further occurred until the weary month's journey ended at Malta on 4th June. Even here, however, through some muddling by the authorities concerned, the headquarters of the regiment were unnecessarily detained for a further full day before being landed, a circumstance which resulted in the loss of two horses. At length the regiment was all disembarked on 6th June, and proceeded to camp at San Antonio, where it formed part of a cavalry brigade under Brigadier-General J. Watson, C.B., V.C., and where it remained throughout its stay on the island.

The Malta Expeditionary Force consisted entirely (excepting two batteries of Field Artillery) of native Indian troops, and comprised two regiments of cavalry, four companies of sappers, and six battalions of infantry. For practical military purposes in an European war its strength was not of great importance. It was as a political demonstration that its despatch created a great and lasting impression far beyond what was warranted by the number of the troops, and it was for the purpose of creating such an impression that the departure of the expedition from Bombay and its stay at Malta was accompanied by a certain display of pomp and circumstance. Thus before the transports sailed, the Viceroy (Lord Lytton) issued to the troops a somewhat florid farewell order, in which he called on them, "the first expedition that has ever left India to strengthen the British forces in the Mediterranean," to uphold by their faithful and devoted performance of their duties, "the honour of the Empire which is now confided to your hands," and assured them that their progress would be watched with pride by the Queen-Empress. Moreover, the Field-Marshal Commanding-in-Chief (H.R.H. the Duke of Cambridge) journeyed from England to Malta and inspected the Expeditionary Force on 17th June, and published a complimentary order in which he expressed his gratification at having had the opportunity of making himself personally acquainted with the troops of the Indian Empire and his admiration of their efficiency.

With these compliments, doubtless well deserved and certainly in the case of the Duke of Cambridge sincere, the Indian troops were obliged to rest content. Even before their arrival Russia had already held her hand when Constantinople seemed to be within her grasp. But she had forced upon her beaten foe a treaty which involved the practical extinction of Turkey as an European power. The British Premier was determined that the terms of this treaty should be modified, and the sudden and dramatic appearance of the Indian contingent in European waters materially assisted the acceptance of his views. The Berlin congress followed, and the conclusion of a peace which lasted for more than a generation was the immediate result of the vigorous British policy. Meanwhile in July all the Indian troops received orders to proceed to Cyprus, which island had just been ceded to Great Britian by Turkey as the price of the good offices of the former in the recent troublous crisis. The Ninth was on the point of starting for its new destination when an extraordinary accident caused all the arrangements to be altered. While at Malta a daily dose of lime juice was administered to the men by direction of the medical authorities. One day, just before they were to embark for Cyprus, by some amazing and criminal carelessness, a bottle of disinfecting fluid was substituted for lime juice, and some of the contents were actually drunk by forty-eight unfortunate men before the mistake was discovered. Three of the victims died in great agony from the effects of the poison, and of those who survived several never fully recovered their health. Who it was who was the direct cause of this deplorable mischance was never discovered. The confusion and dismay created by it were, of course, extreme. There was necessity for a full and careful inquiry, and as a consequence the move to Cyprus by the regiment was altogether abandoned. The Ninth remained at Malta until October, on the 1st and 4th of which month they embarked on two transports, the s.s. *Margaret* and *Scotland*, for Bombay, where the headquarters disembarked on 24th October. They started by rail the same evening for Meerut, where they at length arrived on 1st November after an absence of seven months.

Thus ended for the Ninth this memorable expedition. As has been seen, the fruits of the episode so far as the regiment was concerned were limited to empty compliments, but the fact remained also that the regiment had had the honour of being selected to show in Europe what manner of troops India could produce. Thirty-six years were to pass before the precedent created in 1878 was to be followed in a far greater crisis, but in the meantime its effects were lasting, though quite different in character from what was anticipated by Lord Beaconsfield. Foiled in her cherished ambition to dominate Constantinople, Russia turned her face eastwards, determined that never again should the British in India be free to use their troops against her in Europe. The encroachments of Russia towards the North-West Frontier of India between 1878 and 1904 were the inevitable consequences of the despatch of the Indian expedition to Malta.

Six and a half years now passed without much notable incident in the records of the Ninth. The regiment served at Meerut until November 1881, when it marched to Peshawar, and was quartered there until January 1885. In March 1880, Colonel H. L. Campbell, who had been commandant since 1864, was invalided to England, where he died in August 1881. He was succeeded by Major Thomas Dayrell, who, however, only held the command for a year and two months, and retired from the army in October 1882. Brevet-Colonel T. J. Watson was then appointed commandant. This officer had lately held a similar appointment in the 17th Bengal Cavalry, which regiment had been broken up in the course of reductions following the Afghan war.

Meanwhile there were a few changes in equipment and organisation. On arrival at Malta the regiment was armed with Martini-Henry carbines instead of the Sniders which had been issued just before the departure from Meerut. The former weapons were brought back to India, but were again exchanged for Sniders when the regiment reached Meerut.

There were during these years a good many and contradictory orders relating to the strength of cavalry regiments. It would be tedious to detail these orders and counter-orders, and it suffices to record that in 1882 the strength was fixed at 550 of all ranks, organised in three squadrons, the composition of which remained as laid down in 1863.

A matter of more purely regimental interest was the sanction by the Government of India in June 1883 of the adoption by the Ninth Bengal Cavalry of a special badge, which is described in the following terms:—

A silver, diamond-cut, eight-pointed star, bearing the Royal Crown in gold, raised on a crimson cushion surmounting a garter in gold with silver edging, the garter having upon it in relief in silver the words "Bengal Cavalry." In the centre of the garter the Roman letters IX in silver on a raised crimson cushion. Lying flat under the three lower points of the star, a gold scroll crossed diagonally by two silver ribands bearing the words "Delhi" and "Lucknow" in gold.

It may be added that the regiment desired that the words "Hodson's Horse" might be included in the design, but this was expressly prohibited by the Government of India, the designation in question not being authorised.

Only one other event requires record here—namely, that in 1884 the prize of 100 rupees awarded by the Commander-in-Chief in India for the best shot in the native cavalry was won by Kot Dafadar Sher Singh of the Ninth Bengal Cavalry.

CHAPTER II.

THE SUAKIN EXPEDITION.

On 15th January 1885 the regiment marched out of Peshawar in ordinary course of relief *en route* for Ambala. Colonel Watson was absent on furlough; Lieut.-Colonel A. P. Palmer, the substantive second-in-command, was seconded as Assistant Adjutant-General; and the command of the regiment when it left Peshawar was in the hands of Lieut.-Colonel R. M. Clifford.

On 10th February the Ninth had reached Wazirabad, and was in camp there when a telegram was received by the commanding officer ordering the regiment to move by rail immediately to Cawnpore, there to be supplied with lances and other equipment as a lancer regiment preparatory to proceeding to Egypt on field service. Such an order coming only a few years after the selection of the regiment for service with the expedition to Malta was naturally a source of great satisfaction to all ranks. With the least possible delay arrangements were made for transport by rail, and, moving by squadrons, the regiment arrived at Cawnpore between the 15th and 18th of February, a depot of fifty officers and men being left at Ambala under the command of Lieut.-Colonel H. N. Willis. As each squadron arrived it was at once supplied with lances, as well as with "Mackenzie equipment" and all necessary items for field service, and by 19th February it was ready to begin the journey to Bombay, where it was to be embarked on board ship. Each troop was entrained complete together with its baggage animals, followers, and ammunition, prepared to take the field as a complete unit; and as each train arrived at Bombay its occupants, with animals and baggage, were forthwith embarked the same day on the transport allotted to them.

Meanwhile Lieut.-Colonel A. P. Palmer, having been allowed to relinquish his staff appointment, had rejoined the regiment at Cawnpore, and assumed the command.

The voyage passed without incident, and between the 6th and 10th of March the transports reached Suakin on the Red Sea littoral, where the several troops were at once disembarked, and the regiment went into camp on the outskirts of the town. Here an infantry brigade from India had already in part arrived. This contingent was under the command of Brigadier-General John Hudson, and consisted of the 15th Sikhs, 17th Bengal Infantry, 28th Bombay Infantry, and a company of Madras sappers. The whole force at Suakin, as will be shown below, was under the command of Lieut.-General Sir Gerald Graham.

In order to make the narrative of the next few months intelligible, it is desirable to give here a very brief outline of the events which led up to the despatch of an expedition to Suakin.

The intervention of Great Britain in the affairs of Egypt in 1882, which opened with General Wolseley's well-planned and successfully executed campaign against the nationalist leader Arabi Pasha and the victory of Tel-el-Kebir, was quickly followed by more serious complications in the Sudan. Here an Arab fanatic, Muhammad Ahmad, had a year before proclaimed himself to be the Mahdi, and had started an insurrection against the Egyptian Government. With the assumption of practical control over Egypt, the duty of securing the borders of the country and incidentally of quelling the insurrection of the

Mahdi devolved upon Great Britain. The task proved a troublesome one. The Mahdi's power and influence were increased by various successes. In August 1883 a disgruntled slave-dealer named Osman Digna raised a revolt in the Eastern Sudan in sympathy with that in the Nile valley, and made himself both troublesome and dangerous to the Egyptian garrison at Suakin. Efforts to deal with this side issue at first resulted only in disaster, until in February to March 1884 a force under Sir Gerald Graham inflicted severe defeats on the Arabs at El Teb and Tamai. This quieted the neighbourhood for a time. Meanwhile, however, things were not going well in the Nile provinces. The power of the Mahdi was unbroken, and General Charles Gordon, who had been sent to the Sudan with orders to withdraw the Egyptian garrisons from the country, was practically cut off at Khartoum. Various plans were propounded for his relief, one of which was for an advance across the desert from Suakin to Berber on the Upper Nile, 175 miles below Khartum. Eventually after months of delay and vacillation the proposal of Lord Wolseley for an advance from Egypt up the Nile valley was adopted. But the decision came too late, and in spite of almost superhuman efforts on the part of the troops engaged, Khartum was taken, and Gordon fell before the tardy relief could reach him. This was in January 1885. Then arose further subject for discussion: whether to withdraw entirely from the Sudan and leave it in the power of the Mahdi, or whether merely to fall back and prepare for another effort the following autumn. If the latter course were adopted it was very important that Osman Digna and his following should be finally crushed, so that he might not threaten the left flank of a force advancing up the Nile valley. Moreover, in the event of such an advance being undertaken, the Suakin-Berber road would afford a useful subsidiary line of approach to Khartum. Again the Government in England found itself unable to make a definite decision. While disliking the idea of further entanglement in the Sudan, it yet shirked the responsibility of ordering a complete withdrawal. The question of an autumn campaign was put off for a time, but meanwhile orders were given for operations against Osman Digna. At the same time the scheme of laying a light railway from Suakin to Berber was taken up. Contractors were employed and preparations made for beginning construction. Like all half measures, these decisions were of little value. If operations were to be resumed against the Mahdi in the autumn, much more vigorous preparation was required. If, on the other hand, the Sudan was to be abandoned, then the expedition to Suakin was merely a useless waste of large sums of money and, as it turned out, of many valuable lives.

This brings us to the arrival at Suakin of the Indian contingent in March 1885. The force assembled there was commanded by Lieut.-General Sir Gerald Graham, V.C., K.C.B. It included (besides the Indian contingent already enumerated) a composite regiment of British cavalry (two squadrons each of the 5th Lancers and 20th Hussars) and four companies of Mounted Infantry; G Battery, B Brigade, R.H.A., and two mountain batteries, R.A.; a brigade of three battalions of Guards, a brigade of three battalions of British infantry, and one battalion of Royal Marines; and three companies of R.E. The whole force amounted (with subsidiary services and camp followers) to 10,600 Europeans, 7800 Indians, 1850 horses, 1700 mules and ponies, and 7000 camels. These numbers are worthy of note, in view of the fact that the country in which the force was to operate was entirely destitute of supplies of *any* kind, and that even for water arrangements had to be made by condensing apparatus. For all the British troops and also for the Indian soldiers (except at certain advanced posts) condensed water was supplied for drinking purposes. Wells were sunk for watering the animals, but the water so obtained was brackish and bad. The enormous labour involved in making these arrangements may be readily imagined as well as the discomfort of service in such conditions, especially in the intense heat and in the dust and grime of the Red Sea littoral. It is impossible in this record of the doings of a single regiment to describe in detail the elaborate arrangements and organisation of the whole force. The official history [1] of the campaign is, however, well worthy of study, for it affords a striking example of how difficulties seemingly insuperable can be overcome by forethought, energy, and determination.

The instructions to Sir Gerald Graham at the outset of the operations laid down that their "first and most pressing object" was the destruction of the power of Osman Digna. To effect this he was ordered to attack as early as possible all the positions occupied by the Arab leader and to disperse his troops. The next and secondary object of the campaign was to push on the Suakin-Berber railway.

In March 1885 the enemy was holding an irregular line west of Suakin from Tamai on the south-west through Hashin to Handub on the north-west. The country was for the most part covered with thick scrub of prickly mimosa bushes varying in height from six to eight feet. The positions held by the Arabs were on rocky eminences, the foothills of a volcanic range of mountains farther to the west, while the ground between these heights and amongst the scrub was intersected by numerous dry water-courses. In the immediate neighbourhood of Suakin the country was a fairly open plain. At night the Arabs would issue from the hills and scrub, and crossing this open ground constantly harass the troops of the Field Force by sniping and intermittent firing. No great damage was done, but the dis-

[1] 'History of the Sudan Campaign,' Part II., by Colonel H. E. Colvile (I.D.W.O., 1889).

turbance and annoyance occasioned by this nocturnal firing was serious.

Osman Digna's headquarters were at Tamai, and this, according to Sir Gerald Graham's instructions, was the first objective. But as a preliminary to an attack the General determined to dislodge the enemy from Hashin, where they threatened the right of an advance against Tamai, and whence

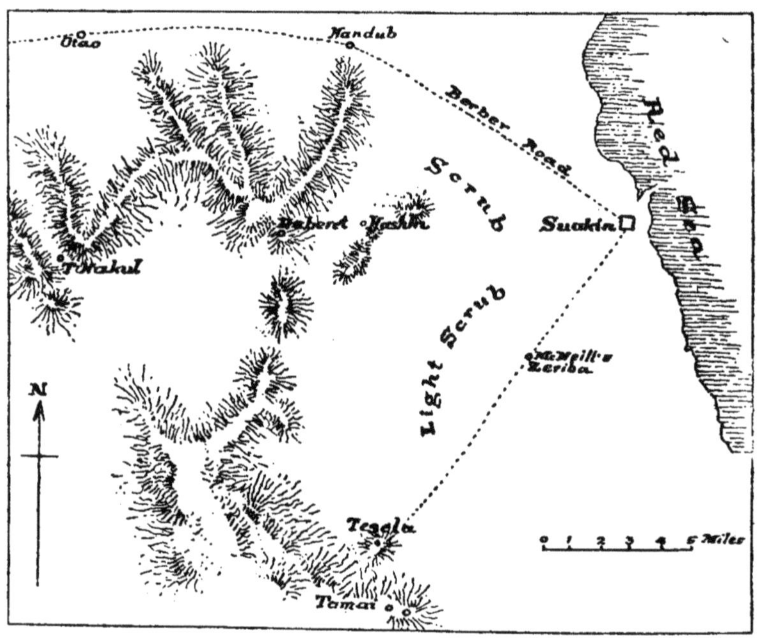

The country round Suakin.

most of the harassing night raids on the British camp generally proceeded. With this object in view a preliminary reconnaissance by the cavalry brigade, supported by the Indian infantry brigade, was carried out on 19th March.

The work of the day began with a parade of the whole field force for the inspection of the Lieut.-General near the West Redoubt, and this having been completed by 7.45 A.M. the cavalry brigade moved off towards Hashin. The front was covered by a screen of scouts, supported by " cossack parties " of one squadron of the 9th Bengal Cavalry, followed by the two other squadrons of the regiment in line of squadron columns. The two squadrons each of the 5th Lancers and 20th Hussars formed the second and third lines, echeloned respectively on the right rear and left rear of the Ninth.

This formation was adopted on every occasion on which the cavalry was employed as a brigade, the 9th Bengal Cavalry invariably furnishing the scouting screen and first line, an important rôle in which the regiment earned a remarkable reputation for coolness, courage, and efficiency.

About 9.30 A.M., after traversing seven miles of open plain from Suakin, the brigade approached a group of low rocky hills, rising abruptly from the surrounding country, and marking the beginning of a region of thick undergrowth which stretched away westwards towards the mountains beyond. A small body of Arabs was occupying the highest of this group of hills (known afterwards as Zeriba Hill), but they fell back as the cavalry approached. A mile and a half farther on a more lofty ridge, known as Dihilbat, or Hashin Hill, rose with bare and precipitous slopes from the low ground, its base surrounded with dense thorny thickets, and a ravine some two or three hundred yards wide separating its northern point from a small eminence known as Beehive Hill. Immediately behind (*i.e.*, north-west) of Beehive Hill is the village of Hashin, a cluster of huts encircled by a jungle of mimosa. The enemy occupying Hashin and the neighbouring heights retreated westwards, leaving only some scattered parties on Dihilbat Hill, who fired occasionally on the advancing cavalry. The leading squadron penetrated the village, which was found to be deserted, but nowhere encountered any resistance.

The object of the reconnaissance having now been accomplished, the retirement was ordered at 10.15

A.M., and by 2 P.M. the brigade reached camp, the 9th Bengal Cavalry acting as rearguard and covering the withdrawal.

On the following day, 20th March, the whole field force except the 1st Battalion Shropshire Light Infantry marched at 6.20 from Suakin under Sir Gerald Graham, with the intention of finally driving the Arabs out of Hashin and of establishing an outpost there on Zeriba Hill. The infantry moved in the formation of three sides of a square, the cavalry covering the front and flanks, the advanced scouts with cossack parties in support being furnished as on the previous day by the Ninth. The infantry reached the detached group of hills seven miles from Suakin at 8.35 A.M., and here the Lieut.-General posted himself on the northernmost of the hills, while the East Surrey Regiment (70th) was detached with the engineers and sappers to form redoubts and a zeriba on the adjacent heights. Meanwhile the rest of the force moved forward towards Dihilbat and Beehive Hills, which were seen to be occupied by considerable numbers of the enemy. The 1st Battalion Berkshire Regiment (49th) and Royal Marines rapidly scaled the heights, driving the Arabs before them, while the rest of the infantry occupied the gorge between the two hills. At this moment Brigadier-General J. Ewart, commanding the cavalry brigade, seeing a number of the enemy making off southwards in the direction of Tamai, took two squadrons of the Ninth to cut them off. As they passed along the south-east face of Dihilbat Hill a large body of Arabs on the hillside threatened their right flank. One squadron, under Major D. H. Robertson, dismounted and opened fire with their carbines. Whilst so occupied they were suddenly charged by a crowd of spearmen, who, creeping through and under the thick bushes with extraordinary speed, threw themselves on the dismounted men almost before the threat of the attack was realised. For some minutes there was a wild hand-to-hand mêlée. The sowars regained their horses as best they could and cleared the front of the Guards Brigade which had been brought up in support; but a good many casualties occurred in the first rush, during which in several instances remarkable coolness and courage were displayed by officers and men. Major Robertson, in saving the life of a sowar who was attacked by two Arabs, received a severe spear wound in the thigh. Captain Garstin was twice hit by spears, but escaped serious injury. In fact all the officers had narrow escapes, including Risaldar Hukm Singh, who, together with nine

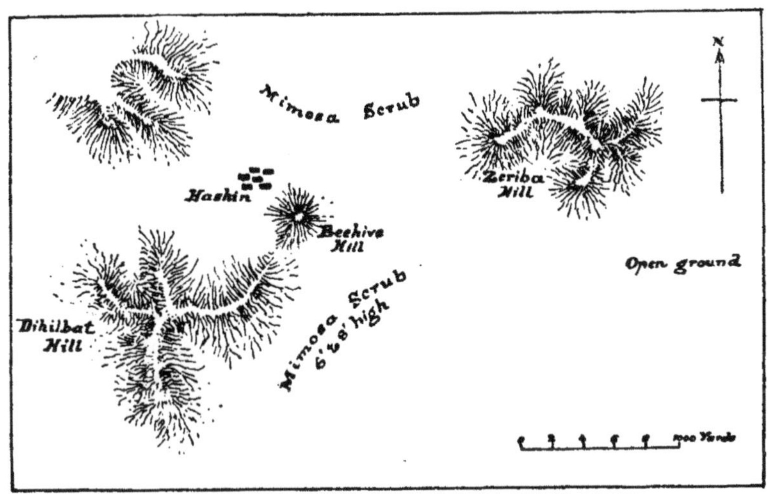

Scene of the action at Hashin.

non-commissioned officers and men, was awarded the Order of Merit for gallantry on this day.

Almost simultaneously another body of the enemy, charging through the bush north of Hashin village, attempted to turn the right flank of the British force and to cut off the Lieut.-General and the troops who were working at the redoubts on Zeriba Hill. This attack, however, was effectually met by a charge of the 5th Lancers (two squadrons) and two squadrons of the Ninth, and the Arabs were driven back with heavy loss.

At 12.45 P.M. orders were given for the force to begin its withdrawal to Suakin. The retirement was covered by the Guards brigade, in which several casualties occurred, for as soon as the troops began to fall back the enemy swarmed over Dihilbat and Beehive Hills, and kept up a hot fire with Remington rifles. Suakin was reached at 6 P.M., the East Surrey Regiment being left to hold the redoubt and zeriba which had been constructed during the action.

The casualties suffered by the Ninth on the 20th March were Ressaidar Shibdeo Singh, eleven rank and file and seven followers killed, Major D. H. Robertson and fourteen rank and file and one follower wounded.

Although this action did not show any very material results, yet it had certainly the effect of putting a stop to the irritating night attacks on Suakin. Moreover, the establishment of a post at Hashin secured the force from the threat of an attack from that quarter when the advance against Tamai was undertaken. It may be added that useful though dearly won experience was gained of the nature of the enemy to be encountered and the manner of his attack, and the vital need for careful scouting was forcibly exemplified. At the same time the events of the day illustrated very clearly the unwisdom of employing cavalry on ground so entirely unsuitable for that arm, where the movement of formed bodies was quite impossible through the thick bush, where even individual men on horseback could only move with difficulty, and where such troops were at a frightful disadvantage against savage spearmen, creeping through the almost impenetrable jungle and able to deliver their attack before their presence was perceived.

It is right to mention here that an entirely distorted and inaccurate account of the Hashin fight appeared a few weeks later in the 'Daily Chronicle,' which attempted to throw discredit on the conduct of the 9th Bengal Cavalry. So great was the indignation created by this false report, and so general the sympathy with the regiment whose gallantry and steadiness had earned praise on all sides, that a strongly worded telegram was sent to the *Central News Agency*, and appeared in the London press early in April, categorically denying the accuracy of the aspersions on the Ninth. This telegram was countersigned before despatch by the Chief of the Staff with the Suakin Field Force.

Having thus by the operations against Hashin secured his right flank, Sir Gerald Graham now set about the more important task of attacking Osman Digna at Tamai. As a preliminary it was necessary for purposes of supply to establish an intermediate post in the desert half-way between Suakin and the intended objective. With this end in view a mixed force, which included the Indian infantry contingent, was despatched on the morning of 22nd March under Major-General Sir John McNeill, V.C., K.C.B., with orders to march out eight miles, where a triple zeriba was to be made. This was to be occupied by two battalions of British infantry, with a space between capable of holding 2000 camels. The Indian contingent was then to return to Suakin, but on the way back another zeriba was to be made, where one battalion was to be left. Only one squadron of cavalry from the 5th Lancers accompanied the force.

It was soon found that the prescribed programme was quite impossible of accomplishment, progress through the bush being very slow for the considerable force employed, accompanied as it was by a great convoy of camels. Accordingly when only some six miles from Suakin Sir John McNeill decided to make the zeriba at a place called Tofrik where there was a clearing in the scrub. It is not necessary to describe in detail what followed. It will suffice to say that when the zeriba was still incomplete and while working parties of infantry were engaged in cutting brushwood and constructing the defences the force was suddenly attacked by a great mass of Arabs, estimated at between 2000 and 3000, who swarming out of the surrounding jungle threw themselves on the working parties, attempted to break into the square, and in one place succeeded in doing so, stampeded the baggage animals and created a disastrous panic among the Indian transport drivers. The situation was saved by the magnificent coolness and courage of the troops, among whom the 15th Sikhs were conspicuous for gallantry and steadiness. But many of the camel drivers and other followers streamed away in frantic terror towards Suakin pursued by the Arab spearmen, who cut them down and killed them as they ran.

At this time a squadron of the Ninth under Lieutenant A. J. Peyton was on the road from Suakin to maintain communications with Sir J. McNeill's force and to relieve a squadron of the 20th Hussars under Major Graves, who had been employed on the same duty. Just as the two squadrons met at about 2 P.M. the sound of heavy firing was heard from the direction of the zeriba. Both squadrons at once hastened towards McNeill's position, and had traversed about a mile when they came in sight of the retreating mob of followers and their pursuers. The ground here being fairly open the cavalry formed line and extending in "rank entire" allowed the fugitives to pass between the horses. The foremost of the Arabs were thus checked, and those behind were then driven back by dismounted fire. On arriving near the zeriba the two squadrons were joined by that of the 5th Lancers and completed the dispersion of the enemy, whose attack had by this time been driven off. The transport animals were now collected as much as possible, a cordon of cavalry was formed round them, and the zeriba having been completed Sir John McNeill's force bivouacked there that night.

Lieutenant Peyton was mentioned in despatches for his conduct on this occasion. His opportune arrival and the promptitude with which he led his squadron to the aid of the defenceless camp followers resulted in many lives being saved.

Several days were now occupied in accumulating supplies of both provisions and water at McNeill's zeriba, the Ninth supplying a squadron every day for scouting duties with the infantry escorting the convoys. On 31st March a reconnaissance to Tamai was carried out by the Mounted Infantry, accompanied by a troop of the Ninth. Only small parties of Arabs were seen, but, so far as could be ascertained, Tamai appeared to be still occupied. Accordingly orders were issued for an advance of the whole field force on the following day against Osman Digna's headquarters. The force marched off from Suakin at 4.30 A.M., the cavalry brigade leading, and the Ninth as usual furnishing the scouting screen and support. At 4.30 P.M. Tesela Hill was reached without opposition being encountered. This point was some two miles from Tamai, and here the force halted, and the infantry having made a zeriba bivouacked therein for the night. The cavalry rode back to McNeill's zeriba to water, rejoining the force early next morning.

At 8 A.M. on the 2nd April the advance was continued. Still no opposition was met with. Only a few stray scouts of the enemy were seen, and these retired towards the hills as the force advanced. Tamai was found to be deserted and was burned to the ground and large quantities of ammunition found there were destroyed. This done the return march was begun at 10.30 A.M., and the cavalry reached Suakin the same evening, having for two days been in the saddle from before dawn until after dark.

Osman Digna's position at Tamai having been thus destroyed and his forces driven back into the hills, Sir Gerald Graham now turned his attention to the second objective given to him in his instructions—namely, the making of the Suakin-Berber railway. With this object in view several posts were established on the Berber road, of which the farthest inland was at Tambuk, about eighteen miles from Suakin. The post there was occupied on 19th April. Meanwhile the Ninth had been employed continuously on protection duties, but on the 18th April a day's intermission was provided by a reconnaissance in force, in which the 20th Hussars, Mounted Infantry, and 15th Sikhs also took part. This force proceeded by way of Hashin to Deberet, where it was met by a reconnaissance from Otao, one of the posts on the Berber road. Having established communication with that place the Suakin force returned, burning Hashin village on the way back. On the following day a troop of the Ninth together with a company of the 15th Sikhs went out again to Hashin and buried or burned the bodies of the men who had fallen in the engagement there on 20th March. From the 19th April the daily patrolling and outpost duties were continued without incident for another fortnight.

The strain of these incessant and fatiguing services in exceptionally trying conditions was now beginning seriously to affect the horses. Hay of very indifferent quality, unwholesome and brackish water and insufficient in quantity, increased the exhaustion resulting from long days in a blazing sun and exposure by night to the sudden chills of the Red Sea littoral. It was with difficulty that the numbers of horses required for orderly and other duties could be found, and in the middle of April the regiment was glad to receive a loan of twenty-five Egyptian horses to assist in these duties.

For the men, too, the daily protection work along the railway and the outpost duties, of which the line extended all round Suakin and the camp from sea to sea, were trying and harassing in the extreme. As a rule every man was on duty at least once in twenty-four hours, and sometimes a long sultry day on outpost duty was followed immediately by a night on piquet. The cheerfulness and willingness displayed by all ranks and the efficiency maintained were an eloquent testimony to the spirit animating the regiment.

Meanwhile little had been heard or seen of the enemy since the destruction of his camp at Tamai, but towards the end of April news was received that a force of Arabs under Sheikh Muhammad Adam Sardun, a trusted lieutenant of Osman Digna, had assembled at a place called T'Hakul, ten miles due west of Hashin, with the intention of harassing our line of communications. Sir Gerald Graham accordingly determined to make a combined attack on the enemy at this place with a mounted column from Suakin and a mixed force from Otao.

The Suakin force consisted of the 9th Bengal Cavalry, 2 companies Mounted Infantry, and the Camel Corps.

Starting at 12.50 A.M. on 7th May under the personal command of the Lieut.-General the column reached Hashin at 2.40 A.M. and at daylight arrived at the south end of the T'Hakul valley. Here the Ninth and half of the Mounted Infantry were sent forward, the latter dismounting and crowning the heights while the cavalry advanced through the valley. Meanwhile the column from Otao reached the northern end of the T'Hakul valley at 5.40 A.M., and succeeded in drawing out the enemy, who was unaware of the approach of the Suakin force. Repulsed by the fire of the Mounted Infantry from Otao, the Arabs rushed back towards their camp, only to find themselves met in that direction by the advancing cavalry, while at the same moment they came under the rifle fire of the troops on the hills. They were thrown into complete confusion, and after rushing backwards and forwards sought safety in flight over the hills towards the south-west, leaving the whole of their sheep, goats, camels and donkeys, besides supplies of grain, clothing, camel saddles, &c., in the hands of their assailants. Their losses in killed and doubtless also in wounded were severe. The casualties in the British forces were no more than one officer and two non-commissioned officers wounded.

At 9 A.M. the Otao column returned to that place, and at 10 A.M. the Suakin column started on its march back to camp. On the way it was fired on by some parties of Arabs, who were afterwards ascertained to have been reinforcements sent by Osman Digna to Muhammad Sardun, but they were quickly dispersed by a few volleys from the Mounted Infantry and Camel Corps, and the column reached Suakin at 4.30 P.M., having covered forty miles since midnight.

The results of this well-planned and well-executed action were excellent. The enemy, already disconcerted by the attack on Tamai, was completely discomfited and disheartened, Osman Digna was a fugitive and his following scattered and reduced, and many of the local tribes were encouraged to make overtures for peace. In short, the object of the Suakin expedition had, for the time at least, been achieved. But now that the successful operations had, after great expense, laid open the road to Berber and Khartum the British Government at length arrived at a tardy decision to evacuate the Sudan and to abandon all idea of further hostilities against the Mahdi. Orders for the break up of the Suakin Expeditionary Force were received early in May and the withdrawal began on 17th May with the departure of the Guards Brigade. While it cannot be said that the prospect of leaving Suakin was unpleasing to the Indian contingent, yet it was not without chagrin that the troops contemplated so futile a termination of a campaign wherein they had endured so much hardship.

Meanwhile early in May Lord Wolseley, who

was in chief command in Egypt and the Sudan, had arrived at Suakin. On the 18th of the month he inspected the Indian contingent, and spoke in terms of high praise of the efficiency and gallantry of its regiments. Lieut.-General Sir Gerald Graham also held a farewell inspection of the contingent, and after some general eulogistic remarks on their good conduct and behaviour in the field he referred particularly to the 9th Bengal Cavalry, complimenting the regiment on the manner in which the men had acquired proficiency in the use of the lance, with which they had been so recently armed, and adding that their conduct in the field and especially their excellent scouting work had excited universal admiration.

In Sir Gerald Graham's and in Lord Wolseley's final despatches the names of Colonel A. P. Palmer, Major D. H. Robertson, and Risaldar Hukm Singh were specially mentioned ('London Gazette,' 25th August 1885), and in the same 'Gazette' Colonel Palmer was made a Companion of the Bath and Major Robertson received a Brevet-Lieut.-Colonelcy. In G.G.O. No. 566, dated 9th October 1885, Risaldar Hukm Singh was admitted to the Order of British India.

The thanks of the Viceroy for their services at Suakin were conveyed to the Indian contingent in Military Department letter dated 27th August 1885; medals and a bronze star (given by the Khedive) were granted to all ranks, and permission to bear the name and date "Suakin, 1885," on badges and appointments. Finally in February 1886 the sanction of the Government of India was received for the regiment in future to be designated the Ninth Bengal Lancers.

The headquarters and two squadrons embarked at Suakin on the 9th and 10th June 1885 in the s.s. *Russian Monarch* and *Italy* respectively, and after a rough passage reached Bombay on the 19th and 21st. Thence they were railed to

Ambala, where they arrived on 26th and 28th June.

The 3rd Squadron, 150 strong, under Captains H. M. Mackenzie and H. L. Dawson, remained with the rest of the Indian contingent at Suakin throughout the summer. The heat was intense, as much as 120° being registered in the shade, and nothing occurred to break the tedious monotony of these trying months. At length on 20th November the squadron was relieved by a similar detachment of the 4th Madras Light Cavalry, and embarked for India in the s.s. *Sirsa*, eventually reaching Ambala on 14th December 1885.

Annexed is a table showing the strength of the regiment and its losses in the Suakin expedition :—

	British Officers.	Indian Officers.	Rank and File.	Total.	Followers.	Horses.	Transport Animals.
Strength on landing at Suakin	10	13	475	498	499	502	192
Killed in action and died of wounds	...	1	11	12	7	10	...
Died of disease	...	1	1	2	...	16	...
Invalided	2	...	22	24	...	...	...
Animals cast	...	...	...	...	...	26	...
Total losses	2	2	34	38	7	52	...
Strength on return to India .	8	11	441	460	492	450	192

Considering the trying conditions of the service the comparatively small number of casualties among the horses and the absence of losses among the transport animals is a remarkable testimony to the efficient stable management in the regiment. But it was long before the horses recovered from the debilitating effects of the campaign, and the 3rd

Squadron naturally suffered the most in this respect as the result of its prolonged detention at Suakin.

This chapter may conclude with some notes on the changes in uniform consequent on the Ninth becoming a lancer regiment. Of these, the most noticeable was the alteration of the facings from red to white. In other respects there was not much modification of the dress of the Indian ranks except that the seams of the kurta were edged with a "light," or piping of the colour of the facings. The pennon adopted for the lances was dark blue and white. The full dress of British officers (except when on duty with the regiment) was entirely altered. The braided tunic worn since 1874 was replaced by a lancer tunic with white lapels, lancer girdle of crimson with gold lace, and lancer cap lines. White gauntlets were worn in mounted dress instead of gloves, and lancer (or "knee") boots took the place of Napoleons. When on duty with the regiment British officers still wore the kurta with Kashmir shawl pattern kammerbund and the blue lunghi, as worn by the Indian ranks.

The several uniforms here briefly described continued in use with but small modifications for the next thirty years, although as time went on khaki took the place of the blue dress with constantly increasing frequency. Among the changes made from time to time in various details may be mentioned the abolition in 1902 of gold stripes on breeches and overalls and the substitution of a double stripe of white facing cloth (nominally in the interests of economy but an economy of doubtful value), the substitution of steel for brass spurs with overalls in full and mess dress, and also in 1902 the final disappearance of Napoleon boots which up to this time had been worn by Indian officers and other ranks, and which were replaced (for use by officers) by lancer boots and (for other ranks) by ankle boots and black puttees.

CHAPTER III.

1886-1898.

SEVERAL years now passed in cantonments without any incident of importance. The regiment was stationed at Ambala till the autumn of 1888, when it marched to Nowshera, being relieved at the former place by its sister regiment the 10th Bengal (D.C.O.) Lancers. At Nowshera it remained for three years, and in November 1891 it moved to Peshawar. Its stay at Nowshera was interrupted once or twice by camps of exercise, and in March 1891 three squadrons and the headquarters marched to Hoti Mardan and formed part of a column called the Buner Field Force of which the object was by a show of force to keep the Bunerwals quiet while a British expedition entered the adjacent Black Mountain country. No disturbance occurred on the Buner border, the object of the assembly of the force was accomplished, and in April the Ninth returned to their cantonments at Nowshera.

Meanwhile in the previous years the only internal change in the regiment had been a rearrangement of its class constitution in 1886, consequent on the establishment being (in common with that of all Bengal Cavalry) increased from three to four squadrons. These were henceforth constituted as follows:—

1st Troop . . Sikhs
2nd Troop . . Sikhs
3rd Troop . . Sikhs
4th Troop . . Dogras

GENERAL SIR ARTHUR POWER PALMER, G.C.B., G.C.I.E.,

Commander-in-Chief in India.

Reproduced by permission of Messrs Elliott & Fry.

5th Troop . .	Punjabi-Muhammadans
6th Troop . .	Punjabi-Muhammadans (Tiwanas)
7th Troop . .	Pathans
8th Troop . .	Hazaras.

Colonel A. P. Palmer, who had been appointed to command the regiment just as it started for Suakin, was promoted to be "Colonel on the Staff," and left the Ninth on 4th December 1888, after being upwards of thirty years in Hodson's Horse, of which nineteen were passed in the Ninth. He was succeeded by Lieut.-Colonel D. H. Robertson, who held the command until February 1893.

During its tour of service at Peshawar the regiment was unfortunate in suffering a good deal from local fever, but in all other respects its stay there was uneventful and the annual reports by inspecting officers show that it fully maintained its unbroken record of efficiency. In 1893 a small party of ten men from the 9th Bengal Lancers under Kot Dafadar Ajub Khan accompanied Sir Mortimer Durand on his diplomatic mission to Kabul, and on their return they and the regiment were complimented by Sir M. Durand and by the Government of India on their good behaviour and useful service.

In August 1894 the Ninth suffered the loss of its Risaldar Major, Mirza Ghulam Ahmad Khan, who died at Rawalpindi, having joined the regiment as a jemadar in 1876. He was succeeded as Risaldar Major by Risaldar Muhammad Akram Khan, but three months later this officer was appointed British Agent at Kabul with the honorary rank of Lieut.-Colonel, and was seconded from the regiment to take up that post in succession to Risaldar Major and Hon. Lieut.-Colonel Mirza Ata-ullah Khan, 10th Bengal Lancers.

In November 1893, after attending a Cavalry Camp of Exercise at Nowshera, the regiment marched in relief to Rawalpindi under the command of Lieut.-

Colonel E. E. Money who had been appointed to the command vacated by Colonel Robertson.

Lieut.-Colonel Money's tenure of the appointment was cut short in a tragic manner only a year later. In November 1894 the Ninth marched to Lahore in order to be present at a durbar and review held there by the Viceroy, the Earl of Elgin. After the durbar the regiment proceeded to Muridki, where an instructional camp was formed under the Inspector-General of Cavalry. The Ninth took part in the manœuvres there, and was about to continue the return journey to Rawalpindi when, on the night of 20th December, a dafadar named Kartar Singh, inflamed by some crazy spite about his loss of promotion, took a carbine from the guard tent where he was on duty and shot a risaldar, Kesar Singh by name, against whom he had a grudge. On the outrage being reported to the British officers who were in the mess tent, they all hurried out to investigate the affair. The murderer, concealed by the darkness of the night, was waiting near the tent and shot the commandant through the head before he had gone a dozen yards. There is little doubt that he intended to shoot others of the officers also, but fortunately he was encountered and promptly seized by Jemadar Ajub Khan before he could do further harm. He was tried at Rawalpindi by a court-martial composed entirely of Indian officers and was condemned to be hanged. The sentence was carried out in public on 31st January 1895.

In the course of the inquiry into this crime it appeared that the actual perpetrator was not alone guilty, and that he had been instigated and suborned by others. As a result of these investigations two Sikh officers, Jemadars Jaswant Singh and Harsa Singh, were arrested and were ultimately handed over to the civil authorities for trial, and on being convicted were sentenced to transportation for life.

After the Muridki manœuvres which had ended in such tragic circumstances the regiment reached

Rawalpindi by rail on 22nd-23rd December, under the command of Major G. L. Garstin, who was appointed to succeed Lieut.-Colonel Ernle Money.

The spring of 1895 was marked by the despatch of the Chitral Relief Force, consequent on the tribal outbreak at Chitral and the investment of that place. The field force under the command of Major-General Sir Robert Low, K.C.B., having assembled near Hoti Mardan crossed the frontier on 1st April, forced the Malakand Pass on the morning of the 3rd and pushed on into and beyond the Swat Valley.

In these operations the 9th Bengal Lancers played a subsidiary part. Their first instructions were to take over garrison duties at Mardan with three squadrons, one squadron remaining at Rawalpindi. Mardan was reached on 1st April. On the 8th April one squadron was ordered to Durgai for duty on the line of communications, with detachments at the Malakand Pass, Shahkot, and Jalala. Three days later the headquarters and two other squadrons moved forward to join the 1st Brigade of the Chitral Relief Force, under Brigadier-General A. A. Kinloch, C.B., at Khar in the Swat Valley. But the regiment was not destined to go beyond this point, or to have any share in the active service of the field force. Chitral was relieved on 20th April, and all serious opposition to Sir Robert Low thereafter ceased. The 11th Bengal Lancers from the field force fell back to Khar and the Ninth was ordered to return to Mardan, where they arrived on 25th April. The squadron at Durgai continued on detachment there for a little while longer, but early in May the whole regiment was once more in quarters at Rawalpindi.

Lieutenant R. B. Low accompanied the Chitral Force as A.D.C. to the Major-General commanding and was awarded the D.S.O. for his work in the operations, and the regiment was allowed to add the name "Chitral" to the list of honours on the appointments.

The close of 1895 was clouded for the Ninth by the shadow of tragedy, as had been its beginning, for in October news was received of the murder at Kabul by one of his servants of Risaldar Major and honorary Lieut.-Colonel Muhammad Akram Khan, who, as has been seen, had only one year before been appointed to the important post of British Agent at the capital of Afghanistan. Lieut.-Colonel Akram Khan had served for sixteen years in the regiment. He was not only a fine specimen of the martial races of India, a thorough gentleman and respected by all ranks, but also an excellent officer, whose untimely death was a great loss to the public service.

A brief mention must be made here of an administrative change which, although it did not materially affect the Ninth Bengal Lancers either at the time or subsequently, was nevertheless of considerable importance. This was the abolition on the 1st April 1895 of the Presidency Army system. In accordance with the orders of Government issued on that date the armies of Bengal, Madras, and Bombay ceased to exist individually and separate from each other, the offices of Commander-in-Chief in Madras and Bombay were abolished, the whole of the forces in the country were brought directly under the Commander-in-Chief in India, and the Bengal Army was organised as the Bengal and Punjab Commands, each under a Lieut.-General. The Ninth, as was natural in view of its composition, was included in the Punjab command. Its title remained unchanged for eight years more, but henceforward the use of the name "Bengal" to describe either regiment of Hodson's Horse became more than ever a misnomer.

There is little to record of the next eighteen months, except the appointment to the regiment as an honorary Captain of H.H. Hamid Ali Khan, Bahadur, Nawab of Rampur, whose connection with the Ninth Bengal Lancers was destined to continue for many years in peace and war. Early in 1897 Risaldar Nadir Khan was selected to form part of a guard of honour of Indian officers to do duty in London on the occasion of the Diamond Jubilee of H.M. Queen Victoria, and he proceeded to England on that service, rejoining the regiment towards the end of August. He was made a member of the Order of British India in recognition of his selection for this honourable duty, besides receiving the Jubilee memorial medal.

Another item to be noted here is the success of the regiment in the spring of 1897 in winning the annual tent-pegging tournament for Bengal cavalry.

This record has now been brought down to a year which was marked by an outburst of fanatical hostility and outrage on the North-West Frontier of India without parallel in the previous quarter of a century. Beginning with a treacherous and quite unexpected attack on an escort in the Tochi country in June 1897, the wave of fanaticism spread to the tribes in the Swat Valley, who in July made violent onslaughts on the garrisons of the Malakand and Chakdara posts. A fortnight later an equally unlooked-for attack was made by the Mohmand tribesmen on Shabkadar fort, only eighteen miles from Peshawar, and the climax was reached when on 23rd August the whole of the forts in the Khaibar were attacked and overwhelmed by a rising of the Afridis, while on the 12th September similarly violent assaults were made by the Orakzais against the fortified posts on the Samana ridge.

Such an accumulation of outrages necessitated punitive measures of exceptional magnitude on the part of the Government of India. The trouble in the Tochi Valley and in Swat was dealt with locally and with promptitude, but more prolonged and deliberate preparations were required before reprisals could be exacted from the Orakzais and especially from the Afridis, a people who excel in all the qualities of ruthless and tireless guerilla warriors as much as their country and its approaches are remarkable for frowning inaccessibility. It was not until October that the main operations against the hostile tribes could be opened, but meanwhile there had been two months of considerable activity all down the frontier, in which the 9th Bengal Lancers had a share.

On the 10th August, two days after the Mohmand attack on Shabkadar, orders were received by the regiment at Rawalpindi to proceed at once by train to Peshawar, there to take over the lines of the 13th Bengal Lancers, who had moved out to the Mohmand border. A third of the rank and file were absent on furlough and leave, and orders were sent to all of these to rejoin at Peshawar. The orders to move were received at 7 A.M. and the regiment left Rawalpindi the same evening, reaching Peshawar on the morning of the 11th August. On the same day a troop was sent out to Shabkadar as escort to guns, and was absent from the regiment for three days. On 17th August another detachment was sent out to Bara Fort, under Captain R. B. Low, D.S.O., and four days later, on the threat of an attack on that post, the headquarters and two squadrons of the regiment, together with a half battalion of the 1st Devons and the 30th Punjab Infantry, were also moved to Bara. Then on 23rd August came the attack on and capture by the Afridis of the Khaibar forts and the danger of an attempt against Jamrud. The force at Bara was hurriedly moved towards the Khaibar, but later in the day it returned to Bara with the exception of the Ninth, which was diverted to Hari-Singh-ki-Burj. On the 28th, however, the regiment was again at Bara but only until 6th September when the headquarters with one squadron rejoined the squadron that had remained at Peshawar, leaving one and a half squadrons at Bara and sending one-half squadron to Jamrud.

This was the position when on 22nd September

orders were received for formal mobilisation for field service. A depot was formed at Rawalpindi, and the regiment waited at Peshawar for instructions regarding its destination. These were received on 5th October when the Ninth was ordered to proceed once more to Bara, there to form part of the Peshawar column of the Tirah Expeditionary Force. The regiment marched from Peshawar on 6th October under the temporary command of Major H. L. Dawson, the commandant Lieut.-Colonel G. L. Garstin being on his way out from England.

The plan of the operations now about to begin was very briefly as follows. The main column, under General Sir William Lockhart, consisting of two divisions, with its base at Kohat was to march direct into the heart of the Afridi fastnesses in Tirah, a region which had never before been entered by a hostile army. Meanwhile a mixed force under Brigadier-General A. G. Hammond, V.C., C.B., D.S.O., A.D.C., termed the Peshawar Column, was to be formed in the neighbourhood of Peshawar, and another similar force, termed the Kurram Column, was to be concentrated in the Kurram Valley. These two latter columns were to support the main advance on the right and left respectively. There was also a reserve brigade at Rawalpindi.

The Peshawar Column consisted of :—

The Ninth Bengal Lancers.
57th Field Battery, R.A.
No. 3 Mountain Battery, R.A.
No. 5 Company, Bengal Sappers and Miners.
2nd Battalion Royal Inniskilling Fusiliers.
2nd Battalion Oxfordshire Light Infantry.
9th Gurkhas.
45th Sikhs.

Although the part assigned to the column to which the Ninth was attached was a comparatively inactive one, yet at the very outset of the operations a duty fell to the lot of a wing of the regiment which involved a severe test of the steadiness and soldierly qualities of all ranks, an ordeal through which they came with conspicuous credit.

On the morning of 18th October Captain G. P. Brasier-Creagh with a force consisting of B and D squadrons, commanded respectively by Lieutenant C. A. Smith and Lieutenant L. L. Maxwell (of the 2nd Bengal Lancers, temporarily attached to the regiment), was ordered by Brigadier-General Hammond to proceed through the Gandao pass (ten miles west of Bara) as far as Mamani at the western outlet of the pass, to reconnoitre and make a sketch of the road to that place, and to find and report on an encamping ground in the vicinity of Mamani. These quite clear instructions, which were in writing, were followed by the rather cryptic warning that the force was not to "get entangled" in the hills. Seeing that the whole road to be reconnoitred and sketched, and the village which was the ultimate objective, were surrounded by hills, it could hardly have been intended that the force should not penetrate into the hills. Probably what the writer of the orders meant was that Captain Brasier-Creagh should avoid being entrapped into an action with the tribesmen among the hills. It was thus at any rate that he interpreted his orders. Accordingly leaving B Squadron at the mouth of the Gandao pass he advanced with the remaining squadron with the utmost caution. The heights on either side were crowned by dismounted men as the headquarters of the squadron and the led horses moved forward through the pass. No signs of any enemy were seen. Mamani was reached and all the orders of the Brigadier were carried out exactly. Captain Brasier-Creagh then started on the return march. In the course of his reconnaissance he had discovered a somewhat better road which followed the banks of the Bara River, and he determined to proceed by this. He accordingly sent a message to Lieutenant

Smith, commanding B Squadron, directing him to cross the river and meet him where the other road debouched from the hills.

Captain Brasier-Creagh with D Squadron had proceeded some little distance along the river road when they were suddenly fired on from the hillside in front, and found their passage barred by a number of tribesmen under cover among the rocks. Part of the squadron was immediately dismounted and the fire of the enemy was returned, and at the same time some men from the rear of the column were ordered to go back and ride down the road through the Gandao pass to ascertain whether that way was open. They had not gone far, however, before they found that here too the road was blocked.

The position of the squadron was now critical. The narrow track between the hills and the river necessitated the men and horses being strung out in single file, peculiarly exposed to hostile fire and at the same time very difficult to protect. As quickly as possible dismounted fire had been brought to bear on the enemy with what appeared to be a good deal of success, but several casualties had occurred and it was quite evident that only by rapid action could the squadron be extricated. The men on foot were therefore remounted and the whole squadron galloped down the road and through the opposing tribesmen. The losses incurred were heavy, but unquestionably the course adopted by Captain Brasier-Creagh, as well as the coolness and gallantry displayed by him and Lieutenant Maxwell and indeed by all ranks, alone saved the squadron from much more serious casualties. As it was, Jemadar Sarwar Khan and five men were killed, four men were wounded, and five were at first reported missing but afterwards made their way into camp, three of them having been kept as prisoners for several days by the enemy. Moreover seventeen horses were killed or missing and ten were wounded of which one was afterwards destroyed. Sixteen non-commissioned officers and men were brought to notice for conspicuous gallantry in this trying ordeal, of whom Kot Dafadar Mir Jafar Khan received the Order of Merit.

After this affair there followed several weeks during which the Ninth, as is usually the case with cavalry in frontier warfare, were split up into numerous detachments and employed incessantly on patrols, escorts, road piquets, reconnaissances, protection of convoys, and all the thousand and one odd jobs that have to be done in such circumstances and for which it is always so easy to call in the aid of a man on a horse. Work of this sort is as trying and exhausting as it is wearisome and thankless. Moreover on the borders of the Afridi country it is also not without its dangers, and in the course of these duties the regiment lost men on more than one occasion during the operations. Meanwhile the headquarters under Lieut.-Colonel Garstin (who rejoined from leave on 27th October) remained at Bara till 19th November, when they moved with two squadrons and the rest of the Peshawar column to Ilam Gadr, about three miles nearer the frontier. Thence a fortnight later the whole force moved through the Gandao pass, past Mamani to Swai Kot, but here it was found that the country by reason of lack of forage and its general character was quite impossible for cavalry, so the Ninth were sent back to Bara, only one field troop remaining with the column.

The headquarters of the regiment were still at Bara when on 17th December the 2nd Division, Tirah Expeditionary Force, under Major-General A. G. Yeatman-Biggs, C.B., having completed the punitive march through the Afridi country, arrived at that place. Thereupon the Peshawar column moved to Jamrud and later to Landi Kotal, but the Ninth did not accompany it. With the exception of the one field troop already mentioned the regiment, or at least the headquarters thereof, remained for the present at Bara and was now attached to the 3rd brigade of the Tirah Force under Brigadier-General Kempster, D.S.O., A.D.C.

Another month now passed in renewed detachment and other duties, one squadron being sent to the 1st Division of the Tirah Expeditionary Force at Jamrud for escort duty with the G.O.C., Major-General Symons, C.B. Some difficult work fell to the lot of that division during this second phase of

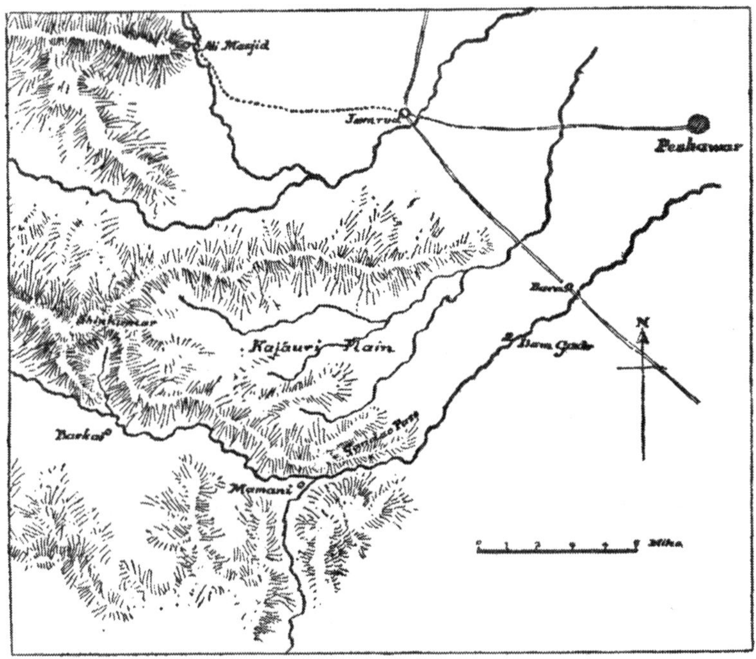

the Tirah operations, but so far as the 2nd Division at Bara was concerned hostilities were practically at an end, or would have been so but for an unfortunate affair on 29th January 1898.

News had been brought to Bara that the tribesmen were daily in the habit of bringing down large numbers of flocks and herds to graze in the Kajauri plain, which lies due west of Bara and is encircled on three sides by the hills of the border. Plans were accordingly laid to cut off and capture the enemy's live stock, and it was arranged that columns should move simultaneously from Bara, Mamani, Jamrud and Ali Masjid, and converge on the Kajauri plain. The headquarters and one and a half squadrons of the Ninth accompanied the Bara column and formed a protective screen in front of the infantry. It would seem, however, that news of the plan had reached the Afridis, for there was no sign of their grazing flocks, nor (so far as the Bara column was concerned) was any resistance encountered beyond a few stray shots at the cavalry screen. Distant firing was heard as the column was withdrawing, and it afterwards appeared that the column from Mamani had been seriously engaged on difficult ground in the Shin-Kamar pass and had suffered severe losses.

Three days later, on 1st February, a squadron under Captain Brasier-Creagh, supported by a second squadron under Captain Low, carried out a more successful raid on the enemy's property. Passing through the Sumghaki pass Captain Creagh with his squadron came suddenly on a large herd of camels. The tribesmen in charge were completely surprised and had no time to offer resistance or to collect to oppose the action of the troops, who drove off the whole herd and brought the animals safely into the camp at Bara to the number of 135.

This exploit may be said to have marked the end of the Tirah campaign so far as the Ninth was concerned. After two more months of continuous detachment and escort duties orders were at length received on 4th April for the regiment to concentrate forthwith at Peshawar and thence to move by rail to Multan where it was to be quartered, and at that station it arrived accordingly on 7th April 1898.

Two years later, by G.G.O. No. 288 of 1900, permission was given to the regiment to bear on appointments the words : " Punjab Frontier " in recognition of its services in the campaign of 1897-98.

CHAPTER IV.

1898-1914.

Almost ten years may now be passed rapidly over, during which few notable events occurred in the history of the Ninth. Of this period the first four years were spent in cantonments at Multan (one wing, however, being in 1899 detached to Dera Ghazi Khan). On 3rd March 1902 the regiment left Multan on relief for Jullundur, arriving there on 28th March, and there it remained until 26th November 1906, but for occasional absences at manœuvres including the concentration of troops at Delhi for the Coronation Durbar in January 1903. In 1906 the Ninth marched to Cawnpore, stopping at Agra by the way, where it took part in a concentration of troops on the occasion of the visit to India of the Amir of Kabul, and where it was inspected by the Commander-in-Chief, Lord Kitchener of Khartum. Arriving at Cawnpore on 8th February 1907, the regiment relieved the 10th D.C.O. Lancers (Hodson's Horse).

In the course of these years the annual reports of inspections show that the regiment invariably maintained its very high reputation. Thus in March 1899 the Inspector-General of Cavalry wrote: "The 9th Bengal Lancers is one of the finest regiments in India . . . fit to go anywhere at twenty-four hours' notice," and similarly satisfactory comments were made on other occasions.

Meanwhile there had been two changes of commandant—namely, in December 1901, when Colonel G. L. Garstin retired and was succeeded by Lieut.-Colonel F. W. P. Angelo; and again six years later, when Lieut.-Colonel A. G. Peyton assumed command on Colonel Angelo being appointed Colonel on the Staff at Faizabad.

An important change in the armament of the regiment was made in April 1903, when the Lee-Enfield rifle, Mark I., was issued in the place of the Martini-Henry carbine, which had been the firearm in use by Indian cavalry for the past thirteen years.

About the same time (in 1902) the use of swords by lancers was abolished, but this order was quickly rescinded and in 1903 swords were restored as part of the armament.

Two very notable changes in the title of the regiment were made during these years. In 1901 the name "Hodson's Horse" was for the first time since 1861 officially added to that of "9th Bengal Lancers," and again by Indian Army Order No. 181 of 1903 the designation was changed once more, the time-honoured but quite misleading title "Bengal" disappeared, and the regiment now became "the Ninth Hodson's Horse."

Although the decade now under review offered no opportunity to the regiment as a whole for active service, yet on several occasions during that time individuals were selected for special duties both in war and peace and all of these earned credit for themselves as well as for their regiment. Particularly was this the case in the course of the South African War from 1899 to 1902. Soon after its outbreak a party of men and syces under Dafadar Wadhawa Singh with thirty horses were sent to South Africa for mounted infantry purposes. The horses were sent with saddlery, clothing and stable kit complete, and were taken over by the Government at an average price of Rs. 600 each, besides the value of the saddlery, &c. Lieut.-Colonel H. L. Dawson was selected to take charge of the horses which were sent to South Africa from a number of regiments, and about the same time Captain G. P. Brasier-Creagh was selected to take some 800 horses from Imperial Service troops which were also despatched to Africa.

Lieut.-Colonel Dawson was later on appointed to command "Roberts's Horse," and subsequently a corps of Mounted Infantry. He rejoined the Ninth in November 1900. Captain Brasier-Creagh was appointed to succeed Lieut.-Colonel Dawson in command of Roberts's Horse, and when serving with that corps he was mortally wounded and died on 27th April 1900.

Dafadar Wadhawa Singh, mentioned above, was on his arrival in South Africa appointed by Field-Marshal Lord Roberts to be his personal orderly. He accompanied the Field-Marshal throughout the campaign, and afterwards to England when Lord Roberts left South Africa in November 1900.

In December 1901 Captain Lukin was selected for service in South Africa, where he commanded a squadron of the 32nd Imperial Yeomanry Hussars. He served with that regiment till the end of the war, and in August 1902 he took the squadron to England.

Early in 1902 two other parties of thirteen and fifteen respectively, consisting of non-commissioned officers and men, proceeded to South Africa, the first for remount duty and the second as a supervising establishment with syces.

Meanwhile in 1900 Major F. W. P. Angelo, Captain R. W. B. Low, D.S.O., and Captain Beatty were selected for field service with the China Expeditionary Force. All of these officers were specially mentioned in despatches, and Major Angelo and Captain Low each received brevet promotion.

Turning to special services in peace, it should be recorded that Captain A. W. Pennington, Ressaidar Deva Singh, Jemadar Ajub Khan, Senior Hospital-Assistant Devi Ditta and Ward Orderly Jaimal

Singh formed part of the Indian contingent which proceeded to England in May 1902, and was present at the coronation of H.M. King Edward VII. The Order of British India was awarded to Ressaidar Deva Singh and Jemadar Hospital-Assistant Devi Ditta in commemoration of this event.

In the following January, when the whole regiment took part in the military concentration at the Coronation Durbar at Delhi, Trumpeter Kala Singh was selected to be one of the trumpeters accompanying the Chief Herald at the Durbar, and the silver trumpet with embroidered bannerol, which he used on the occasion, was presented to the regiment.

In March 1906 Risaldar Major Muhammad Ali Beg, Sardar Bahadur, proceeded to England to be one of the Indian orderly officers on duty with H.M. the King-Emperor during that season. The Risaldar Major, who was accompanied by Dafadar Nur Ahmad, rejoined the regiment on 27th August 1906.

This list may be closed with the note that in 1904 the Ninth was called upon to supply a party of six men to serve as consular guard and escort to the British Minister in Persia. Dafadar Rajah Khan, with five sowars, was deputed for the duty, and went to Teheran in September 1904. Rajah Khan unfortunately died the following year, and his place in Persia was taken by Sowar Muhammad Khan.

The year 1907, at which we have now arrived, brought the fiftieth anniversary of the raising of Hodson's Horse, and the Ninth determined to celebrate the occasion in a suitable manner. It was decided that the week from 2nd to 8th December would be the most suitable time, and the fact that this coincided with the Cawnpore polo week facilitated the entertainment of visitors. One hundred pensioners of all ranks were invited to take part in the festivities as the guests of the regiment, free railway tickets being provided for them with the generous assistance of the railway authorities, who granted return tickets for single fares. The regiment received handsome donations from Hon. Major H.H. the Nawab of Rampur (Rs. 2000) and Hon. Lieutenant Malik Mobaraz Khan (Rs. 500) towards defraying the cost of these journeys and of the entertainments, which was necessarily heavy.

The proceedings opened on 2nd December with a reception by the British officers of all the pensioners, among whom were eight veterans of the Mutiny Campaign. These, who had all served under Hodson, were as follows :—

Risaldar Hukm Singh.
,, Jamal Khan.
Dafadar Khushal Singh.
,, Kishn Singh.
,, Indar Singh.
,, Sultan Khan.
,, Ditta Khan.
Sowar Santokh Singh.

The events of the succeeding days included a torch-light tattoo, a gymkhana, ceremonies in the Masjid and Mandir, regimental sports, a regimental ball, and on Saturday evening a dinner to the whole regiment and the guests, followed by a nautch.

On the following day, Sunday, 8th December, the Mutiny veterans and a deputation from both regiments of Hodson's Horse, headed by Brigadier-General F. W. P. Angelo, went over to Lucknow and deposited on Hodson's grave in the grounds of La Martinière College a wreath of white flowers, tied with ribbons of the regimental colours, and a card bearing these words :—

MAJOR W. S. R. HODSON

IN MEMORIAM

FROM HIS COMRADES IN ARMS

OF

IX. AND X. HODSON'S HORSE.

On 9th December the pensioners and the party from the 10th Lancers, who had been the guests of the Ninth during the celebrations, left Cawnpore, having thoroughly enjoyed the welcome which they had received from the regiment. On the same day Brigadier-General Angelo returned to his new command at Faizabad, bidding a final farewell to the 9th Hodson's Horse, much to the regret of all ranks.

There now follows another quiet period of service in cantonments with very little of interest to record. During the years 1907 to 1911 the regiment remained at Cawnpore. A squadron was in 1907 detached temporarily to Calcutta and once or twice there were brief absences at manœuvres.

In September 1909 Risaldar Mir Jafar Khan was selected to be A.D.C. to the Commander-in-Chief in India, and two years later the same officer together with Lieut.-Colonel R. B. Low, D.S.O., and Risaldar Major Ram Singh were chosen for the more signal honour of forming part of the contingent which represented the Indian Army at the coronation of their Majesties King George V. and Queen Mary.

In October 1911 there was once more a change of commandant, Lieut.-Colonel R. B. Low succeeding Colonel A. G. Peyton, whose tenure of the appointment expired.

At the end of the same month the stay of the regiment at Cawnpore came to an end and the Ninth left that station *en route* for Ambala. On the way, however, they halted for a month at Delhi, where they took part in the great Coronation Durbar at which their Majesties the King and Queen were present in person. At this ceremonial the regiment again supplied one of the trumpeters (Trumpeter Sadr Din), who attended the Chief Herald, and as in 1903 the silver trumpet used by him was presented to the officers' mess as a memento. Moreover pensioned Dafadar Wadhawa Singh (the same who had been Lord Roberts's orderly eleven years before in South Africa) was selected to be State mace-bearer at the Durbar. In honour of the occasion Hon. Colonel H.H. the Nawab of Rampur, G.C.I.E., A.D.C., was made a Knight Grand Cross of the Royal Victorian Order, and Risaldar Major Ram Singh received the Order of British India, while several officers and men were awarded the commemorative medal.

The regiment left Delhi on 19th December on the conclusion of the Durbar ceremonies and reached Ambala on the 1st of January 1912.

The last item to be recorded in this uneventful period of peace is the appointment on 10th December 1912 of Major-General F. W. P. Angelo to be Honorary Colonel of the regiment, the first officer to be selected for this distinction.

CHAPTER V.

SERVICE IN FRANCE, 1914-1918.

THE month of August 1914, memorable for the outbreak of the Great War, found the Ninth Hodson's Horse still stationed at Ambala where it formed part of the Ambala Cavalry Brigade under the command of Major-General C. P. W. Pirie, the other units being A Battery, Royal Horse Artillery (the "Chestnut Troop"), the 8th (King's) Royal Hussars, and the 30th Lancers (Gordon's Horse).

The hot weather being at its height, a large proportion of the men were on leave and furlough when on 31st August orders were received for the regiment to mobilise for service in France. At such a season the strain of rapid preparation was naturally trying, but all worked cheerfully, buoyed up as they were with the prospect of active service in Europe, a distinction which not even the most sanguine had ever anticipated.

One of the first duties was the thankless task of selecting a staff for the depot to be left behind at Ambala, and for this the lot fell on Lieutenant E. V. Seymour [1] as commandant, with Lieutenant F. W. Messervy [1] as Adjutant, and two Indian officers, Risaldar Mir Jafar Khan, Sardar Bahadur, and Jemadar Bhola Singh.

All unfit men and horses were weeded out and posted to the depot, and by 12th September the

[1] Both of these officers subsequently joined the regiment on field service

mobilisation was complete and the regiment ready to start. It was not, however, until 9th October that the journey was begun, the regiment moving in three troop trains to Bombay, where on arrival it was encamped near the Bombay Gymkhana Club. No noteworthy incident occurred *en route* from Ambala except the escapade of a horse from A Squadron which managed to jump out of the train by the way. It was, however, recovered by the next troop train a few miles farther down the line, quite uninjured and seemingly making its way to Bombay.

On 13th October the regiment under Lieut.-Colonel R. B. Low, D.S.O., embarked for France on board two vessels, the *Clan Macphee* and the *Aratoon Apcar*, its strength being 8 British and 18 Indian officers, 521 rank and file, and 58 followers. The rest of the Ambala brigade and other units made up a convoy of sixteen ships in all, which set out from Bombay on the evening of 16th October.

The voyage occupied twenty-two days, including a stop of three days at Port Said. The time was passed in lectures, drills, and exercising the horses as much as possible, a few spare stalls enabling this to be done. Three horses died during the voyage and one sowar had to be left sick at Port Said. Marseilles was reached on 7th November and the regiment disembarked next day and went into camp at La Valentine. Thence a few days later it was moved by squadrons to Orleans, where it was concentrated on 17th November and encamped at Les Grues. Here those British officers who had been at home on leave when the war broke out rejoined the regiment, and a busy but very uncomfortable ten days were spent in equipping the men with warm clothing, bayonets, &c., and in getting the horses into condition. The latter stood the change of climate wonderfully well, in spite of the fact that the weather was very cold, with much snow, and that the camp was deep in slush and mud. But for the men the severe weather and damp meant great discomfort and they frequently put on so much clothing that they could only mount their horses with difficulty.

On 26th and 27th November a move was made for the Flanders front. The regiment was entrained by squadrons and, detraining on the 28th at Lillers, marched to Le Reveillon and Allouagne some five miles west of Bethune, where they went into billets. This was the regiment's first experience of billeting, but it was not long before all ranks learned how many men or horses could be got into a barn, and also realised that many an unprepossessing place will furnish quite good cover from the elements.

Here some more time was spent in exercising the horses and in getting all ranks acquainted with the surrounding country as well as in training the men in the use of the bayonet and of hand grenades. Here too the regiment saw something of that dreadful unpreparedness for a big war which, in small ways and in great, was to be the bane of the British army in France for many months to come. Bombs and hand grenades were almost unobtainable, and make-shifts of all sorts—"hair-brush," "tobacco-tin," &c. —had to be improvised for lack of something better. Almost more serious and certainly less excusable was the complete lack of horse-shoes. Before being disembarked all horses had been shod with the light shoe of Indian pattern, but four weeks' work on French roads, many of them *pavé* or cobbled, had so worn these shoes that many horses were by now unshod. It was therefore deplorable that the Ordnance Department could not supply any shoes at all and continued incapable of making good this deficiency until the new year. The result was that a great number of horses were laid up with laminitis and that many had to be destroyed, while others were rendered useless for a long period.

Another defect in the organisation was the fact that, although waggons had been provided for the transport of the Indian units, it had not occurred to any one that Indian drivers ought to accompany the troops. Throughout the year 1915 none but English drivers were available, and the difficulties inevitably arising from such an arrangement were not decreased by the further fact that these men were untrained civilians and unused to discipline.

On 9th December orders were received for squadrons to be sent by turns into the trenches for instructional purposes, and during the ensuing fortnight all squadrons in rotation carried out this duty for twenty-four hours at a time, gaining much useful experience thereby.

During this period there was heavy fighting about La Bassée and consequently it was not unexpected when at 1.25 P.M. on 21st December orders were received for the brigade to assemble at 2.20 P.M. and move forward. Arriving at Béthune the brigade was ordered to go into billets for the night, but at 1.15 on the morning of the 22nd sudden orders were issued for it to fall in at once, leaving transport and led horses at Béthune. The brigade moved off at a rapid trot along the road to Festubert, the Ninth leading. At Gorres a halt was made, the men dismounted, the horses were sent back to Béthune and the advance continued on foot, a wearisome trudge over rough ground in the dark. Festubert was reached a little before dawn, and here the brigade was ordered to attack at once and to retake some trenches that had been lost. After some delay, due to part of the brigade having missed connection in the darkness, the attack was about to begin, led by the Ninth, when the remnants of two previous attacks came back to Festubert with the information that they had been ordered to abandon the trenches and retire to that place. The brigade attack was accordingly countermanded, and A and D Squadrons at once took over some trenches south of Festubert. Subsequently B and C Squadrons relieved the 15th Sikhs in some trenches between that regiment and the 9th Gurkhas. This part of the line was particularly bad. The trenches were breast deep in water and the parapets were thin and made of wet sticky clay. It was the first experience of the regiment in the terrible conditions of trench warfare in Flanders and the men suffered severely. They had previously known nothing of "trench feet" and in any case the rapidity of their advance to the line and the urgency of the occasion would have prevented them from taking any precautions against the evil. By 2 P.M., when the regiment was relieved in the line and the brigade was concentrated at the cross-roads west of Festubert, the feet of many of the men were in a very bad condition from standing for hours in freezing water, and when a further move was made later in the afternoon most of these were quite unable to keep up.

At 4 P.M. the brigade fell in and marched about one and a half miles northwards to a cross-roads close to Rue de l'Epinette, where some reserve trenches were occupied, and here again C Squadron was unfortunate in having to take over a wet trench. Meanwhile before the brigade moved off the German advance had enabled some of their snipers to get into the village of Festubert, with the result that a few casualties occurred, among these being Ressaidar Sultan Muhammad Beg, who was severely wounded and paralysed for life.

The night of the 22nd was bitterly cold and snow began to fall in the morning, and it was therefore with much relief that at 1 P.M. on the 23rd orders were received for the brigade to fall in and march back to Béthune. By this time the men of C Squadron had suffered so severely that only twenty-five were able to parade with the regiment.

The horses were met about a mile out of Béthune and on reaching that town the brigade went into billets for the night. Next morning they moved off at 8.30 and marched to Laires, distant about twenty miles to the west, where the regiment, with the rest of the Ambala Brigade, was lodged in comfortable billets for some three months, the headquarters of the 1st Indian Cavalry Division (Major-General H. D. Fanshawe) being at Norrent Fontes, south of Aire, and Brigade Headquarters at Bomy.

In this brief but trying experience of trench warfare the regiment lost one man killed and Lieutenant T. W. Corbett, Ressaidar Sultan Muhammad Beg and eight men wounded, besides a considerable number incapacitated by frost-bite. Among the latter was the Commandant, Lieut.-Colonel R. B. Low, D.S.O., who was invalided to England and was never able to rejoin the regiment.[1]

During the time spent at Laires systematic squadron, regimental, brigade and divisional training proceeded constantly both by day and night, but mounted work was much handicapped by the condition of the horses' feet owing to lack of shoes. On the 19th January 1915 the Cavalry Corps was inspected by the Commander-in-Chief, Sir John French, who expressed his admiration of both men and horses.

At length on the afternoon of 7th March 1915, orders were suddenly received for the brigade to move forward, the Ninth to go into billets at Rely, some five miles towards Béthune. Although the distance to be traversed was but short, yet the roads were found to be so blocked by the 2nd Indian Cavalry Division, which was also moving forward, that billets were not reached until 9 P.M. This advance was in anticipation of the British attack at Neuve Chapelle, where it was fondly hoped that the German line might be broken and that an opportunity might be afforded for the use of cavalry. The event, however, proved that such hopes were vain, and after a further advance on 11th March to Marles and thence to the Bois des Dames, the brigade without having been employed retired to Cauchy, and, on the night of the 15th-16th, to

[1] The regiment was subsequently awarded the battle honour of "Givenchy, 1914," for the services here described.

their old billeting area south of Aire, where the Ninth went into billets in the villages of Nédon and Nédonchelles. Here five weeks were passed in comparative inaction, except for constant training work and for the employment of detachments in digging reserve lines in the neighbourhood of St Venant, and here too Lieut.-Colonel A. W. Pennington, who had up to now been employed with Imperial Service troops, rejoined the regiment and took over the command in place of Lieut.-Colonel Low.

This temporary quiet was rudely disturbed by the news on 22nd April of the German gas attack near Ypres, and by consequent orders on 24th April for the brigade to move northwards at once. The rendezvous was at 6.45 P.M., and after a cold, wet march of twenty-one miles through the night, Staple, five miles south of Cassel, was reached at 4.15 on the morning of 25th April. Thence at noon on the same day the brigade moved on to the neighbourhood of the Cassel-Poperinghe road, a little east and south of Watou. Once more however the services of the division were not required. On the morning of 2nd May the brigade marched back to Staple, and on the night of 4th-5th May to Racquinghem, about midway between St Omer and Aire, where the regiment was billeted until 27th May, except for a brief move to Allouagne on 17th-19th May.

On 27th May the Ambala Brigade, with the rest of the 1st Indian Cavalry Division, moved up for another turn in support of the Ypres front. Arriving at Steenbecque at 11 A.M. on the 28th they left the horses and moved on by 'bus to Brandhook where the Ninth went into huts one mile south-west of Vlamertinghe. Here gas masks were received for the first time (but of a very primitive kind) and the regiment practised the adjustment and use of them. All the officers now did duty for various periods with units in the line, and the machine-gun section was also employed on several occasions, being brigaded with sections from other units. In the course of these duties Captain Atkinson, Lieutenant Morris, and Jemadar Tek Singh were all slightly wounded. Meanwhile the headquarters of the regiment were none too peaceful. The huts were heavily shelled daily, and sometimes twice in a day, with extreme accuracy, but good shelter trenches were dug and the men got into them whenever necessary and casualties were thus avoided.

After about a fortnight at Vlamertinghe the division was sent back on 14th June to the neighbourhood of Aire, appreciative orders being published by Generals Plumer and Allenby, commanding respectively the 2nd Army and 5th Corps, in which they acknowledged the help and support received from the 1st Indian Cavalry Division.

The regiment was now billeted first at Ecques for about three weeks and then until the end of July at Cléty, about five miles still farther west. While at the first-named place the Indian Cavalry Corps was inspected by Lord Kitchener, and here too a visit was received by the Ninth from Major-General F. W. P. Angelo, their honorary colonel. Training work during these weeks was carried out with difficulty, the country being almost all under standing crops, and on two occasions large detachments numbering 200 men were sent forward to the front line and were employed for several days in digging reserve trenches.

A complete change of scene came on 1st August 1915, when after completing eight months' service in the Flanders area the Indian Cavalry Corps was transferred to the Somme and posted to the 3rd Army. Moving by way of Renty, Marles, Vauchelles and Condie-Folie the Ninth reached Long on 10th August, and here the regiment was billeted for a fortnight. The crops had now been harvested so that there was room for useful mounted training, and the time was thus passed profitably and not without amusement with fishing, partridge riding (now practised for the first time) and regimental sports, the latter much appreciated by the inhabitants.

On 22nd August however these occupations were interrupted by the receipt of orders for a trench party to go up to the line. Eleven British and eight Indian officers and three hundred of other ranks started from Long and marched to Forceville Wood, distant about thirty-two miles. From here the horses were sent back and at 10.30 P.M. on the 23rd the dismounted party proceeded through Martinsart to the trenches east of Authuille (on the banks of the Ancre about three miles north of Albert). These trenches were taken over from the 2nd Indian Cavalry Division and were held until the night of 2nd-3rd September, when the regiment was relieved by the 2nd Lancers and went back to billets at Long. A week later they went up to the line again and took another turn in the same trenches until finally relieved on the night of the 16th-17th.

During this trench duty the regiment lost four men killed and twenty-one wounded, while Captain Moody was awarded the Military Cross and four men received the Indian Order of Merit.

The Ambala Brigade was now transferred from the 1st to the 2nd Indian Cavalry Division, and on leaving the trenches the Ninth accordingly went into new billets at Picquigny for a few days, and thence on 23rd September moved to Bonneville in the Candas area. Here the brigade was located for a month, during which time the operations about Loos were in progress. All the rest of the Indian Cavalry Corps was placed during this period under the direct control of the British Army Headquarters, the Ambala Brigade alone being retained as reserve to the Third Army.

Whilst at Bonneville the regiment lost their Risaldar Major, Ram Singh, Sardar Bahadur, who retired on pension. A splendid type of Sikh soldier and gentleman, whose father and grandfather had assisted in raising a squadron for Hodson, and who himself was respected and beloved by all, he carried with him the good wishes of every one, and it was with the greatest regret that only a year later the regiment received the tidings of his death.

On the 21st October 1915 the Ninth moved to the neighbourhood of Oisemont to rejoin the 2nd Division, and here it went into billets for the winter, distributed comfortably in the following villages :—

Headquarters and D Squadron at Faucoucourt.
A Squadron at Nesle-Hôpital.
B Squadron and Machine Gun Section at Lignières.
C Squadron at Mouflières.

Several months now passed without incident, but fully occupied with training work of every description. Meanwhile among the British officers Captains J. C. Russell and F. St J. Atkinson left the regiment temporarily to do duty with infantry units, of which they were later appointed to the command; and in June 1916, Lieut.-Colonel A. W. Pennington, M.V.O., left for Egypt to take up command of the Patiala Lancers and was succeeded in the command of the Ninth by Lieut.-Colonel G. A. H. Beatty. The commands both of the Ambala Brigade and the 2nd Indian Cavalry Division had also changed hands and were at this time held respectively by Brigadier-General C. H. Rankin and Major-General H. A. MacAndrew.

On 10th June 1916 the division moved to the neighbourhood of Abbeville for training, but only until the 27th of the same month when, in preparation for the great infantry attack on the Somme front, the cavalry was brought forward in the hope that it might be possible to employ mounted troops. Any expectation of this sort proved however to be unfounded. One brigade (Secunderabad) did indeed break through the German line on the afternoon of 14th July, but the advance only reached as far as three-quarters of a mile when dismounted action and trench warfare reasserted their grip of the operations.

Meanwhile and for some weeks about this time the regiment frequently supplied working parties of 100 to 300 men to dig trenches round Mametz Wood, Longueval, Bécourt-Château and Contalmaison. These parties suffered considerable casualties, Jemadar Samand Singh and three men being killed, Lieutenant Douetil and thirteen men wounded. For the same operations Sowars Hayat Muhammad and Abdullah Khan were awarded the Indian Order of Merit.[1]

After several weeks at and near Querrieu, some eight miles east of Amiens, the regiment moved to Inval-Boiron and thence on 27th August to Lanchères near Cayeux-sur-Mer at the mouth of the Somme, where for ten days field firing and other training was carried out and every one enjoyed the luxury of sea bathing. From this pleasant interlude the next move was on 6th September through Inval to Molliens-Vidame, some ten miles west of Amiens, and thence to Bussy-les-Daours, where the cavalry concentrated on 7th September. On the 14th the 2nd Indian Cavalry Division advanced to Dernaucourt, just south of Albert, and on the following day to Montaubon, where for two days it waited behind the infantry line in constant hope of a chance to break through. Once more, however, these hopes were disappointed, and on 17th September the division fell back to Bussy and the Ambala Brigade on the 26th to Molliens-Vidame. Here five weeks were spent in combined manœuvres and other training, while recreation was found by the officers in occasional pig-sticking. On 1st November the Ninth moved to the valley of the Bresle, where in the area between Blangy and Gamaches the regi-

[1] For services during this period the regiment subsequently was given the battle honours "Somme, 1916," "Bazentin," and "Flers-Courcelette."

ment was billeted for the winter months in the villages of Rieux, Montières and Enfer.

During these autumn weeks all dismounted men were at first attached to the 14th Corps at Montauban and were used for consolidating the line. Later a Pioneer Battalion was formed, two hundred strong, commanded first by Major Rowcroft and afterwards by Major Dyce, and was attached to various corps in the course of the winter.

At the headquarters of the regiment the winter was spent in training of all sorts, staff rides, bombing practice, musketry, &c., in the course of which latter work an unfortunate accident occurred when Lieutenant Reeves, a very promising young officer, was accidentally killed, owing to a live cartridge having got among dummy rounds at a musketry parade.

On 1st November the regiment provided an escort for H.R.H. the Duke of Connaught when he inspected the Guards Division, and on 7th February 1917 a party under Major Rowcroft formed part of His Majesty's escort at the opening of Parliament.

In March 1917 there came the German general retirement, and in consequence of this the regiment moved forward until, on 24th March, it took over an outpost line in touch with the German rearguards from Etreillers to Beauvois, a few miles west of St Quentin. During the next four days a number of reconnaissances were carried out by day and night, both mounted and dismounted, of the Bois d'Holnon, Attilly, Marteville and Vermand. Much valuable information was obtained and the regiment received the thanks of the divisional commander for its useful work. These operations were not unaccompanied with loss, Kot Dafadar Sultan Muhammad being killed, four men missing, and afterwards reported as prisoners, and seven wounded. At the same time three dafadars earned the Indian Order of Merit and three sowars the Indian Distinguished Service Medal.

Six weeks followed without incident in billets, first at Warfusée-Abancourt (near Corbie-sur-Somme), and afterwards between Tertry and Coulaincourt (west of Vermand). Then on 15th May, the regiment being at Coulaincourt, a trench regiment 300 strong was formed under Major Rowcroft and attached to the Secunderabad Brigade, at first in support and afterwards in the trenches east of Le Verguier. Here it was employed until 29th May, in the course of which time six men were wounded and two taken prisoners. After a week's rest at Tertry a trench regiment of the same strength as before, commanded by Lieut.-Colonel Beatty, again moved up to the front, and was sent to Vadencourt as brigade reserve. Hence on 12th June B and C Squadrons relieved two squadrons of the 18th Lancers in the line, Squadrons A and D remaining in reserve at Vadencourt.

The two latter squadrons now devoted themselves for six days to careful and detailed practice and rehearsal of a raid, which it was proposed to carry out on the German trenches in and about the village of St Hélène, and the details of which it will be of interest to describe. Photographs of the trenches to be attacked and their surroundings having been obtained from aeroplanes, an exact plan was plotted out with tapes on somewhat similar ground a few miles behind the lines, and all noticeable features were marked, as well as enemy dug-outs and machine-gun posts so far as they were known. Here the whole plan of the raid was rehearsed both by day and night until everybody concerned was thoroughly familiar with his duties. Moreover all officers and non-commissioned officers and some of the men were taken on several occasions to an observation post in the French lines, which adjoined those occupied by Hodson's Horse, and from which the trench system to be attacked was commanded. Very careful measurements were taken of the distances to be traversed in the approach and these were checked by an officer's patrol, so that the time required could be accurately calculated.

The programme of the raid was as follows. There was to be no preliminary bombardment. The attack was to depend for success on complete surprise. The raiders were to advance in two columns, "A" and "B." The enemy's wire was to be cut by explosion and the trenches were then to be rushed. "A" column, leaving a party to block the trench towards the north, was to turn southwards, bayoneting the garrison and bombing the dug-outs as they went, until they reached the point of retirement, where another gap was to be blown in the wire. Meanwhile "B" column was to cross the enemy's front trench and proceed eastwards as far as a sunk road running north and south through St Hélène; here they were to turn south, pass through the village and so back to the gap of exit.

The raiders were to trust as far as possible only to the bayonet, but a proportion of bombs were to be carried, filled with phosphorus with a small bursting charge. These were to be ignited and hurled into the dug-outs. The explosion of such bombs scatters the burning phosphorus which sticks and burns wherever it falls, emitting also a thick suffocating smoke.

The wire was to be cut with "Bangalore torpedoes," 3-inch pipes filled with ammonal and exploded with a fuze and detonator. Each torpedo would cut a gap five yards wide, and as the raiding party was a large one two torpedoes were to be exploded so as to make a gap of ten yards. The German wire was some ten feet across, so each torpedo had to be twelve feet long in order to ensure cutting right through the wire. They were therefore awkward things to carry and it was particularly difficult to prevent the long metal tube from grating against the wire as it was pushed underneath. Specially good men were selected for this task and an Indian officer and non-commissioned officer were detailed for the duty of lighting each torpedo.

The explosion of the torpedoes, followed by a red and a green Verey light and a red and a green rocket, were to be the signals for a box barrage of artillery and machine-gun fire to come down round the threatened sector, and this was to continue for forty-five minutes. Five minutes before the barrage was to cease a rocket was to be fired to warn the raiding party.

The raid was carried out on the night of the 18th-19th June, and thanks to the previous careful practice everything worked exactly as arranged. The columns moved off at 10.10 P.M., " A " column under Captain Vigors (who also commanded the raid), and " B " column under Captain Stevens (10th Lancers), the men in jerseys and breeches, the officers without any regimental badges, all ranks with white bands on each arm. The magazines of rifles were charged, but no round was in the chamber and only ten rounds of ammunition were carried by each man, together with two Mills' grenades. At 11.20 the raiding force moved forward from the " assembly point," which was calculated to be five hundred yards from the German wire. Thence the raiders advanced slowly and very cautiously until, rather less than four hundred yards from the assembly point, the wire was seen in front. The columns now lay down while the torpedo parties crawled forward with their burdens. At 11.48 the torpedoes exploded and Verey lights were sent up. A moment later the raiding columns dashed forward into the German trench, and simultaneously down came the box barrage. All the Germans who were met were bayoneted and bombs were exploded in all the dug-outs by " A " column, while " B " column made their way through the village. The noise of the shells and machine-gun bullets close overhead was so deafening that it was almost impossible to make commands heard, while the men were so excited that in any case it would have been difficult to communicate orders; but they had all learnt their parts so thoroughly that no hitch occurred, and the blaze of the bursting shells lit up the scene as brightly as daylight.

Punctual to the prearranged time both columns reached the point of exit where another gap had been made in the wire, and by thirty minutes after midnight the German trenches were evacuated. The whole affair had been executed with perfect precision and the only thing which reduced the value of the success was the fact, learnt from a prisoner, that on the very night of the raid the garrison of the sector attacked had been reduced to a single platoon of eighteen men. The whole or almost the whole of these were accounted for, but it was a source of great disappointment to the squadrons concerned that a larger garrison had not been encountered.

Our casualties were one British officer (Lieutenant G. Wilson) and six rank and file wounded.

For their conduct in this well-planned and gallantly executed undertaking Captains Vigors and Stevens were awarded immediately the Military Cross by the Field-Marshal Commanding-in-Chief, while Risaldar Nur Ahmad Khan and Dafadar Fateh Khan (11th Lancers) received the Indian Order of Merit.

The regiment remained for six days more in the trenches about Le Verguier, during which time much valuable work was done in improving the defences, as well as in constant patrolling. On the 23rd June they were relieved by the Poona Horse and returned to tent bivouacs near Tertry. Thence on 13th July they marched north to Ostreville near St Pol where the brigade remained for nearly three months.

In the autumn of this year it was hoped that the long-planned and costly attack in the Ypres salient which culminated on the Passchendaele Ridge might at last give the cavalry their opportunity, and in this expectation the Ambala Brigade with the rest of the division moved on 6th October to Tennay and thence on the 11th to the area behind Ypres, where the Ninth found themselves in their old haunts and for three days were held in readiness at Watou. But it soon became evident that these hopes were no more likely to be realised than on previous occasions. On 14th October therefore the Ambala Brigade withdrew to Wardrecques and thence to the neighbourhood of Hesdin, twenty miles west of St Pol, where the regiment was billeted until 8th November. On that date the brigade again moved to the Somme valley, and passing through Bray reached Brusle on 12th November, where the whole Cavalry Corps was concentrated.

We now come to the Cambrai battle of the last ten days of November 1917, in which the Third Army under Sir Julian Byng sought, with the aid of tanks—the first occasion of their use in a considerable number—to break the German line and to make a gap wide enough for the cavalry to pass through and, supported by infantry, to roll up the enemy's line on either side of the aperture. The surprise by the tanks was complete and a considerable local success was achieved, but the scheme as a whole was a failure because the tanks were not in sufficient numbers to do more than pierce the line on a narrow front. Nevertheless the main object of the attack was so nearly attained that some of the Cavalry Corps actually succeeded in getting through the gap made by the tanks and inflicted considerable loss on the enemy. The subsequent phases of the fighting in the Cambrai area, when the Germans endeavoured by counter-attacks to recover the ground which they had lost, afforded to the Ambala Brigade and notably to the 9th Hodson's Horse an opportunity for mounted action which was one of the most brilliant pieces of cavalry work in the whole war.

In anticipation of a break-through on 20th November the brigade moved forward on that date as far as Villers-Pluich, and on the following day to Marcoing, but when it became evident that success had not been achieved it fell back to Equancourt and on 27th November to Tertry.

Here orders were received by the regiment to go into the trenches dismounted on 30th November. Accordingly on the morning of that day all arrangements had been made for a spell of duty in the trenches about Le Verguier. The saddles were stripped and packed away, the men were equipped for dismounted work, and the horses were out at exercise. Meanwhile continuous and heavy drum fire had been heard to the north, and although this was at first attributed to an artillery "strafe" arranged to interfere with the periodical German reliefs, yet little surprise was felt when at 8.30 A.M. urgent orders were received for the regiment to fall in mounted and to rendezvous at the cross-roads east of Estrées-en-Chaussée, the transport being ordered to follow later.

Messengers were sent out immediately to call in the horses and to countermand all previous arrangements. Every effort was made by all ranks to carry out the orders with the least possible delay, and with such good success that at 11 A.M. the regiment was at the rendezvous, the first unit of the whole division to arrive there.

At 11.30 the Ambala Brigade moved off followed by the rest of the division, and proceeded at a rapid trot without a check for eleven miles through Roisel and Villers-Faucon to a point about three-quarters of a mile north-west of Épéhy, where it halted for upwards of an hour.

Here some sort of information was obtained of the events of the morning. It appeared that a sudden and heavy attack had been delivered against the Seventh Corps about Honnecourt and Gouzeaucourt. The blow was quite unexpected and had been so energetically followed up that the Germans had overwhelmed our front line and had reached the bivouacs almost before any warning had been given. The troops were in every stage of unpreparedness, getting their breakfasts, shaving and the like. For a little while the retirement was a case of *sauve qui peut.* Senior commanders escaped in various sorts of undress from " slacks " to pyjamas and one officer who was in his bath only avoided capture by retreating precipitately wrapped in a bath towel.

Stories of this kind however were gathered later. Meanwhile all that could be ascertained for certain was that the line had been forced back and that the Germans were still advancing, that the Guards Division had been brought up to stem the enemy's progress and was making a counter-attack on Gouzeaucourt. About 1 P.M. the divisional commander (Major-General MacAndrew) gave orders verbally to the Ambala Brigade, with the 8th Hussars leading, to push on to Gauche Wood and to get into touch with the Guards on the left. The rest of the division, as it arrived, was to carry on the line to the right. Accordingly the 8th Hussars at once moved forward towards Gauche Wood, but at Chapel Crossing on the railway line they were checked by the wire about our own second line trenches, and while passing this obstacle they came under heavy fire from the north-west of Villers-Guislain and suffered a good many casualties. For a time their further progress was impossible, but one squadron had already managed to get through and had gained the shelter of a sunken road about 400 yards west of Gauche Wood. In this emergency the 9th Hodson's Horse was ordered to go up on the left of the 8th Hussars, fill the gap between them and the Guards Division, who had by this time recaptured Gouzeaucourt, and attack Gauche Wood. The regiment trotted forward as far as Revelon Farm where it halted for a few moments to make sure of the right direction, a precaution which was necessary in order to avoid being held up unexpectedly as had already happened to the 8th Hussars. From here the two leading squadrons (C and D, commanded respectively by Major A. I. Fraser and Captain M. D. Vigors) advanced rapidly, the other two and the headquarters remaining a few minutes more at Revelon. The country was difficult to traverse as the whole of our second line trench system had been dug since the Cambrai operations of the previous week, and most of it had been wired. After proceeding a little distance a trench was reached where an ordnance officer with a party of clerks and orderlies were found in possession, doing their best to hold the line intact, but visibly relieved at the appearance of the regiment. Near here a gap was found in the wire, where it crossed a road running north-east to Gouzeaucourt. Through this gap C Squadron passed and breaking into a gallop made straight across the intervening stretch of perfectly open country to where, nearly a mile away, the sunken road, already occupied by the squadron of the 8th Hussars, ran north and south, just west of the railway line. At this moment the German infantry were seen emerging from Gauche Wood and dropping into the sunken road, but when they saw Major Fraser's squadron galloping towards them they ran back to the shelter of the wood, covered by a few machine-guns on the railway line.

The passage of C Squadron through the gap in the wire and their advance across the open ground beyond had been so rapid that the enemy's artillery had not had time to get on to the troops; but by the time that D Squadron under Captain Vigors began to pass through the defile they met with a very different reception. The leading troop was almost blown to pieces by the concentrated fire on the narrow opening, but the remainder of the squadron never wavered nor did they change their pace. Advancing with the utmost steadiness through the gap, the troops spread out in diamond formation forty yards apart, and breaking into a gallop crossed the intervening open ground under heavy shell and machine-gun fire and joined the leading squadron in the hollow road.

This gallant episode which was watched with admiration by onlookers, was one of which any army in the world might be proud. For mounted troops to advance in the face of modern firearms is in any circumstances an undertaking requiring the greatest pluck and daring. To make such an advance in orderly formation, first through a narrow defile and then across open ground exposed throughout to hostile artillery and infantry fire, betokens not only dash and gallantry but also coolness and nerve on the part of the leaders and steady discipline and *moral* of a very high order in all ranks. Nor was this merely a vain display of reckless bravery. The opportune arrival of Hodson's Horse at a point where the defence was very seriously weakened, and at a moment when the single squadron of the 8th Hussars that had established itself there was in danger of being overwhelmed, was successful in definitely checking the German advance and materially assisted in keeping the line unbroken.

To return to the narrative of events. C Squadron, having reached the sunk road and dismounted, found their further progress checked by a German machine-gun on the railway immediately in front. Major Fraser with another officer and four men immediately left the cover of the road and dashing forward endeavoured to rush the gun, but Fraser fell shot through the head, and the others seeing the attempt useless got back under cover, most of them wounded. At this moment D Squadron came up and dismounting a little behind the road, where a fold in the ground gave cover for the horses, advanced and prolonged the line to the left of C Squadron until they got into touch with the 20th Hussars, from the 1st Cavalry Division, who had come up on the right of the Guards at Gouzeaucourt.

Meanwhile the regimental headquarters and the two other squadrons had also advanced and finding another gap in the wire managed to escape many casualties, but suffered the very serious loss of Major F. St J. Atkinson who was mortally wounded by a shell at the head of A Squadron. B Squadron under Major Dyce now prolonged the line to the right, and A was held in reserve. After a time when the hostile barrage slackened the led horses were sent back to Revelon Farm and later to Heudecourt. They were got away only just in time for the sunk road and ground behind were heavily shelled by the enemy, but the position held there was a strong one and not many casualties occurred among the troops. The rest of the 8th Hussars came up on the right of Hodson's Horse, the 20th Hussars being on the left, and the line of the sunken road was held in this order through the night, the men being able to make themselves comfortable with the bedding, food, &c., left behind by our infantry in their hurried retreat that morning.

Early on the 1st December the 18th Lancers relieved the 8th Hussars, and at 7 A.M. moved forward in support of the Guards, who now attacked Gauche Wood. At the same time the Ninth, extending to the right, took over the part of the line vacated by the 18th. Thus they remained until 4.15 P.M., when A Squadron was sent forward to prolong the line held by the Guards and 18th Lancers southwards from Gauche Wood. Twice during this day the regiment suffered several casualties, the first occasion being in the early morning when some tanks, moving up to support the attack on Gauche Wood, lost their bearings and taking the Ninth for Germans sprayed our unfortunate men with machine-gun bullets. A very fine non-commissioned officer who had won the Order of Merit and Distinguished Service Medal was killed, an Indian officer wounded, and the Commanding Officer narrowly escaped with his life. Again between 6 and

6.30 P.M. in the evening the sunken road and dressing-station immediately in rear of it were heavily shelled, Jemadar Sardar Khan being severely, and Captain Dudding slightly wounded.

At length on the morning of 2nd December the regiment was relieved by the 7th Dragoon Guards and 20th Deccan Horse, and marched back under heavy shell-fire to a bivouac one mile south of Heudecourt.

The losses in Hodson's Horse on these two memorable days were two British officers (Major A. I. Fraser, D.S.O., and Major F. St J. Atkinson, D.S.O.) and eight non-commissioned officers and men killed, two British and three Indian officers (Captain Dudding, Lieutenant Murphy, Risaldar Harbant Singh, Jemadar Mir Alam and Jemadar Sardar Khan) and forty-three of other ranks wounded (of whom two afterwards died and one was missing). Of these losses twenty-five killed and wounded were in D Squadron, and almost all of these were hit in the few minutes occupied in passing through the gap in the wire near Revelon.

The loss of Majors Fraser and Atkinson was a specially heavy one. The latter had only recently rejoined from the command of a British infantry battalion to which, as has been related, he was posted two years before. Both were particularly brilliant officers and squadron leaders, and Major Atkinson was also famous as one of the finest polo players of his day. Their deaths were deeply deplored by the whole regiment.[1]

Among many deeds of devotion and gallantry the splendid work of Captain Dutt of the Indian Medical Service is deserving of record here. Through-

[1] A remarkable and perhaps unique testimony to the high esteem and affection with which these and other officers of the Ninth were regarded by their squadrons is supplied by the fact that, on his return to India after the war, Ressaidar Sardar Khan (Awan), of Major Fraser's squadron, erected, at his own expense and of his own initiative, a monument to his commander's memory at Ara, in the Jhelum district of the Salt Range.

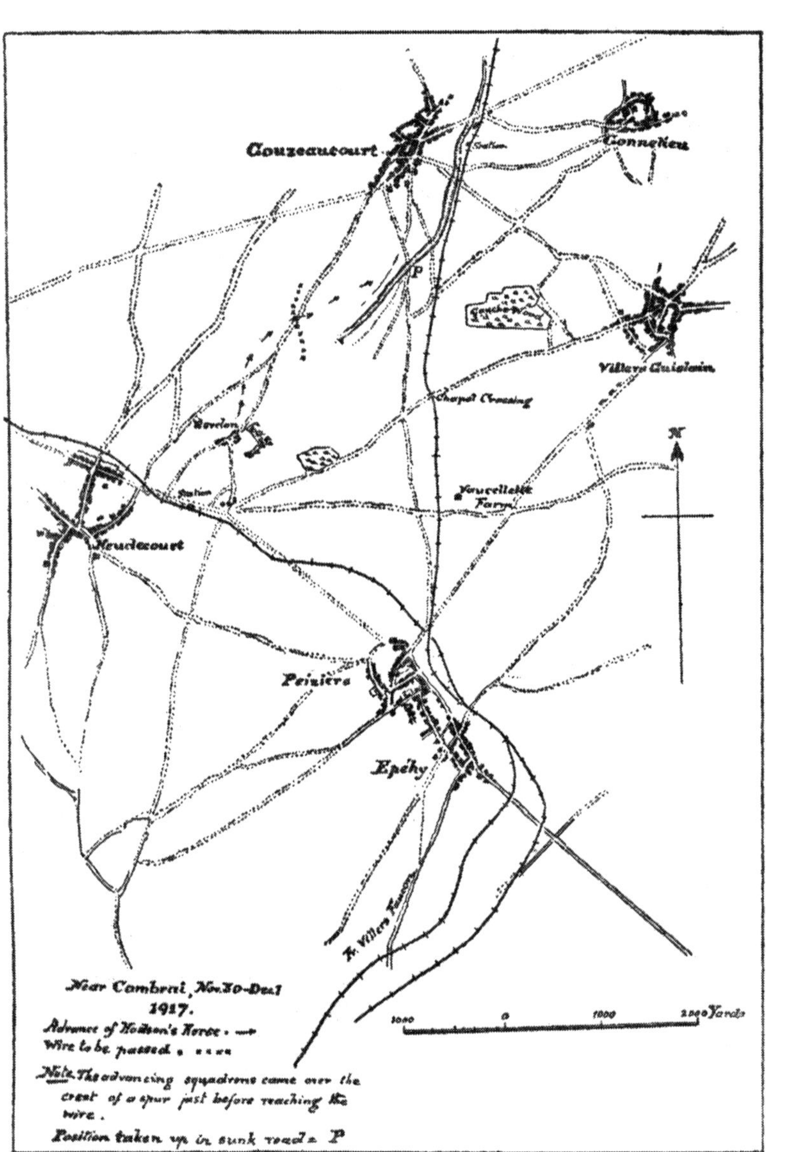
Gouzeaucourt
Station
Gonnelieu
P
Gauche Wood
Villers Guislain
Chapel Crossing
Revelon
Station
Vaucellette Farm
Heudecourt
Peizière
Epéhy
N
Near Cambrai, Nov. 30-Dec. 1
1917.
Advance of Hodson's Horse. →
Wire to be passed.
Note. The advancing squadrons came over the crest of a spur just before reaching the wire.
Position taken up in sunk road = P
1000
0
1000
2000 Yards

out the whole action and under the heaviest fire he carried on his duties coolly and indefatigably and well earned the Military Cross with which his conduct was rewarded. It may be related that a German Colonel, severely wounded and a prisoner, whose wounds were dressed by Captain Dutt under heavy shell fire, pulled off his Iron Cross from his own breast and handed it to his benefactor with heart-felt expressions of gratitude.

Other rewards earned in this action were a bar to the Distinguished Service Order by Lieut.-Colonel Beatty, the Military Cross by Captain and Adjutant Graham, the First Class of the Indian Order of Merit by Ressaidar Nur Ahmad Khan, the Second Class of the same order by Jemadar and Wordi Major Sardar Khan and by three other ranks, besides twelve Indian Distinguished Service Medals and seven Indian Meritorious Service Medals.

After spending the night of 2nd December at Heudecourt the regiment moved to Villers-Faucon where it bivouacked on the 3rd. The next day it was again in the trenches, this time near Vaucelette Farm, and here Ressaidar Harditt Singh [1] was mortally wounded. On 5th December it was once more relieved and finally left this part of the trench line. It moved back again to Villers-Faucon, with the loss of one man, and a few days later marched to Brusle (four miles east of Péronne) where it remained until the 20th.

While at Brusle Lieut.-Colonel Beatty, D.S.O., left the regiment, being transferred to the command of the Lucknow Cavalry Brigade, and the command of the Ninth devolved upon Lieut.-Colonel C. H. Rowcroft.

The weather at this time was very severe with hard frosts and heavy snow, and work in the trenches was proportionately trying. During the time spent at Brusle the regiment was daily employed in strength-

[1] Ress. Harditt Singh was the son of Hon. Capt. and Risaldar Major Ram Singh (see p. 166 and App. III.).

ening the support trenches north of Le Verguier, but on 21st December it moved to its old billeting area about Tertry and from here a trench regiment 270 strong under Major Dyce, M.C., went into the trenches in front of Vadencourt on the Oise, the led horses being sent back to Mouflières, L'Étoile-Brucamp and Villers-sous-Ailly in the Abbeville area. During this tour of duty the regiment performed the usual trench patrol work and on the night of 12th-13th January 1918 carried out a small raid. It was at length relieved by the 15th Hussars and left the trenches for the last time on 15th February. It marched on that day to Vermand, whence it was railed to Roisel. On the 16th it moved by rail to Saleux and thence proceeded in lorries to the billets near Abbeville where the horses were waiting.

Thus ended the active service of the 9th Hodson's Horse in France, for orders were now received for the break up of the Ambala Brigade and for the Indian regiments to proceed to Egypt. On 26th February 1918 the Ninth marched to the Prousel area and entrained next day at Saleux for Marseilles where they arrived on 1st and 2nd March. The embarkation of the horses and mounted men was effected on 5th March, while all dismounted men were sent by rail to Taranto whence they were shipped across to Alexandria.

For nearly three and a half years the regiment had been employed in the greatest war in history and against some of the most efficient troops in the world. The opportunity so long desired of showing their value in the legitimate rôle of cavalry had been almost wholly denied them, but in the far severer test of the trenches, in long nights and days of cold and wet and misery, in face of artillery bombardments such as had not before been dreamed of, and in despite of all the other horrors of modern scientific warfare, their fortitude was not shaken nor did their cheerful endurance falter.

In a special order of the day the Commander-in-

Chief of the British Forces in France, Field-Marshal Sir Douglas Haig, wrote as follows :—

"As the Indian Cavalry Regiments are now leaving France I wish to record my great appreciation of the valour, determination and devotion to duty shown by all ranks in the field. Indian officers, non-commissioned officers and men have been absent for more than three years in a foreign country, thousands of miles from their homes and families, in a climate to which they are totally unaccustomed, and have by their gallant deeds added even greater lustre to the already glorious names of their respective regiments."

"I am very sorry indeed," wrote Brigadier-General C. H. Rankin, commanding the Ambala Cavalry Brigade, "that our old community is being broken up, but I am glad and thankful to have had the pleasure and the honour of such close association with the 9th Hodson's Horse for a year and a half.

"I should like you and your officers to understand how much I have appreciated the feeling of absolute confidence that no matter what the circumstances might be the 9th Hodson's Horse would most surely do the right thing at the right time."

Map of North-Eastern France . . . *at end.*

CHAPTER VI.

PALESTINE AND SYRIA, 1918.

THE men and horses of the Ninth embarked as has been seen on 5th March 1918 at Marseilles for Alexandria. The horses together with those of the 19th Lancers (Fane's Horse) were on the transport *Hydaspes*, with one man to every five horses. The rest of the personnel embarked on the transport *Ellenga*. Other units occupied three other ships and the whole convoy escorted by four British destroyers left Marseilles on 7th March.

At this time enemy submarines were very active in the Mediterranean and all possible precautions had to be taken against attack, especially by night when lights were carefully extinguished. In these unwonted conditions it was surprising to landsmen to see with what speed and steadiness these huge ships could be navigated in pitch darkness, though often not more than three hundred yards apart.

Malta was reached without incident on 12th March, and on leaving there the convoy was increased by two French ships, while the escort now consisted of two British and four Japanese destroyers. Setting forth at night and steaming at high speed one of the transports, when two hours out from Malta, ran down and sunk in the darkness a patrol boat with a crew of nine men of whom only one was saved. In the confusion resulting from this accident the *Hydaspes*, which was the next ship to port of the transport concerned, was herself nearly rammed by the latter and was only saved by the readiness of the nearest destroyer. The incident is described in graphic words in a very interesting little volume by Lieut.-Colonel Rowcroft, 'With the 9th Hodson's Horse in Palestine'[1]:—

"How the splendid sailor commanding the Japanese destroyer on our left grasped the situation was a marvel; but grasp it he did, and rushing at lightning speed across the bows of the *Hydaspes* he turned sharp right, then left about, and laying his ship against the bows of the large transport he regularly rode her off, like a well-trained polo pony. It was a fine example of quick thinking, dash and resource."

This danger past, the convoy reached Alexandria without further mishap on 16th March, and the regiment disembarking next day was railed to Tel el Kebir, where it went into camp and found the dismounted men from Taranto already arrived.

The climate of Tel el Kebir in March and early April is bright and fresh and the country around proved a good training ground, although rather dusty. Both men and horses quickly improved in condition and much useful training was done, especially mounted work, as well as some much-needed refitting. Leave to India, long looked for, was at last opened to all Indian ranks, and every one was able to get leave to visit Cairo and Ismailia.

On 19th April the regiment started for Palestine, accompanied by the 18th Lancers, both regiments being posted to the 5th Mounted Brigade, of which the third unit was the Gloucestershire Hussars. The commander was Brigadier-General P. V. Kelly, C.M.G., D.S.O., a fine cavalry soldier who had served for ten years in the Egyptian Army and was in command of the brilliantly successful Darfur Campaign of 1916. Marching by way of Kassassin, Ismailia, and el Ferdan to Kantara, the regiment there entrained on 22nd April and, after an exceed-

[1] 'With the 9th Hodson's Horse in Palestine,' by C. H. R. (Bombay: Thacker & Co., Ltd., Publishers.)

ingly dusty and dirty journey in half-open trucks, arrived at Ludd on the following day. Here they went into bivouac near Surafend, where in ideal cavalry country and a good climate five days were spent in valuable field practices.

At this moment the whole of the force under General Allenby in Palestine was in process of reorganisation. The violent German attacks in France had necessitated the call to the Western front of almost all the Yeomanry regiments in Palestine as well as many battalions of infantry, siege batteries and machine-gun companies. The Yeomanry were replaced by the Indian cavalry lately withdrawn from France, the infantry by Indian regiments from Mesopotamia and India. The military situation at the end of April 1918 was as follows. The British, having taken Jerusalem, had established their line from the Mediterranean to the Jordan sufficiently far north of Jaffa and Jerusalem to secure those places from all but long-range gun fire from the enemy. They had seized and maintained an important bridge-head across the Jordan at Ghoraniyah, a few miles east of Jericho, and they were about to make for the second time a raid across the Jordan, with the object of destroying an entrenched position at Es Salt and of co-operating with the Arab forces under Emir Feisal. In this attempt the Ninth and other Indian cavalry units did not arrive in time to take part.[1] On its conclusion on 5th May the Commander-in-Chief decided, for several strong reasons, to continue through the summer to hold the positions which had been secured in the Jordan Valley, notwithstanding the serious objections on climatic grounds to such a course. It had hitherto been considered to be impossible for Europeans to exist in that region during the hot weather, when even the native Arabs move to higher ground. Not only

[1] Capt. T. W. Corbett, 9th Hodson's Horse, was employed in these operations, and was recommended for the Military Cross for his services there.

is the heat intense in this low-lying marshy tract, but besides this malaria of a virulent type prevails at that season and renders the country almost uninhabitable. On the other hand it was of great importance to deny to the enemy the passages of the Jordan immediately north of the Dead Sea, and still more desirable, in view of General Allenby's plans for his autumn campaign, to persuade the Turkish leaders that our next big advance would be east of the river. The decision to hold the Valley involved serious demands on the endurance of both men and horses, but it was fully justified by the results.

At the beginning of May, then, the 9th Hodson's Horse and 18th Lancers were ordered to march to the Jordan Valley, and leaving Surafend on 2nd May they reached Jerusalem on the following day. The 5th Mounted Brigade to which they belonged was for the present posted to the Australian Mounted Division commanded by Major-General H. W. Hodgson. The whole cavalry force in General Allenby's army (at first organised in three and later in four divisions) was named the Desert Mounted Corps, and was commanded by Lieut.-General Sir Harry Chauvel.

After a night at Jerusalem the regiment moved down to Jericho, a very trying march by a bad road and a continuous descent from 2500 feet above sea level to 1000 feet below. The bivouac at Jericho on 4th May was oppressively hot, nor was that at the Wadi Obedieh on the next night much better. Thence on 6th May the Ninth marched to the Mellahah Wadi where they took over a section of the line from the Imperial Camel Corps. This section was now organised in seven posts, all of which were named in monosyllables beginning with T, and, reading from right to left, were Tea, Tail, Tick, Thin, Tart, Tame, and Tool. The Gloucestershire Yeomanry prolonged the line to the right in a succession of posts of which the names began with S (subsequently occupied by the 18th Lancers). Each post was dug and wired for all-round defence after the fashion in which the regiment had had much practice in France on the St Quentin–Le Verguier sector, and the line was afterwards strengthened by a continuous belt of wire along the front, connecting the posts. In addition to this defence work, large parties were sent out daily to drain the Mellahah Wadi and neighbouring swamps, in order to reduce the swarms of mosquitos which were as bad at night as were the flies by day. The Turkish line was at a varying distance of from 800 to 1500 yards away, the whole *terrain* being very open, but much cut up by *wadis* filled with brackish water. When the regiment first took over the line the Turkish sniping was rather troublesome, but our patrols were sent out nightly and our snipers were placed in concealed positions in front of the line during the day, and by these means the annoyance was entirely stopped. The led horses and transport were all sent back to the Obedieh Wadi, except a few horses kept for pack work and despatch riding. One man remained with every four horses, and this left between fifty and sixty rifles to each squadron in the line at first, but these numbers soon dwindled until, towards the end of the time in the Mellahah Wadi, a squadron was thought lucky if it could muster thirty strong.

Rations and water were brought up every night after dark, a distance of six miles each way. The supply of water presented a serious difficulty. The water of the Mellahah was so salt that it blistered the lips; that of the Jordan, which was sweet and swift-running, was inaccessible, being some distance away and under fairly close fire and observation from the Turkish posts. Drinking water had therefore to be brought up with the rations in water-carts and in "fantassies," large copper tanks, usually carried on camels but now packed on waggons.

As will be seen from the above there was much work to be done, most of it by night, both for the sake of coolness and to avoid observation. The conditions of the service were hard and trying, the heat was great and the air sultry and oppressive, the flies and mosquitos made sleep difficult even when occasion for it offered, and water was scarce and rations none too plentiful.

Patrols, as has been mentioned, were sent out at night to locate the enemy's posts. It was noticed that the Turks were very jumpy at night and would break out into noisy singing apparently to encourage themselves whenever they fancied they heard one of our patrols approaching. It was determined therefore to take advantage of this nervousness and to arrange a night raid of the kind which had been so successful in France. A preliminary reconnaissance was made on the night of 17th May by Captain T. W. Corbett and on subsequent nights by Kot Dafadar Gujar Singh and Dafadars Ram Singh, Dhalip Singh and Serain Singh, by whom full reports were obtained of the Turkish posts selected for attack. These were opposite the part of the line held by A and B Squadrons, and the first of these Squadrons under Captain Corbett, with Lieutenant I. B. F. Pierce and three Indian officers, was detailed for the duty.

The raid took place on the night of 22nd-23rd May and was accomplished with complete success, a low estimate of the enemy casualties being twenty killed and thirty wounded.

It will be of interest here to quote verbatim the regimental order for the operation[1]:—

"9TH HODSON'S HORSE OPERATION ORDER NO. 1.

1. A raiding party (A Sq.) will attack two enemy posts located in X 30c to-night, 22nd and 23rd. The party will be supported by one Troop D Sq. and 2 Vickers guns under

[1] From 'With the 9th Hodson's Horse in Palestine.'

Lieut. G. W. Ninis. All to report to Capt. Corbett at Tick Post at 10 P.M. to-night.

2. 1 Verey Light will be fired from Tick Post at 03.20, 03.25, 03.30 to guide return of party. (O.C. B Sq. will arrange this.)
3. M.O. to send 4 stretcher-bearers, who will carry arms, and 2 stretchers with raiding party.
4. Raiding party will take a proportion of wire-cutters and bombs, but all work to be done with the bayonet as far as possible.
5. Password is "Russell Sahib." [1]
6. On a green Verey light being fired by the raiding party, followed by a red light immediately, the Ayrshire Battery will put a barrage on the two posts in X 30c.
7. O.C. B Sq. will arrange—
   (*a*) to fire the Verey lights ordered in 2,
   (*b*) to be ready to cover the retirement of the whole party from his battle positions.
   (*c*) to have tea ready for the raiders.
8. O.C. Raiding Party will hold a Roll Call on return to small wadi in X 28c."

A preliminary covering party with one Hotchkiss rifle under Lieutenant Pierce went on at 7.30 P.M. to make certain of the route. The raiding party, 47 strong besides officers, started at 10.15. The night happened to be a particularly bright one, with a brilliant moon. The initial advance was therefore very slow and gradual, the men dribbling forward, not more than six being on the move at any one time, and 200 yards being kept between troops. At 12.45 some shots were fired by the Turks, and there were signs of alarm in their line; further advance was therefore stopped, and the

[1] Major J. C. Russell, 9th Hodson's Horse, who had been killed at Passchendaele six months before, while commanding the 6th Batt. Cameron Highlanders, had for several years commanded A squadron and had been much beloved by his men.

raiders lay quiet until the moon went down. But unfortunately whilst they were lying thus a Turk passed right through their lines and doubtless reported their presence, so that when the attack took place the posts had been manned with all available reserves.

As soon as the moon set at 2.15 A.M. the advance was continued and the distance between troops was closed to 100 yards. Having arrived at 150 yards from the Turkish posts the two leading troops at 2.45 A.M. deployed one against each post, while the two other troops halted and lay down ready to support either attack as need might arise. Immediately the deployment and attack began the enemy opened a very heavy fire from the front and right flank with rifles and machine-guns. The attacking troops broke into a double and were on the Turks in a few moments. As already reported by the reconnoitring parties, there was no wire in front of either post, but they were entrenched and were held by about 50 men in one and 75 or 100 men in the other, besides several outlying points with four or five men in each. The Turks put up a considerable resistance and the attacking troops were therefore immediately reinforced by their supports. Only two prisoners were taken, but the raiders remained for twenty minutes in the enemy posts, killing and wounding with the bayonet or with point-blank fire quite fifty or more. The raiding party then withdrew, bringing in all their own wounded, the retirement being covered by the fire of two Hotchkiss rifles and of two sections which Captain Corbett collected round him for the purpose. Artillery fire as arranged was called for at 3.12 A.M. and at 4 A.M. the whole raiding party, including the covering party, was back at Tick Post, after a most gallant and successful exploit.

The casualties on our side were Captain Corbett, Ressaidar Bur Singh, Jemadars Indar Singh and Bhagwan Singh and nine of other ranks wounded.

Very warm congratulations on this brilliant little affair were conveyed to the regiment by the Corps and Divisional Commanders and by Brigadier-General Kelly, who wrote :—

"The Turks were met in considerable strength in an entrenched position with a good field of fire, and it is only due to the determined leading and the soldierly spirit and dash of leaders and men that success was obtained."

For gallant conduct in this raid Captain Corbett was awarded the Military Cross (he subsequently received a bar to the same decoration for work in the operations of April 1918 in Es Salt, where he was present on the staff). The Indian Order of Merit was awarded to Jemadar Bhagwan Singh, and the Indian Distinguished Service Medal to Kot Dafadar Gujar Singh, Dafadar Ram Singh and Dafadar Dhalip Singh.

On 26th May the regiment was relieved by the Gloucestershire Hussars and went into brigade reserve in the Wadi Obedieh, where it was found that the horses had been heavily shelled on the last two mornings and had suffered some loss. The bivouac was therefore moved to the Wadi el Aujah, and here the regiment remained until 5th June. Working parties were meanwhile supplied daily and strong patrols were sent out at night into a gap in the line to the right of the S posts, a sector which was only lightly held because the difficult nature of the ground about the precipitous gorge of the Jordan obviated any danger of surprise attack at this point.

From 6th to 18th June the Ninth were again in the Mellahah Wadi, occupying the S posts, which they took over from the 18th Lancers. No unusual incident occurred in the trenches during this period, and an attempted raid on the Turkish posts disclosed the fact that they were unoccupied by night and held only by snipers by day. On the 12th June the led horses, which had been brought out of the *wadi* and picketed on the open plain, were heavily shelled, and suffered serious casualties, thirty men being wounded, and six horses killed and twelve wounded. The horses and their personnel were thereupon taken back into the Wadi Obedieh.

On relief by the 5th Australian Light Horse the regiment moved back to Talaat el Dumm, the headquarters of the Desert Corps, with the further prospect of going for a three weeks' rest to the Bethlehem area. The latter move was delayed for a few days owing to a threat of a heavy assault on the Aujah posts, which caused the Ninth to be sent down again to Jericho; but the anticipated attack not having developed they returned to Talaat el Dumm on the night of 30th June and next day marched through Jerusalem to Beit Sawir, three miles south of Bethlehem on the Hebron road, where they remained in a comfortable bivouac until 17th July. From the nature of the country mounted training was difficult, but the time here afforded a useful opportunity for some musketry practice and staff rides, as well as for cleaning up kit and accoutrements and improving the condition of the horses, who had gone through a trying two months in the Jordan Valley. Here too the regiment was inspected by the Divisional Commander (Major-General Hodgson) who was most flattering in his praise of its conduct and appearance.

On the 8th July, while the Ninth were still at Beit Sawir, a large party of Muhammadans from the regiment as well as similar parties from the 18th and from other regiments went to Jerusalem to celebrate the Id. Six days later almost all of those who had paid this visit were attacked with some unusual form of fever, which was called influenza but which was quite strange to the medical officers. It was suggested that the outbreak might have been caused by enemy action while the men were at Jerusalem, such as by the infection of the praying mats. It was certainly a curious coincidence that just when the men so infected were all incapacitated by illness the anticipated attack on the Aujah defences at length took place. It was probably known to the Turks that some thousands of Muhammadan soldiers would go to Jerusalem for the festival, and it was at any rate a suspicious circumstance that the day fixed for the enemy's attempt coincided exactly with the time when the disease contracted on 8th July had rendered its victims *hors de combat.*

However this may have been, the attack on the Aujah line was heavily repulsed by the Australian and New Zealand Division so that the need for support by the Indian regiments did not arise; nevertheless the 5th Mounted Brigade was warned to hold itself in readiness, and owing to the sickness among the Muhammadans it was necessary to make C and D temporarily into one squadron.

Many of the sufferers from this mysterious outbreak never recovered, or at least for several months afterwards did not thoroughly recover, their health.

On 18th July the brigade left the pleasant bivouac at Beit Sawir and after camping that night at Talaat el Dumm proceeded on the 19th to relieve the New Zealand Horse on the extreme left of the line in the Jordan Valley. Beyond (westwards) there was a gap of some two miles of rocky and precipitous ground which was patrolled by Arab auxiliaries, and then, on the top of the Judæan hills, was the right of the infantry.

Until the 27th July the Ninth was in divisional reserve. On that date they relieved the 18th Lancers in the W posts of the Aujah line. Here the ground was much higher than in the Mellahah Wadi, being only 600 feet below sea level, and the climate was consequently better. The regimental headquarters and reserve squadron encamped by the Ain ed Duq, a beautifully cool clear stream, which furnished a plentiful supply of pure water for both men and horses; two squadrons were in the front line, and one squadron in close support at night. The Aujah provided very good sport for leisure moments, there being plenty of seesee partridge and chikor. The Turkish posts were much farther away than in the other parts of the line and were located on the hills in front. Patrols as usual went out daily, a prominent point of call being a ruined house on a small knoll known as Tel el Truni, and this place was also held at night by a strong troop. One day at about 1 P.M. a Turkish patrol was seen to ascend this knoll from the neighbouring foothills. Captain Carr-White, 31st Lancers (attached to Hodson's Horse), to whom this was reported at once, arranged with the artillery to put a blocking barrage round Tel el Truni. Then taking a troop of Dogras (B squadron) who were on duty at Wick post he went forward as far as possible under cover in a small *wadi*, whence charging across the open ground he captured seven of the Turks before they had time to escape. He and his party were exposed to a good deal of fire from rear and left flank as they retired, but they brought their prisoners in without loss. Captain Carr-White was afterwards awarded the Military Cross for his bold initiative in this affair. The incident was a small one and in itself comparatively unimportant, but, like the exploit before Gauche Wood in the previous November, it had a special value in showing that, notwithstanding the long months of trench warfare in France, the Indian Cavalry had preserved undimmed the brilliance in action and quickness of resource which was to stand them in such good stead in the near future.

After ten days in the line the regiment was relieved by the Worcestershire Hussars and on 10th August moved further down the Wadi el Aujah, near the old bivouac in the Obedieh Wadi, where it became divisional reserve to the troops in the Mellahah section of the line. Here the health of both officers and men, which up to now had been surprisingly good on the whole, began seriously to suffer from the effects of the pestilent climate in the "Valley of Desolation." It was therefore a welcome relief when on the 15th August the regiment was ordered to Tel el Sultan, on the high ground above Jericho, where the brigade concentrated preparatory to moving westwards to the coast. This move indicated that the anticipated autumn offensive was approaching. The reorganisation of the Desert Mounted Corps was now completed. The 5th Mounted Brigade was withdrawn from the Australian Division and became the 13th Cavalry Brigade in the 5th Cavalry Division, of which the commander was Major-General Harry MacAndrew, C.B., D.S.O., the same under whom the 9th and 18th had served during their last year in France. The other brigades in the division were the 14th (Sherwood Rangers Yeomanry, 20th Deccan Horse and 34th Poona Horse) and the 15th (Imperial Service) Brigade (Jodhpur, Mysore and 1st Hyderabad Lancers).

The 13th Brigade left Tel el Sultan on 19th August and, marching only by night in order to avoid observation, on the morning of 21st August it went into bivouac at Wadi Hannein, two miles south of Richon le Zion and about eight miles south-east of Jaffa.

It is impossible in this record of a single unit to discuss at any length the details of the campaign which was now about to begin. In its daring conception it will ever rank high among decisive military operations. The dash and determination which carried it with lightning speed to an overwhelmingly successful conclusion place it in the very forefront of such annals. The conception was the work of the Commander-in-Chief, a great general and a brilliant cavalry commander. The conclusion was due equally to the forceful determination of the high command and to the tireless enthusiasm of the troops, whose endurance and ardour knew no bounds until victory, unsurpassed and splendid, had been completely achieved.

In order to make what follows an intelligible

narrative it may be said briefly that General Allenby's plan was to attack the Turkish right near the coast with an overwhelming force of infantry, which, having carried the defences, was then to wheel to the right and crumple up the whole enemy line against the Judæan hills. Through the wide gap thus made three divisions of the Desert Mounted Corps were to dash forward. Riding up the coast the cavalry were to pour over the hills of Samaria into the Plain of Esdraelon, seize the railway junction at Afule, the Turco-German Headquarters at Nazareth, and later the port of Haifa. From Afule part of the force was to ride down the Valley of Jezreel to Beisan and the Jordan and seize the important railway and road junctions there, while an Arab force was to accomplish a like work east of the Jordan at Deraa. Simultaneously our infantry in the Jordan Valley was to advance and capture the bridge at El Damieh. The whole of the Turkish Armies (VII and VIII) west of the Jordan would thus be practically surrounded and their communications entirely cut, while the Fourth Army east of the river would be isolated and exposed to the converging attacks of our forces from the west and the Arabs from the east.

It will be seen that the scheme depended almost entirely for success on the celerity and determination of the cavalry attack. None but a commander thoroughly imbued with the cavalry spirit would have conceived such a plan, nor would even he have ventured to play so bold a game had his resolution not been fortified by complete and well-founded confidence in his troops.

Scarcely less essential for success was the element of surprise, and in this respect also the details of General Allenby's plans displayed admirable foresight and care. The two raids across the Jordan in the spring of 1918 had given to the enemy the impression that our next advance would be made in that direction. As has already been mentioned it was largely to confirm this impression that General Allenby decided to hold the positions in the Jordan Valley all through the summer. When the time approached for the great attack every sort of precaution was taken not only to conceal the transfer of troops to the coastal sector but even to simulate the increase of our forces on the Jordan. There is no space here in which to describe the ingenious measures taken with these objects in view, and it must suffice merely to say that they entirely achieved their purpose. A German report dated 17th September which was found at the headquarters at Nazareth stated that "far from there being any diminution in the cavalry in the Jordan Valley, there are evidences of twenty-three more squadrons there," and another report declared: "They [the English] will use most of their forces—attack troops and cavalry in this case—against the country nearest the Jordan."

We left the 13th Cavalry Brigade at Wadi Hannein rejoicing in the fresh air of the Mediterranean littoral after the sultry heat of the Valley of Desolation. Men and horses quickly picked up in health and spirits in the improved surroundings, and for more than three weeks every advantage was taken of the open, rolling downs amid which the bivouac was situated to prepare for the coming advance.

At length at 7 P.M. on 17th September the brigade left the training area and moved up to Sommeil just east of Jaffa, where the 5th Division concentrated. Here they remained all next day carefully concealed among the orange groves, from the shelter of which no one was allowed to issue forth. At 7 P.M. on the 18th a further advance was made along the sea-shore to the neighbourhood of El Jelil where, close behind our infantry line, the division dismounted, watered and fed the horses and waited for the dawn.

This silent ride in the bright moonlight along the level sands by the sea was a memorable experience.

The long column of horsemen moved almost noiselessly, the only sounds being the occasional jingle of a bridle or clink of a stirrup iron. Water at the halting-place was plentiful and good, for wells dug anywhere along the shore, though only a few yards from the sea, gave a generous supply of perfectly pure water, free from any taste of salt.

At 4.30 on the morning of 19th September the artillery massed against the Turkish positions opened a tremendous bombardment. Five infantry divisions at the same moment dashed forward to the attack. The enemy taken completely by surprise were driven from their trenches almost without resistance. By eight minutes past five the front line was taken, and thirty-five minutes later the whole position here was in our hands and the infantry began to wheel to the right. The door for the cavalry advance had been opened.

The 5th Cavalry Division, to which was allotted the left of the advance, being screened from view by the cliffs above the sea-shore, was able to ride almost on the heels of the infantry. The 13th Brigade acted as advance guard with the 9th Hodson's Horse leading, and the regiment moved off at 6 o'clock. For two hours the going along the sand of the shore was very heavy and when the Nahr el Falik was reached at 8 A.M. the long trot had taken a good deal out of the horses.

Up to this point no opposition had been encountered, indeed the extreme left of our infantry attack extended as far as the Nahr el Falik when the Ninth arrived there, but the leading squadron (D under Major Vigors) had not advanced more than a few hundred yards beyond that stream-line when fire was opened on them by some Turkish cavalry dismounted. Two machine guns returned the fire while Major Vigors moved round amid the sand dunes to turn the position; at the same time the Commanding Officer ordered C Squadron (Captain Stevens) to attack direct in column of troops widely extended and at increased distances. The enemy did not await the attack but made off in haste pursued for some distance by C Squadron. Major Vigors now received a warning message dropped from an aeroplane that about 200 Turks with two guns and some transport were in an orchard and farm some 400 yards to the right front. The advanced troop under Risaldar Nur Ahmad had already come under the fire of this party. Major Vigors at once brought up the remainder of the squadron and without any delay delivered a mounted attack with complete success, capturing three officers, fifty or sixty men, two guns and twelve waggons with teams.

D Squadron was now out on the open plain over which many scattered parties of Turks and transport were retreating, and a certain amount of hostile fire was encountered. Fire was also opened on the regiment by some guns in rear, but the squadrons being opened out and the pace increased, little harm was done. Meanwhile the advance proceeded steadily and the village of Murkhalid was soon reached. Near here the right advanced patrol of D Squadron under Dafadar Mehtab Singh charged and captured or killed four men working a Lewis gun, and secured the gun. These men stuck to their gun to the last and wounded one of the patrol besides killing a horse before they were disposed of. As the squadron approached Murkhalid a troop of Turkish cavalry opened fire from some trenches on a rising ground about three hundred yards north of the village, but they were promptly charged by a troop under Jemadar Nawab Ali Khan and forthwith surrendered. Some miles farther on another Turkish troop was encountered covering the crossing of the Nahr Iskanderuneh, but they too were charged in front and their retreat threatened by another troop on their flank, and they made haste to retreat.

All this time the main body had been advancing at a steady trot so that the leading squadron, which had to clear the front on these several occasions, while maintaining the advance unchecked, had been severely tried, and that too through a day of great heat. The horses were now much exhausted, and accordingly soon after the Iskanderuneh had been crossed B Squadron took the place of D at the head of the column. The advance was continued without a check across the plain of Sharon until at 11.40 A.M. the Nahr el Hudeira was crossed and the regiment rode into Liktera, where the Turkish garrison surrendered without striking a blow. Here many prisoners were taken, mostly Germans of the Transport and Medical Corps, as well as some lorries and a hospital sumptuously equipped.

At Liktera a halt was called. The horses were watered and fed and the whole brigade got what rest they could for six hours. Since the start they had ridden twenty-six miles, the first ten of which were over very heavy ground. About five hundred prisoners had been taken by the Ninth, who had been fighting more or less for the last sixteen miles.

Quite early on the 19th Lieutenant J. E. Walker had been sent out with a small patrol with orders to reconnoitre the crossings of the Falik and Iskanderuneh wadis and the villages of Murkhalid and Liktera, and to send back reports to the regiment of any opposition likely to be encountered at those places. His progress at first was checked through no fault of his own, so that at the outset he was not able to get a sufficient start of the brigade, while the rapidity of the general advance made forward reconnoitring and transmission of reports almost impossible. Nevertheless the patrol pushed on with much dash and resource and actually reached Liktera well in advance of the regiment, a performance which was not achieved by either of the patrols sent forward by the two other brigades of the Fifth Division.

At 6 P.M. on the 19th the division started again on the second part of their great ride, more arduous than that which they had accomplished, for it involved a long night march of nearly thirty miles through difficult mountainous country, and by the roughest of tracks. Before moving off the Divisional Commander received information from an advanced patrol that the road which was to be followed, the more westerly of two roads over the Carmel range, passing by way of Sindiane and Jarak to Abu Shusheh in the Plain of Esdraelon, was unfit for wheel traffic. The divisional transport was therefore directed to cross the hills by the eastern, or Musmus Pass, in rear of the Australian Mounted Division, while the 15th Brigade was ordered to remain at Liktera till the following day and then cross by the Sindiane track by daylight with the artillery of the division. The 13th and 14th Brigades left Liktera as related above at 6 P.M., the former still leading the march. The place at the head of the 13th Brigade was now taken by the 18th Lancers. Arriving at Sindiane the column turned north-eastwards and at once became involved in a tangle of rocky hills. The track was so bad that for miles the men had to march on foot leading their horses, and in the uncertain light it was difficult to keep to the road or indeed to distinguish it at all. In these trying circumstances the situation was saved by the special qualifications and aptitude of Brigadier-General Kelly, whose thorough knowledge of Arabic and experience in traversing almost trackless country enabled him to obtain all possible information from the stray Arabs encountered during the night, and to follow their vague directions with success. At Jarak on the crest of the hills A and D Squadrons of the Ninth were left to hold the road and to serve as a flank guard for the remainder of the Corps, which was crossing the Carmel range by the Musmus defile. The rest of the column pushed on down the pass into the great Plain of Esdraelon, or Megiddo, reached Abu Shusheh at about 2.30 A.M. on the 20th September, crossed and cut the Afule–Haifa railway near Warakani, and at length reached the foothills below Nazareth. Here the 14th Brigade halted till daylight, but the 13th still continued the advance up the hills *viâ* Jebata and El Mujeidil. At the latter village some delay was caused by the Arab guide declaring that the place was Nazareth, in consequence of which it was surrounded and searched before the mistake was discovered. Two hundred Turkish soldiers were captured, but the delay was unfortunate. Again just outside Nazareth a similar mistake occurred with regard to the village of Yafa, and although more prisoners were taken yet the Nazareth garrison was now alarmed. At 5 A.M. the Gloucestershire Hussars galloped into and round the town and about 900 prisoners were taken, most of them Germans, as well as many lorries and motor-cars and the whole of the documents of the Turco-German G.H.Q., but Marshal Liman von Sanders himself, who it was hoped would be captured, dashed out of his house in his pyjamas, jumped into a car and got away along the Tiberias road just in time to avoid being taken.

Meanwhile a considerable number of the enemy established themselves in houses north of and above the town, on both sides of the road fork to Tiberias and Acre, and in particular in a convent on high ground a little east of the road. An attempt was made to carry these positions, a squadron of the 18th attacking dismounted, supported by B Squadron of the Ninth. But the Turks were strongly posted and were well supplied with machine guns, and little progress was made. Moreover the 13th Brigade was now seriously reduced in strength owing to the detachment left at Jarak, to escorts having been detached with prisoners, and to the fact that no small number of men had lost touch during the rapid advance over difficult roads in the darkness. At 10 A.M. therefore Brigadier-General Kelly decided to withdraw his exhausted men and horses from

Nazareth and to hold the hills to the south of the town. Before retiring the troops put out of action all the motor-cars of the enemy G.H.Q. and the lorries in the German lorry park, and the town was then evacuated. A few hours later in consequence of orders received from the divisional commander the brigade withdrew to Afule and bivouacked on the plain a little to the north of that place, B and C Squadrons of the Ninth being on outpost.

Since the start from El Jelil on the previous morning the brigade had traversed fifty-six miles. The casualties in the 9th Hodson's Horse from enemy action were two men killed and Lieutenant W. S. Shepherd and eight men wounded.

At 10 A.M. on 21st September A and D Squadrons rejoined the regiment from Jarak. On the same day the 13th Brigade moved forward again to occupy Nazareth, the Ninth being ordered to advance by the direct Afule-Nazareth road and to attack the town from the south, while the rest of the brigade were to make their way through the hills to the Nazareth-Tiberias road and to attack from the north-east. The Ninth moved off at 12.15 P.M. A patrol sent forward under Lieutenant Ninis found no signs of the enemy, and as the place appeared to be unoccupied D and C Squadrons were at once pushed through and round the town to the hills on the north-east outskirts and the convent on that side, while A and B covered the advance from the heights on the south. A few shots were fired from the convent, but a white flag was soon run up and all opposition ceased. It was found that all the enemy except about a dozen men had retreated during the night. Unfortunately the large stores of arms, ammunition, &c., which had been accumulated here at the Headquarters of the Turco-German armies had, on our withdrawal on 20th September and the subsequent retreat of the enemy, been to a great extent looted by the Arabs. The rest of the 21st and part of the 22nd September was occupied in collecting as much as possible of what remained.

On the night of the 21st-22nd the brigade outposts were attacked by a battalion of Turks marching from Haifa, but the attempt was brilliantly defeated and the assailants put to flight with very heavy loss by the 18th Lancers, in whose sector the attack was made.

On 22nd September the whole of the 5th Cavalry Division concentrated about Nazareth preparatory to an attack on Haifa and Acre on the 23rd. The 13th Brigade was detailed for the latter duty, which proved an easy task, the place surrendering after firing a few harmless shells. Most of the Turkish garrison bolted, but three guns were captured and about 120 prisoners. The two other brigades had a harder job at Haifa where the Turks and Germans put up a good fight, but the town was in our hands by 2.30 P.M.

The first part of General Allenby's plan had thus been executed with absolute and startling success, and the achievement of the troops was briefly and fittingly acknowledged by the Commander-in-Chief in the following order :—

"I desire to convey to all ranks and all arms of the Force under my command my admiration and thanks for their great deeds of the past week, and my appreciation of their gallantry and determination, which have resulted in the total destruction of the VIIth and VIIIth Turkish Armies opposed to us.

"Such a complete success has seldom been known in all the history of war.

E. W. ALLENBY,
General,
C.-in-C."

26th September 1918.

The defeat of the Turks west of the Jordan had laid open the road to Damascus, and the Commander-in-Chief now determined to complete his victory by the capture of the great Arab city, the capital of Syria. The IV Turkish Army east of the Jordan was in full retreat northwards, endeavouring to reach Damascus where they might make a stand on the difficult ground south of the town, and give time for reinforcements to arrive from Aleppo. But General Allenby intended again to employ his cavalry to forestall this attempt, and the operations of the next few days thus resolved themselves into a race for Damascus between the Desert Mounted Corps and the retreating Turks.

The British advance was to be made in two columns. The 4th Cavalry Division was to cross the Jordan at Jisr Mejamie, join hands with the Arab army about Deraa and follow hard on the heels of the IV Turkish Army in its retreat. At the same time the Australian Mounted Division and 5th Cavalry Division were to move by the Nazareth-Tiberias road, cross the Jordan just south of Lake Huleh and push on to Damascus over the spurs of Mount Hermon.

Orders for the advance were issued on 25th September. The distance from Nazareth, where the Headquarters of the Desert Corps were then located, to Damascus is about 90 miles. Two days' food and forage was to be carried by the troops. When this was exhausted the cavalry was to live on the country. All transport, even the regimental water carts, was to be left behind, the only wheels with the force being the guns, ammunition waggons, and a few light ambulances with each division.

After a welcome rest at Acre on 24th and 25th September the 13th Brigade left that place at 5 A.M. on the 26th and marched by Shefr Amr to Kefr Kenna (Cana of Galilee), a distance of 26 miles by very bad and stony tracks over the Galilean hills, almost all of which had to be traversed in single file. Camp was reached at 3 P.M. but the water supply was so scanty that it was 7 o'clock before the horses had all been watered and the men were able to settle down to get some rest.

Next morning (27th September) an early start was made at 3 A.M., but this proved to be quite unnecessary and might have been avoided by more foresight and better arrangements on the part of the staff. The 13th Brigade was the rearmost of the 5th Division and was kept so long waiting before the units ahead of it had cleared the road that it was 10 A.M. when it reached Tiberias—eleven miles in seven hours. Here the horses were off-saddled, watered and fed, and there was time for all ranks to get a refreshing bathe in the lake. Then at 1 P.M. the march was resumed to Kusr Atra just south of Lake Huleh, a further twenty miles, which was reached at 11 P.M. after a very long and wearisome day. At this point the Australian Mounted Division, which was leading the advance, had been held up for some hours by a rearguard of the Turks, and it was not until dusk that the passage of the river at Jisr Benat Yakub (the Bridge of the Daughters of Jacob) was secured, and some of the Australians were able to move on up the steep slopes of the left bank. These circumstances accounted to some extent for the block on the road which had so much delayed the progress of the 5th Division.

On the morning of 28th September the bridge over the Jordan, which had been seriously damaged by the Turks, was not yet repaired, nor was the Australian Division clear of the Jordan ravine, the ascent of the eastern slopes being very precipitous and extremely difficult for the guns. It was not until 1.30 P.M. therefore that the march of the 13th Brigade began, and on reaching the Jordan there was a further halt until the bridge was at length passable, during which time the horses were well watered and fed. At 4 P.M. the brigade started again and at midnight reached El Kuneitra, a Circassian village on the Hermon range, where at an altitude of some 3000 feet above the sea level a very cold night was passed in bivouac. This day's march had been about eighteen miles.

After a good rest on the night of the 28th and morning of the 29th the Desert Mounted Corps started to ride the last forty miles to Damascus. The Australian Division moved off at 2 P.M. followed at 5 P.M. by the 5th Division and it was hoped that by marching all night the two divisions would be able to surround the city on the following morning. But again the advance was delayed by the Turkish rearguards. The Australians encountered considerable opposition near the village of Sasa, and when dawn broke on the 30th September the leading troops were still twenty miles from Damascus. At the same hour the 13th Brigade of the 5th Division had only got some five miles from the starting-point.

But the enemy was now completely broken and the column was able to push on without further opposition. At Sasa all transport was left and the brigade trotted on to Khan Shiha, about a quarter of a mile north of which place the horses were off-saddled, watered and fed. At 1 P.M. orders were received for the village of Kaukab to be occupied, but as a matter of fact this had been done by the Australians some hours before, when the Turks made their final stand. Meanwhile, in consequence of a message received from an aeroplane that a large body of Turks (in fact the head of the retreating IV Army) was approaching Kiswe on the Deraa-Damascus road, the 14th Brigade had been sent to try and intercept this force, and at 2.30 P.M. the 13th Brigade, with the 9th Hodson's Horse leading, was also despatched on the same duty.

The country eastwards from Kaukab was cut up with numerous deep and almost impassable water-courses and the going was very difficult. D Squadron under Major Vigors with two machine-guns formed the advance guard, and at about 4.30 P.M. when a mile south-west of Kiswe the scouts reported that a body of some seventy Turks was retiring towards that village. Major Vigors pushed on to engage the enemy, who thereupon opened fire. The machine-guns and one troop were halted to give covering fire while Major Vigors with two other troops moved some distance up the Wadi Zabirani and then wheeling to the right charged down on the Turks, who at once ceased fire and surrendered. There proved to be as many as three hundred of them, all in a most exhausted condition. Meanwhile Risaldar Nur Ahmad alone with his orderly had penetrated into Kiswe, and now sent back word that the place was full of Turks. The greater part of two troops were sent to his assistance and a number of prisoners were collected, who with those already taken were sent off under an escort towards Kaukab.

While this was going on some shells came over from the east and a column of about a thousand Turks with transport was seen approaching Kiswe along the slopes of the Jebel el Mania, evidently part of the IV Army. Major Vigors at once determined to head off this column and hold it until the 13th Brigade could come up and cut the enemy off. In the meantime however a message had reached him from the regiment telling him that the brigade was retiring to bivouac at Kaukab and ordering him accordingly to withdraw his squadron and rejoin the regiment. But in view of the approach of the Turkish column, Major Vigors sent the messenger back to the regiment, explaining the situation and reporting that he was about to engage the enemy in anticipation of the arrival of the brigade. By great ill luck this message was not received by the regiment. The squadron was so weak that not more than one man could be spared to go with it. The messenger's horse broke down on the road, and the brigade moved back to Kaukab followed by the Ninth, in ignorance of the chance that was being lost and of the predicament in which the advanced squadron was left. This was indeed not without peril. As soon as he had despatched his message Major Vigors, with the one troop that remained at hand and with two Hotchkiss rifles, galloped across the open to some low hills on his left front, whence he thought that he could best hold up the advancing Turks. The latter, seeing how small was the party that barred their way, opened fire with two mountain guns and six machine-guns and continued to advance. The fire was returned by our Hotchkiss rifles and by dismounted men, but the numbers with Major Vigors were too few for it to be possible for him to close with the enemy, while the hostile fire, although fortunately inaccurate, yet caused several casualties among the horses including that of the troop leader, Risaldar Dost Muhammad Khan. Darkness was now coming on, there was no sign of the looked-for support from the brigade, and the Turkish force was only about two hundred yards from the position taken up by our troop. Moreover the squadron was hampered by a large number of prisoners, and finally it found itself being fired on by some parties of the enemy on the left front. Reluctantly therefore Major Vigors decided to break off the engagement and retire. The Hotchkiss rifles were left to cover the withdrawal and the men in charge of them behaved with great gallantry under heavy machine-gun fire. Unfortunately as these in turn were retiring one of the horses was killed and the Hotchkiss rifle which it was carrying had to be abandoned. Some six or seven horses in all were killed but no men were hit, and as there were several Turkish ponies in the vicinity the men were mounted on them. The squadron was collected under cover of the Wadi Zabirani and Risaldar Nur Ahmad's party from Kiswe was picked up here together with a large batch of prisoners.

The enforced abandonment of the attack on the Turkish column was rendered all the more disappointing because, just as the troop retired, the advance guard artillery of the pursuing 4th Cavalry

Division opened fire from the south-east and the Turks began to break up and to straggle away over the hills. A little while later a patrol of the 4th Division was met with, and contact between the two wings of the Desert Mounted Corps was thus achieved after a separation of five days.

D Squadron then started for the bivouac at Kaukab, where it arrived at 8.40 P.M. It was found to be impossible to get all the prisoners in that night, and one troop was left about half a mile from camp and brought the last batch in next morning.

The events of the afternoon reflected great credit on the whole squadron for their gallantry and dash and especially on the fine leading, coolness and determination of Major Vigors.

The distance covered by D Squadron on the 30th September was about 45 miles, while Risaldar Nur Ahmad's troop had gone some 18 miles more. Nearly seven hundred prisoners were taken as well as two field and two mountain guns.

At 5 A.M. on 1st October the 5th Division started on the last march of its great advance, and passing by way of Kiswe reached the eastern outskirts of Damascus. Here the 13th Brigade was sent on and got into touch with the 3rd Australian Light Horse Brigade, which had passed through the northern quarter of the city in the early morning. C Squadron at the same time was sent to the Maidan station to assist in sorting out and endeavouring to get into some order the thousands of prisoners in every stage of disease and exhaustion who crowded the streets. Little could be done for them, the less so because our own troops were without transport or supplies except the merest essentials.

The next day a composite squadron from each regiment took part in a formal and official march through Damascus. It was not intended to be a triumphal entry, but merely as a display of force to overawe the turbulent elements in the population. Meanwhile those of the division who were fortunate enough to escape participating in this parade marched to El Judeideh, 12 miles south-east of Damascus, where in a very pleasant camp, under shady trees and with excellent water at hand, men and horses had a welcome and well-earned rest during the next two days.

Thus ended the second phase of General Allenby's campaign. In the brief thirteen days that had elapsed between the start on 18th September and the occupation of Damascus on 1st October the whole of the Turkish positions had been overrun and captured, the Desert Mounted Corps had traversed the enemy country almost from end to end —the 9th Hodson's Horse itself had marched two hundred and fifty miles,—all the forward bases, depots, and headquarters of the enemy had been taken, his three armies in the field had been not merely defeated but utterly annihilated. Nothing remained but to give the Turkish power in Asia the *coup-de-grâce*.

Unfortunately the forces available for any further movement were sadly depleted. Scanty food and the long and trying marches, mostly by night, had much exhausted the horses. The men were not only worn out with fatigue but were suffering in large numbers from exposure to the climate and to malarial infection in the swamps and marshes of the Jordan Valley. In this respect the 5th Division was not in so bad a case as the Australian and the 4th Division, both of which, within a few days of reaching Damascus, were so prostrated with malaria as to be quite incapable for a time of further effort. Even the 5th Division could scarcely number more than 1500 men, and with so small a force it seemed almost impossible to make a further advance or to achieve yet more conquests. Nevertheless the advantages to be gained by, at least, securing the port of Beirut and the railway between there and Damascus were so great that the Commander-in-Chief determined to make the attempt.

The 5th Division therefore was ordered to march on 5th October, while the 4th Division also moved on the same date to Zabdani, two days' march from Damascus on the Beirut railway. After two long marches the 5th Division on 6th October occupied Rayak, an important railway junction, whence a broad-gauge line runs northward through Homs to Aleppo and thence joins the Baghdad line at Muslimie. Here the division halted for two days while the armoured cars attached to the force went on to Beirut which they entered without opposition, the inhabitants receiving them with rejoicing and handing over to them the small Turkish garrison of 600 men who had been disarmed. On the 9th October, the Beirut line having been secured, the division moved a few miles up the Homs-Aleppo road as far as Tel es Sharif, and the armoured cars reconnoitred as far as Baalbek. Still no opposition was encountered, nor was there any sign of the enemy except parties of Turks disarmed by the Arabs and, on one occasion, a German aeroplane which bombed the camp, wounding one man of the Ninth and killing four horses.

In these circumstances General Allenby decided that the 5th Division supported by armoured cars should make a further advance to Homs, under Major-General MacAndrew. That officer divided his force into two columns. Column A, which led the advance and was accompanied by divisional headquarters, consisted of three batteries of armoured cars and three light car patrols, with the 15th (Imperial Service) Brigade, two regiments strong. Column B, which followed generally a day's march behind, included the rest of the division.

The 13th Brigade moved to Baalbek on 12th October and the march was continued without incident to Homs which was reached on 16th October. Everywhere the force was greeted with enthusiastic demonstrations of joy, while on the other hand enemy stragglers got short shrift from the inhabitants; indeed in several places the camping grounds and roads were made most unpleasant by the stench arising from the bodies of Turks and Germans, who had been killed as they passed through in retreat.

At Homs once more the question for or against continuing the advance had to be decided, and once more the Commander-in-Chief determined to push on, and to round off his campaign with the capture of Aleppo. It was known that there was a considerable force of Germans and Turks in and about that city, amounting it was said to some 20,000, but about 8,000 of these were already beaten and demoralised, and General Allenby believed that the victorious 5th Division, even though greatly reduced in strength, would be equal to the task of overcoming whatever resistance they might meet with.

The advance was continued by A Column on 20th October, B Column including the 13th Brigade following next day. But now it was found necessary to leave A squadron of the Ninth at Homs owing to an outbreak among the men of some sort of acute influenza, and this squadron only rejoined the regiment a fortnight later at Aleppo; moreover there was so much sickness among the British soldiers that the Ninth had to lend men to the Horse Artillery to help in driving the guns. Four days of varied marches brought the brigade to Ma'arit el Naaman on 24th October. Meanwhile the divisional commander had pushed on with the armoured cars to within six miles of Aleppo on the 23rd and had thence sent to the Turks a summons to surrender. The Turkish commander, Mustapha Kemal, refused to consider such a proposal, but he was so much impressed by Major-General MacAndrew's boldness that he began at once to withdraw his troops from the town. On the 24th a force of Sherifian Arabs arrived and in the course of the following night these were admitted by friends into Aleppo. The Turks who still remained were thereupon driven out and on 26th October Major-General MacAndrew motored in with the armoured cars. A sharp fight was put up by a Turkish rearguard on the Alexandretta road north of the town in which the 15th Brigade (Jodhpur and Mysore Lancers) behaved with great gallantry, and after this all resistance was at an end.

Column B reached the neighbourhood of Aleppo on 26th October. On the 27th this column passed round the west of the city to a point about four miles north-west on the Alexandretta road. Here the 5th Division was once more concentrated and here on the same day the 13th Brigade relieved the 15th on outpost duty. At noon on the 31st October the armistice with Turkey came into force and Sir Edmund Allenby's campaign in Palestine was at an end.

The achievement of the 5th Cavalry Division, and indeed of the whole Desert Mounted Corps, in these memorable operations was a splendid one. Since leaving the concentration area near Jaffa thirty-eight days before the division had marched 567 miles (the distance covered by the 9th Hodson's Horse was 509 miles), and had taken eleven thousand prisoners and fifty-eight guns. The total captures of the Desert Mounted Corps were 83,700 prisoners and about 160 guns. In the four weeks since the fall of Damascus the 5th Division had encountered but little opposition. The Turks were demoralised and beaten and the boldness with which the victors pressed forward was exactly calculated to complete the despair of the enemy. But our troops had to contend against foes more relentless and insidious than the Turks. We have seen how even at Damascus extreme fatigue and exposure had already made them a prey to the attacks of disease. As the further march of two hundred miles proceeded their ranks were thinned in daily increasing numbers, and daily the physical capacity of those who remained was reduced. Men and horses had reached almost to the limits of their powers. It was only the consciousness of victory and their grim determination to complete their task that carried these gallant horsemen through to the final goal.

The following officers served with the 9th Hodson's Horse in General Allenby's final campaign :—

Lieut.-Colonel C. H. Rowcroft .	Commanding
Captain L. C. T. Graham . .	A Squadron commander
Captain F. W. Messervy . .	B Squadron commander
Lieutenant G. W. Ninis . .	B Squadron officer
Captain C. W. L. Stevens . .	C Squadron commander
Lieutenant D. T. Long . . .	C Squadron officer
Major M. D. Vigors . . .	D Squadron commander
Lieutenant J. E. Walker . .	D Squadron officer
Lieutenant N. N. Morris . .	Adjutant
Captain W. S. Shepherd . .	Signalling officer
Lieutenant I. G. F. Pierce . .	Quartermaster
Lieutenant C. M. Hutchings . .	C.O.'s galloper
Lieutenant A. C. Chilton . .	Commanding Divisional Transport
Captain S. Dutt	Medical officer

The following immediate honours were conferred for gallant conduct in the operations :—

Major M. D. Vigors, M.C. .	Distinguished Service Order
Risaldar Nur Ahmad Khan, I.O.M.	Military Cross
Risaldar Dost Muhammad Khan	Indian Order of Merit
Jemadar Nawab Ali Khan .	,,
Risaldar Major Muhammad Akram Khan, Bahadur	Indian Distinguished Service Medal
2792 Dafadar Mehtab Singh .	,,
3331 Sowar Mir Badshah . .	,,
3253 Sowar Kabul Khan . .	,,

CHAPTER VII.

NOVEMBER 1918-1922.

THE occupation of Aleppo and the armistice with Turkey were followed almost immediately by the cessation of hostilities with Germany and the end of the great war. But a year was yet to elapse before the return of the Ninth to India. The defeat of the Turks involved the abolition of their rule in Syria, and the administration of the country was thereupon undertaken temporarily by the British, whose principal agents for several months were the 5th Cavalry Division.

After some days on outpost duty on the Alexandretta road north-west of Aleppo the Ninth were relieved on 2nd November and moved to a bivouac near the city, where A Squadron rejoined the regiment from Homs. Here nearly two months were passed, and as men and horses recovered from the strain of the late operations the usual routine of training was resumed, while recreation was found in regimental sports, racing and a little polo. As the season advanced and the weather broke up, some new Turkish barracks were handed over to the regiment, but they were very dirty, and although they were cleaned up with the help of Turkish prisoners, yet most of the men still preferred to sleep in their bivouac shelters in the open.

Meanwhile on the receipt of the news of the armistice in France the following congratulatory telegrams were exchanged between the 5th Cavalry Division and Sir Douglas Haig :—

" *To* F.M. Sir DOUGLAS HAIG.

" Nov. 20. The 5th Cavalry Division, Aleppo, send you their heartiest congratulations on your glorious victories in France. They will always regret that they were not there to play their part under your command. MACANDREW."

" *To* G.O.C. 5th Cavalry Division, Aleppo.

" Very sincere thanks for your telegram of congratulations from myself and from all ranks of the armies under my command. We in our turn heartily congratulate you and the 5th Cavalry Division on your brilliant work in Palestine. We have never ceased to regret the departure of yourself and your splendid squadrons from my command in France during the critical battles of this year.

D. HAIG."

On 3rd December Sir Harry Chauvel, the commander of the Desert Mounted Corps, inspected the regiment and expressed his appreciation of the condition of the horses and mules. He added that he had inspected the greater part of the Desert Corps since the armistice but that he had seen no unit that could compare with the 9th Hodson's Horse in the condition of the horses or in general smartness.

A week later the Commander-in-Chief, Sir Edmund Allenby, came to Aleppo, and on 12th December he passed in procession through the town, on which occasion his escort was commanded by Lieut.-Colonel Rowcroft and was composed of B Battery, Honble. Artillery Company, and two squadrons each of the 9th Hodson's Horse and the 18th Lancers.

At the end of the year the regiment moved in four marches to Killis in Cilicia, about forty miles due north of Aleppo, where it was employed on police and administrative work, with a detachment of B Squadron performing similar duties at Katma, and another from A Squadron at Azaz. The labours involved were by no means light or easy. The country was in a state little short of chaos. The peaceable elements of the population were at the mercy of wandering bandits who murdered and pillaged indiscriminately, Armenians being their peculiar prey. Brigandage of this sort had to be put down, local governors were appointed and their administration had to be supported, and the awards of a reparations committee, which was promptly set up, had to be executed. In all these duties the troops had a prominent share, the officers commanding the several centres lending their assistance to the local authorities, non-commissioned officers and men superintending the work of Turkish gendarmes in investigating reports of murder and violence, and patrols and piquets in outlying villages maintaining order and protecting the inhabitants from outrages. Before long general confidence was restored, the normal life of the people was resumed, and in spite of great difficulties much progress was made in restoring the property of those who had suffered spoliation during the recent turmoil.

After some six weeks of this work at and around Killis orders were received by the regiment to move to Marash, a large town in the extreme north of Cilicia, about a hundred and twenty-five miles distant. Leaving Killis on 13th February, on being relieved there by the Gloucestershire Hussars, the Ninth marched through Aintab (fifty miles), where the headquarters of the 13th Brigade were located, and where a halt was made for a few days. Thence they proceeded across the Aksu (one hundred miles) and reached Marash on 22nd February.

The occasion of this move was the receipt of reports that a wholesale massacre of the Armenians in the town was contemplated by the Turks, who formed a large proportion of the population, and whose troops were not many miles away. In view of this threat it was foreseen that a long occupation of the place would be necessary, and the regiment was therefore accompanied by a considerable column of subsidiary services, including a Ford car motor section, a wireless section, a hospital section, and food supplies for several weeks. The regiment was greeted on its arrival with the wildest enthusiasm by the Armenian population, and was accompanied by them to the place selected for its bivouac near the American Mission buildings, where it quickly settled down. Comfortable quarters were found for all ranks, the water supply was plentiful and good, and excellent grazing was to be had by the river some two miles from the town. It had been anticipated that the Turks might attempt to oppose the march of the regiment and dispute the passage of the river Aksu, but nothing of the sort happened, and although our relations with the Turkish authorities at Marash were at first strained and difficult yet matters soon mended, and tact and discretion secured friendly feelings on all sides. The duties that had to be performed here were similar to those at Killis, but they were naturally more difficult in consequence of the stronger Turkish element and of the proximity of purely Turkish territory and Turkish troops. Throughout the time spent at Marash precautions were taken against hostile disturbances, patrols being sent out daily, and protective measures adopted at night. Indeed in October 1919, towards the close of our occupation, the presence of Mustapha Kemal's troops in the neighbourhood necessitated special precautions. Hotchkiss rifle posts were placed on the approaches to Marash, and the non-commissioned officers in charge of these were furnished with written notices in Turkish to the effect that no one was allowed to pass. Orders were issued, however, by Mustapha Kemal warning his patrols against coming into collision with the British forces, and no trouble occurred.

In other respects the months of 1919 passed without much notable incident. In addition to the special duties in aid of the civil power which have been described, the regiment carried out as usual constant training practices. Amusement was found in regular polo and a certain amount of pig-sticking, and after a few weeks at Marash a very successful race meeting was organised, at which the acting Corps Commander, Sir Harry Hodgson, was present, as well as the Mutaserrif and all the local officials, and which was largely attended by the populace.

Meanwhile the New Zealand and Australian Mounted Divisions left Palestine in the early spring, and on 7th June 1919 the famous Desert Mounted Corps ceased to exist. The occupation of the conquered territory was continued for five months more by the so-called " North force," consisting of the 4th and 5th Cavalry Divisions and two divisions of infantry, the whole under the command of Major-General G. de S. Barrow, until in November the administration of northern Syria was taken over by the French and the British forces of occupation were withdrawn. The first parties of the French relieving troops reached Marash on 28th October, and on 1st November the Ninth left the town on the first stage of their homeward journey. For more than eight months the regiment had kept the peace and maintained friendly relations with all sections of the population, consisting of Turks, Armenians, Circassians and Kurds. How successful they were in this delicate and difficult task may be gathered from the following expressions of appreciation which were addressed to them on their departure :—

" *To* Lieutenant-Colonel Rowcroft

Commander of English troops of occupation in Marash.

The grateful Armenian community of Marash expresses her cordial thanks to the 9th Hodson's Horse,

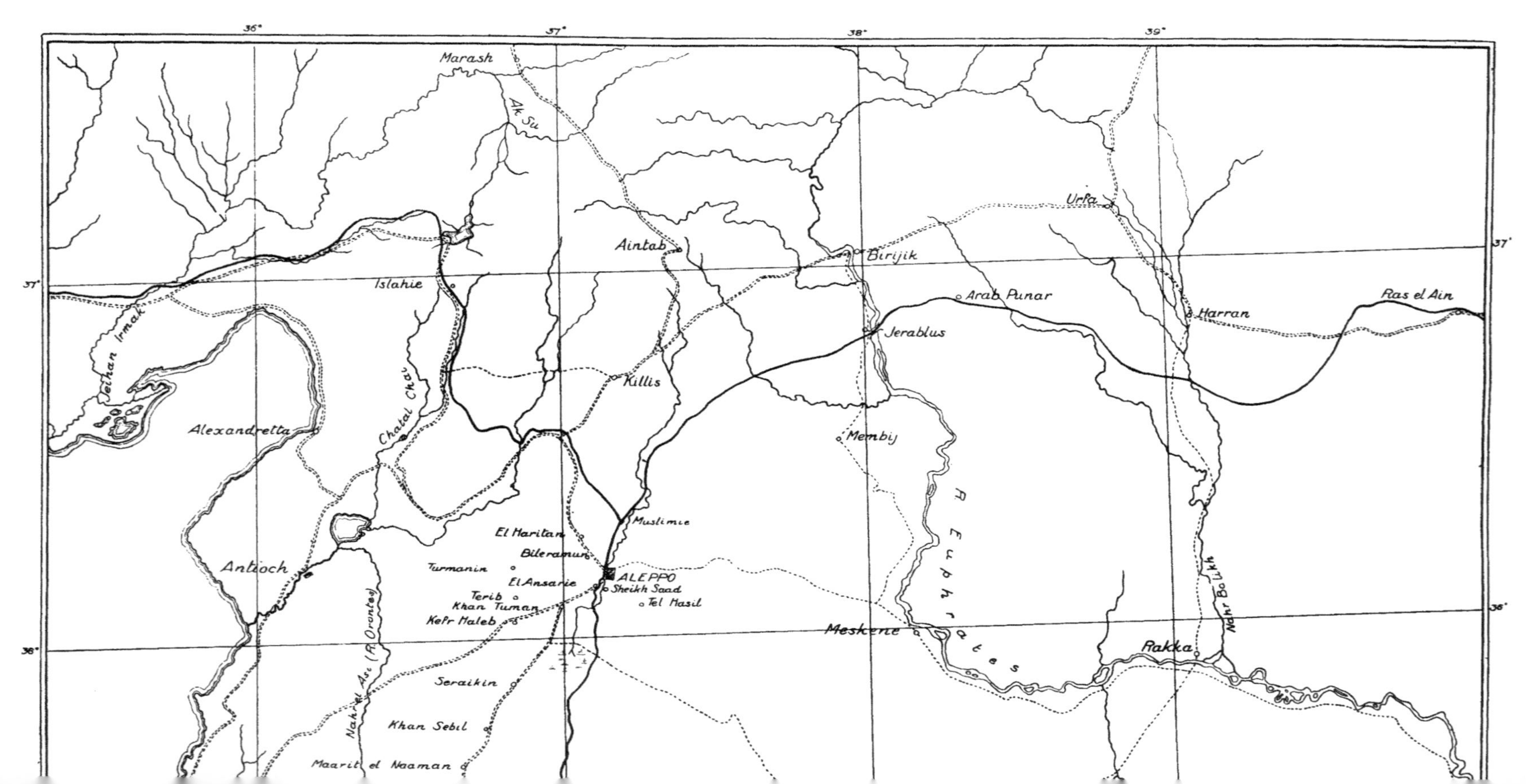
36°
37°
38°
39°
Marash
Ak Su
Urfa
Birijik
Aintab
Islahie
Arab Punar
Ras el Ain
Harran
Jerablus
Killis
Jeihan Irmak
Alexandretta
Chatal Chai
Membij
Muslimie
El Maritan
Bileramun
Turmanin
El Ansarie
ALEPPO
Sheikh Saad
Tel Hasil
Terib
Khan Tuman
Kefr Maleb
Antioch
Nahr el Asi (R. Orontes)
Seraikin
Khan Sebil
Maarit el Naaman
Meskene
R. Euphrates
Rakka
Nahr Bolikh

Banias
Zor Defai
HAMA
El Rastan
Telbise
HOMS
Tadmor
Kusseir
Jendar
Hasie
Rible
El Mineh
TRIPOLI
Hermil
Ras Baalbek
Lebwe
Nebk
Baalbek
BEIRŪT
Zahle
Shtora
Kuteife
Zebdani
Khan Ayash
Khan Kuseir
Duma
Harista el Basal
Khan Meizelun
Saida
DAMASCUS
Kalana
Roads
Tracks
Railway
miles
miles
35°
34°

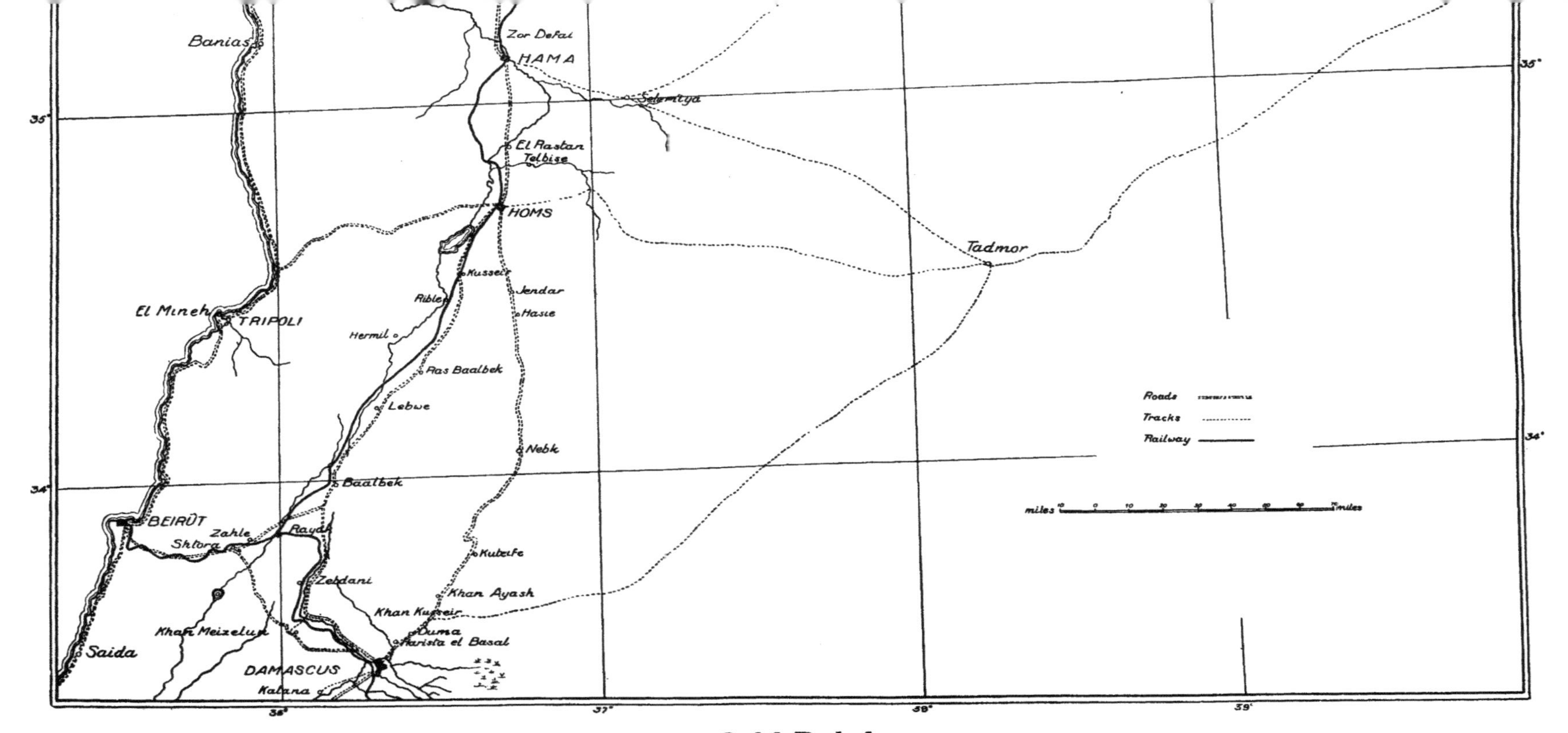

SYRIA

whose protection she has enjoyed for eight months. Together with our gratitude we feel a great sorrow that we have not been able to entertain you as we would like, on account of political conditions.

On the occasion of the departure of the said brave regiment we wish your excellency and your soldiers to be sure that your sacred memory is engraved deeply and will remain so for ever in the heart of each Armenian individual.

LONG LIVE ENGLAND."

"*To* His Excellency the O.C. British Troops about to depart from Marash.

I beg to inform you that the whole Muhammadan population of our town is grateful and thankful to you for your just and neutral actions that you carried out during your stay here and for the proved good conduct of your soldiers, and I beg you also if you would kindly inform all high concerned about this population.

Accept please the respects of our population.

(Signed) Chief of Municipality of Marash
HASSAN."

1/11/19.

"*From* the Mayor and Municipality (Turkish) of Marash.

On your departure we wish to thank you for good government, impartiality, just treatment and the proved good conduct of the troops, and we beg that you will suggest your methods to your successors."

There is little to record of the march southwards from Marash except the peculiarly abominable weather which made the roads almost impassable and caused the whole experience to be very trying both to men and horses. At Ma'arit el Naaman, the third march out from Aleppo, some thieves raided the camp at night, but a volley from the stable pickets brought down three of the marauders, and a threat to blow up the house of the village sheikh resulted in the recovery of all the purloined property. Tripoli was reached on 23rd November after a terrible march across the Lebanon, and here the regiment halted for upwards of a fortnight, the horses being much exhausted with three weeks' continuous marching in appalling weather. Resuming the march on 10th December the regiment arrived at Beirut on the 13th and remained there till the last day of the year, during which interval the horses rapidly regained condition. On 31st December the Ninth embarked at Beirut in two parties, arrived at Kantara on 3rd January 1920, and after a fortnight to re-equip moved by rail to Alexandria, where they went into a camp near the sea at Chatby on 17th January.

Thus ended the service of the regiment in Palestine and Syria. During the year and eight months which the Ninth had passed in those countries they had played a distinguished and honourable part in one of the most remarkable campaigns of modern times; they had helped to restore the credit of the mounted arm which had been freely and ignorantly assailed as a result of the experiences in France; they had proved, as the Officer Commanding the 13th Cavalry Brigade wrote in his farewell order on their departure, "that bold and well-trained cavalry can overcome all opposition"; their actions in the great advance to Nazareth and Damascus had furnished (to quote again from the same order) "historical examples to all soldiers of what can be accomplished by determined horsemen imbued with the true cavalry spirit"; above all they had both in war and peace not only well maintained but also added fresh lustre to the reputation of the Indian cavalry and of their regiment.

The Ninth had now been upwards of five years on foreign service and it was but natural that all ranks should look for a speedy return to India. Instead of this, on their arrival in Egypt they found themselves faced with the prospect of an indefinite delay consequent on the political conditions in that country. It is unnecessary to comment on the causes of the trouble in question. So far as the Ninth was concerned the material fact remained that, after a long period of varied and trying active operations, the officers and men were perforce required to restrain yet further their legitimate desire to return to their own country. The cheerfulness with which they met this new demand on their patience was a characteristic proof of their loyal spirit.

The disturbed situation in Egypt and the hostile attitude of a section of the population, which were the occasion of the retention of the Ninth at Alexandria, necessitated the regiment being located close to the town. Only this fact could account for the choice of the camp ground. It was indeed as unsuitable for cavalry as any place could be; there was however no alternative but to make the best of it.

For many weeks after the arrival of the regiment at Alexandria it was frequently called upon to assist the civil authorities in keeping order in the town. But these irritating and uncongenial duties never developed into open hostilities, for the mobs of turbulent Egyptians invariably melted away as soon as the cavalry appeared. In other respects the time was passed usefully and not without amusement. So far as the limited size of the parade ground permitted all sorts of training were constantly practised, while the officers were able to get a great deal of polo, and won the Cairo Open Cup, the Alexandria American tournament, the Alexandria Open Handicap Cup and the Public Schools Cup. All ranks fully appreciated the amenities of civilisation at Alexandria after so many months of hardship and privation, and the horses were got into such excellent condition as to elicit warm commendation from inspecting officers.

On 2nd February, 1920, Lieut.-Colonel C. H. Rowcroft, D.S.O., who had commanded the regiment through the last two eventful years, left for India under orders to assume command of the 26th Light Cavalry, and Captain (temporary Major) Vigors, D.S.O., M.C., took command of the Ninth. Within a few months, however, this arrangement was cancelled and Lieut.-Colonel Rowcroft rejoined and resumed command on the 19th June.

At last in the autumn of the year orders were received for the regiment to return to India. Unfortunately these orders involved leaving all the horses in Egypt where they were to be handed over to the Remount Department. The Ninth had everywhere while on foreign service been conspicuous for the excellence of their stable management and general horsemastership, and the condition of their horses when they handed them over in 1920 was described by the Brigadier, under whom they had been serving, as wonderful. It was therefore with all the more regret that they were obliged to leave them behind.

The regiment embarked at Suez on 16th December, 1920, and disembarked at Bombay on 30th December, and this record of its service during the last six years may be fittingly closed with the farewell order issued by Field-Marshal Lord Allenby, who had been in chief command of the campaign in Palestine and who was now Commander-in-Chief in Egypt :—

"General Headquarters,
Egyptian Expeditionary Force,
30th Nov. 1920.

" Officer Commanding
9th Hodson's Horse.

On your departure from Egypt and the E.E.F. please express to all ranks my high appreciation of the services they have rendered and their admirable spirit and conduct in all circumstances.

Your regiment has upheld with distinction the best fighting traditions of the Indian Cavalry.

I thank you and wish you good luck.

ALLENBY, F.M.
C.-in-C.—E.E.F."

In the course of their service out of India the Ninth had suffered the following losses :—

	Killed.	Wounded.	Missing.	Died of disease.
British officers	4	5	...	...
Indian officers	4	7	...	...
Other ranks	20	130	3	15

Further details of the casualties among officers will be found in Appendix V.

A list of honours and rewards earned by the regiment is given in Appendix IV. This has been made as complete as possible, but it is feared that there may be some inaccuracies in it and perhaps some omissions.

The following battle honours were awarded the regiment for service in the Great War :—

France and Flanders, 1914-18.
Givenchy, 1914.
Somme, 1916.
Bazentin.
Flers-Courcelette.
Cambrai, 1917.
Palestine, 1918.
Megiddo.
Sharon.
Damascus.

A complete list of the British officers who served

with the regiment in the field is given at the end of this chapter.

The 9th Hodson's Horse reached Ambala on 1st January, 1921, having been longer absent on service overseas than any other unit of the Indian Army. They were greeted on arrival with a congratulatory message from the Commander-in-Chief in India (General Lord Rawlinson) to the following effect :—

"On your return to India from field service overseas, His Excellency the Commander-in-Chief extends to you and all ranks under your command his heartiest welcome, and congratulates all on the gallantry and devotion with which they have mainained the high traditions of the army.

H. Hudson,
Lieut.-General,
Adjutant-General in India."

The Officer Commanding,
9th Hodson's Horse.

Leave was at once opened for all ranks and was given to as many as possible. For those who remained with the regiment there was plenty of work to be done in settling down in the new conditions of service and in preparing for the further changes which were now about to be introduced.

During the previous year the Government of India had arrived at the decision, as the result of the experience of the war and after prolonged and exhaustive inquiry, that the silladar system, which ever since the Mutiny had been maintained in all but three of the Indian cavalry regiments, must be abolished. Simultaneously with this great change it had also been decided that a drastic reduction should be made in the strength of the cavalry arm, the number of regiments being reduced from thirty-nine (as it stood before the war) to twenty-one. The need for economy and the increased dependence on mechanical substitutes for mounted troops were the main arguments to support this sweeping change. In order to give effect to the decision, all but three of the existing units of cavalry were arranged in pairs, and from each pair one regiment was organised. It was claimed that this method of reduction preserved the existing regiments from complete extinction, but, practically, many fine corps ceased to exist. The 9th Hodson's Horse were more fortunate than these. It was a matter of course that in such a scheme they would be re-united with the second regiment of Hodson's Horse, the 10th Duke of Cambridge's Own Lancers. Each naturally regretted the loss of a separate individuality which had lasted so long, but both still retained the name which was most honoured by them, nor did the re-union bring any break in the traditions which still surround the memory of their common founder. The Ninth and Tenth disappeared but Hodson's Horse still survives unchanged in spirit and doubly enriched with the achievements of more than sixty years.

The Ninth moved from Ambala to Lahore cantonments on 3rd September, 1921. There they joined the 10th Lancers, who had arrived some time before, and the work of amalgamating the two units proceeded rapidly. The separate titles of the two were retained in the army lists for another year, but to all intents and purposes the union was effected in the autumn of 1921, under the command of Lieut.-Colonel C. H. Rowcroft, D.S.O., who had led the 9th for so many eventful months in France and Palestine.

Note.—A Squadron and R/B (Sikhs) are now represented by A Squadron of the present regiment.

C Squadron and a few of D (Punjabi Muhammadans) are represented by the present B Squadron.

L/B Squadron (Dogras) is merged in the present C Squadron.

D Squadron (Pathans) has disappeared.

A List of the British Officers who served with the Ninth Hodson's Horse during the Great War.

Lieut.-Col. R. B. Low, D.S.O.
Lieut.-Col. A. W. Pennington, M.V.O.
Lieut.-Col. G. A. H. Beatty, C.M.G., D.S.O.
Lieut.-Col. C. H. Rowcroft, D.S.O.
Major H. L. Dyce, M.C.
Major O. F. Smith.
Major J. C. Russell, D.S.O. (killed in action, 1917).
Major A. I. Fraser, D.S.O. (killed in action, Cambrai, 1917).
Major F. St J. Atkinson, D.S.O. (killed in action, Cambrai, 1917).
Major E. de Burgh, D.S.O., Brev.-Lieut.-Colonel.
Major W. S. Craster (8th Cavalry).
Major G. R. P. Wheatley (27th Light Cavalry).
Capt. M. D. Vigors, D.S.O., M.C.
Capt. G. de la P. Beresford, M.C. (10th Lancers)
Capt. T. W. Corbett, M.C.
Capt. E. V. F. Seymour.
Capt. L. C. T. Graham, M.C.
Capt. C. F. L. Stevens, M.C. (10th Lancers).
Capt. G. B. Reeves (accidentally killed).
Capt. H. W. Luttman-Johnson.
Capt. F. W. Messervy.
Capt. R. R. M. Porter, I.M.S.
Capt. S. Dutt, I.M.S., M.C.
Assist.-Surg. J. Parkinson, D.C.M.
Capt. Kirkwood (17th Lancers).
Capt. J. A. C. May-Sommerville (11th Lancers).
Capt. F. K. Moody, M.C. (13th Lancers).
Capt. R. A. Carr-White, M.C. (31st Lancers).
Lieut. G. W. Ninis.
Lieut. J. R. K. Murphy.
Lieut. J. Y. Tollemache.

Lieut. D. T. Long.
Lieut. E. H. Gastrell.
Lieut. J. E. Walker.
Lieut. M. S. Bendle.
Lieut. R. H. R. Cumming.
Lieut. I. G. F. Pierce.
Lieut. G. S. Hurst.
Lieut. C. M. Hutchings.
Lieut. R. A. Oswald.
Lieut. J. R. M. Stephen.
Lieut. E. S. M. Prinsep (11th Lancers).

Indian Army Reserve of Officers.

Capt. S. P. Davis.
Capt. C. F. Birley (Duke of Lancaster's Own Yeomanry).
Capt. C. Shepherd-Cross (Duke of Lancaster's Own Yeomanry) (killed in action with M.G.C.)
Capt. M. N. Morris.
Capt. J. A. Ewart, M.C.
Capt. W. S. Shepherd.
Capt. M. Dudding.
Lieut. R. Morgan (Westminster Dragoons).
Lieut. K. S. Stewart-Martin.
Lieut. T. Stallibrass.
Lieut. Lord Monkswell.
Lieut. P. V. Douetil.
Lieut. R. P. Hawkins.
Lieut. R. W. Fremlin.
Lieut. W. D. Woellworth.
Lieut. J. M. Wilson.
Lieut. J. A. Brown.
Lieut. Kinloch.
Lieut. G. Wilson.
Lieut. A. M. Wallace.
Lieut. A. C. Chilton (R.H.G.)

Note.—As has been mentioned in the narrative, Majors Russell and Atkinson were employed during 1916-18 with infantry battalions in France, and the former was killed at Passchendaele at the end of 1917 while commanding the 6th Battalion Cameron Highlanders. Several other officers were similarly absent from the regiment from time to time. Among these Brev.-Lieut.-Colonel E. de Burgh after the arrival of the regiment in France served on the staff as Brigade-Major, 3rd Cavalry Brigade; G.S.O.2, 3rd Cavalry Division; and finally G.S.O.1, 2nd Cavalry Division (2nd battle of Ypres, Loos, Arras, Cambrai, Somme, 1918, &c. Despatches, 1916 and 1919, D.S.O., brevets of Major and Lieut.-Colonel).

PART III.

THE TENTH DUKE OF CAMBRIDGE'S OWN LANCERS (HODSON'S HORSE)

CHAPTER I.

1859-1878.

The separate record of the 2nd Regiment of Hodson's Horse opens at Gonda where, as related in Part I., the regiment arrived on 25th May, 1859.

Ten days previously Capt. C. H. Palliser had been appointed to the command of the regiment, a post which he continued to fill with credit to himself and with advantage to the corps for no less than twenty-two years. It is difficult in these days to imagine conditions in which so prolonged a tenure of command could be either beneficial to a regiment or even tolerable to the individual, but whatever may have been the merits and defects of the system, the fact remains indisputable that to Capt. Palliser the 10th Bengal Lancers (as the regiment soon became) owed more than to any other individual with the single exception of Hodson himself. His was the fostering care and sympathetic guidance which gave the regiment cohesion, character, and all the qualities which are meant by *esprit de corps*. He found it little better than a half-trained and half-disciplined crowd of fighting men, and he left it, after twenty-two years of his wise and capable control, one of the most efficient as well as one of the most celebrated cavalry regiments in India.

Naturally enough, after the final operations in the Mutiny Campaign, the attention of the regiment was centred for the most part on perfecting its organisation, drill and equipment. In the latter respect the work had to start almost from the beginning. Thus we find from the manuscript record prepared by the commandant that in 1859 a complete outfit was provided for the men, including brown leather sword-belts, black leather pouches and pouch-belts, Napoleon boots, swan-neck brass spurs, dark blue alkalaks with scarlet facings and pipings (which were obtained from England), and cloaks of country blanketing. A year later khaki cotton blouses were obtained for summer wear instead of the blue alkalaks, and in 1861 the uniform was further supplemented by the provision of white blouses to be worn in the hot weather by men on orderly duties.

The armament of the regiment at this time was a dual one, sixteen men in each troop receiving, in accordance with G.O.G.G. of 27th April, 1860, the so-called "Victoria" carbine, while the rest of the regiment was armed with lances, with shafts 9 feet long and having pennons 26 inches long by 9 inches wide, the upper half scarlet, the lower blue.

It was not until 1862 that steps were taken to equip the regiment with saddlery of a uniform pattern. In that year "Nolan" saddles with high brass-bound cantles were obtained, and at the same time headstalls, reins, breastplates and cruppers of black leather with steel mounts. Shabracks also were issued to all ranks, of blue cloth edged with scarlet; and steel spurs were substituted for those of brass previously in use.

Meanwhile in May, 1861, the 2nd Hodson's Horse received the title of 10th Regiment of Bengal Cavalry, and about the same time its establishment was fixed as follows as regards Indian ranks:—

3 Risaldars.
3 Ressaidars.
1 Wordi Major.
6 Jemadars.
6 Kot Dafadars.
48 Dafadars.
6 Nishanburdars.
6 Trumpeters.
420 Sowars.

MAJOR GENERAL SIR CHARLES PALLISER, G.C.B.

Two years later, by Adj.-Gen's. letter 152 of 11th January, 1863, the composition of the regiment was determined, namely :—

1 troop of Trans-Indus border tribes.
1 troop of Punjabi Mussulmans.
2 troops of Sikhs.
1 troop of Dogras and other hillmen.
1 troop of Jats.

The stay of the regiment at Gonda, its first cantonment, lasted from May, 1859, till June, 1861, when it marched to Bareilly, detaching a troop each to Fatehgarh and Moradabad.

Eighteen months later, on 18th January, 1863, it was again on the move, in the first place to Agra where the Viceroy (Lord Elgin) was holding a durbar, and thence as part of His Excellency's escort by way of Muttra, Delhi, Meerut, and Saharanpur, to Ambala, where the regiment arrived on 27th March.

Its stay at Ambala was interrupted by troubles on the North-West Frontier, resulting in the Ambela Campaign of 1863 and operations against the Mohmands in 1863-64, but although the regiment was moved up to Peshawar in November-December 1864, with detachments on the disturbed border at the forts of Michni, Abazai and Shabkadar, it took no part in the operations, and after attending a cavalry camp of exercise under the Commander-in-Chief (Sir Hugh Rose) at Chamkanni in February 1864, it marched back to Ambala, arriving there on 7th April.

The next year and a half were uneventful but for certain noticeable changes in the equipment of the regiment and of its title. On 3rd May, 1864, a General Order by the Commander-in-Chief directed that the 10th, 11th, 13th and 14th Regiments of Bengal Cavalry were to be designated "Lancers," these being the first regiments of Indian Cavalry on which that honourable and coveted title was ever conferred.

At the same time the "Victoria" carbines, with which a proportion of the men was armed, were withdrawn, and the armament thenceforward consisted of a bamboo lance 10 feet long, with bayonet-shaped head, a sword and a pistol. Sabres of light cavalry pattern were obtained from Government stores at half cost price, and wooden scabbards covered with black leather and with metal shoes were made regimentally.

Some few other changes in clothing and equipment about this time included the adoption of *posteens* for use by all ranks in cold weather, the garments being obtained at Peshawar while the regiment was on duty there in the winter of 1863-64; the issue of new lunghis both full dress and undress, the former being red, the latter dark blue; and finally in 1865 the introduction of "Mameluke" bits, which were manufactured at Kotli near Sialkot and issued to all native ranks. In these humanitarian days, and (if one may say so) days of more intelligent horsemastership, the use of such a bit in a cavalry regiment is unthinkable. To a rider with perfect hands it certainly gave complete control over his horse. Used by a clumsy man it was liable to cause intense pain, and sometimes the injuries inflicted by it on horses' mouths were horrible. Nevertheless it remained in use until 1888, when, though not at once discontinued, it was modified in such a way as to remove its most objectionable features.

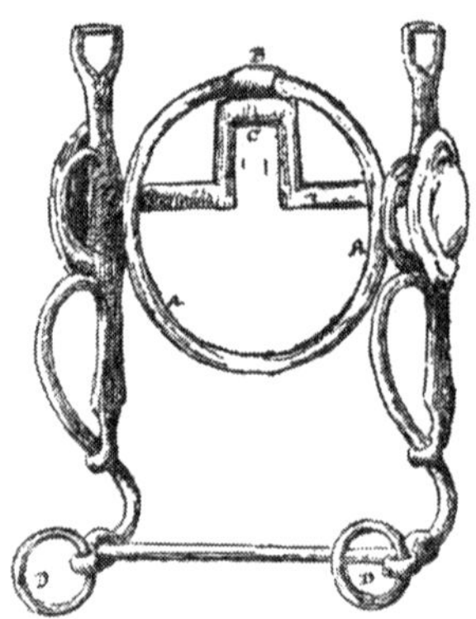

The large ring A A hangs loosely from the sleeve B which is fixed to the top of the port C. When the port is placed in the horse's mouth the lower part of the ring passes round the horse's under jaw. When the curb rein is used from D D this ring acts as a rigid and very severe curb.

Towards the end of 1865 the regiment marched to Jhansi, where however it did not remain more than a year, for in December, 1866, it moved to Saugor, where it relieved the 4th Madras Cavalry.

Here again its stay was destined to be short, but the time was made both toilsome and expensive by the uncongenial task of building new lines. The old lines vacated by the Madras Cavalry (a " regular " regiment) were ruinous, and the Tenth, being on the silladar system, could not look to the Government to provide new accommodation. All the assistance that the regiment could obtain at first was permission to use the materials of the old lines, in default of which the provision of materials would at Saugor have been extremely difficult. Later on, the commanding officer, Major Palliser, appealed to the Government for the grant of a hutting allowance, such as would be admissible to an infantry regiment, pleading that the considerable outlay involved in the building of the lines had fallen on the regiment only a few months after it had been obliged to incur heavy expense in repairing the accommodation at Jhansi. This appeal received favourable consideration, and a hutting allowance was eventually granted.

Meanwhile all through the spring and early hot weather the men worked hard at the task before them. The whole work, not only of building the huts but also of drainage and of collecting stones with which to make the raised horse-standings, had to be carried out by the regiment. The labour was arduous and prolonged, but it was undertaken with energy on the part of all ranks, with the satisfactory result that the new lines were completed by the time that the monsoon broke.

The regiment did not long enjoy, however, the fruits of this hard toil, for in the autumn of the same year, 1867, an opportunity occurred for more congenial employment on field service.

Early in the summer serious tension had arisen between Great Britain and the African kingdom of Abyssinia, the British representative at the Abyssinian capital being insulted and British subjects imprisoned by King Theodore. At length the situation necessitated a resort to hostilities, and it was decided to send an expedition from India under the command of Hodson's old friend, Lieut.-General Sir Robert Napier, then Commander-in-Chief of the Bombay Army. The undertaking was one of considerable difficulty. Over and above the initial risks of transporting a large force of all arms from India to the African coast, the expedition entailed a march of some hundreds of miles from the coast to the capital of Abyssinia, through an almost unknown country, deficient in supplies of food or water. At the end of the march the invading force would be required to meet an enemy who, though uncivilized, had a considerable fighting reputation, and it would probably be necessary to attack a rock fortress of great natural strength. For such a task the employment of an ample force was essential, and prolonged and careful preparation was required in respect of every detail.

The troops selected for the expedition amounted to nearly 14,000 men and consisted of four and a half regiments of cavalry, seven batteries and one Indian company of artillery, eight companies of engineers and sappers, and four regiments of British and ten of Indian infantry. In this force two regiments of Bengal Cavalry were included, namely the Tenth and the Twelfth.

Some months before the force was actually detailed the Tenth volunteered *en masse* for service in Abyssinia in case of need, a circumstance which was the more notable and praiseworthy because none of the men had ever served overseas, and in those days such a service was a much more serious, or at least more unusual adventure then it became in later years.

At length on 9th October orders were received at Saugor for the regiment to hold itself in readiness to join the Expeditionary Force, and on the next day Lieutenant England accompanied by Jemadar Isar Singh proceeded to Calcutta, where the regiment was to embark, to supervise the preparation of shipping and the provision of food supplies and water for the different classes and castes.

On 12th October the Tenth marched from Saugor towards Allahabad *en route* for Calcutta. The start was made in unfavourable circumstances. Heavy rain, unusual at that season, impeded the march, making the rough roads almost impassable for wheel carriage, and filling the water-courses so that five separate rivers had to be swum by the horses between Saugor and Nagode, which was reached on 25th October. The regiment marched into Allahabad on 5th November, and here it was halted until 23rd December when, the transports being ready for its reception, it was moved by rail to Calcutta and encamped on the *maidan* opposite Prinsep's Ghat.

The embarkation began on 28th December as follows :—

Date.	Unit.	Strength— Rank and File.	Strength— Horses and Baggage Ponies.	Commander.	Name of Transport.
1867 Dec. 28	1st Troop	78	135	Lieut. D. M. Strong	Callirrhoë
1868 Jan. 1	3rd Troop and part of 4th	94	160	Major A. T. Armstrong	Dallam Tower
" 2	2nd Troop	76	130	Lieut. A. England	Ellen Stewart
" 2	6th Troop	56	106	Lieut. H. C. Greenaway	Water Witch
" 2	5th Troop and part of 6th	86	146	Capt. O. Barnes	Challenge
" 3	Hd.quarters and part of 4th Troop	60	125	Major C. H. Palliser Lt. and Adj. A. P. Palmer Lieut. H. Wyllie	Winchester

Each of the transports also carried the followers attached to the unit on board, and the numbers of these somewhat exceeded those of the fighting men. The transports, hired for the purpose, were all sailing ships, and in order to accelerate the voyage each was towed by a steamer. The destined port of disembarkation was Zula, in Annesley Bay, but the transports were directed to call *en route* at Aden.

The voyage was not effected without mishap. When only so far as the Sand Heads at the mouth of the Hughli the steamer towing the *Dallam Tower* broke down, and Major Armstrong, commanding the troops, having telegraphed to Calcutta for instructions, was ordered to make the best of his way under sail to his destination.

Still more serious was the misadventure to the detachment under Captain Barnes on board the *Challenge*. When off the Madras coast cholera broke out among the men and a trumpeter and three sowars died. The vessel was put into Madras and was detained there until all fear of a recurrence of the disease had disappeared, when, having meanwhile been thoroughly disinfected, she proceeded on her way, leaving however twenty men, with their horses, in quarantine at Madras.

Meanwhile the rest of the transports had made a good voyage to Aden, where the headquarters of the regiment arrived on 31st January. There however a great disappointment awaited them. Instead of hastening on to Annesley Bay, orders were received from Sir Robert Napier that the regiment was to disembark at Aden in consequence of the scarcity of water at Zula, and as each transport arrived the men and horses were disembarked at Steamer Point and marched to Sheikh Ottiman, an Arab village eleven miles away, where sufficient water was obtainable for its requirements.

As things eventually turned out this detention at Aden was not of very long duration. The *Challenge*, which was the last transport to reach that port, arrived on 16th February. On the 24th the regiment received orders to re-embark and proceed to Zula, a summons which was obeyed with all possible alacrity, only the unlucky 5th troop remaining at Aden because it was considered that the cholera-stricken *Challenge* was unfit to carry it farther. Even this troop was enabled to embark a few days later on another transport, the *Legion of*

Honour, and reached Annesley Bay a fortnight after the rest of the regiment.

But the delay in the arrival of the Tenth at the base of operations was fatal to its hopes of taking part in the attack on Magdala, the Abyssinian capital. Sir Robert Napier's main infantry force started from the coast on 25th January, 1868. The 12th Bengal Cavalry which, although it left India after the 10th, was not detained at Aden but proceeded at once to Zula, was able to overtake the main column at Antalo, half-way between the coast and Magdala. The Tenth, on the other hand, in spite of the anxiety of the regiment to take part in the operations, arrived too late to be present at the capture of the stronghold.

The whole regiment had reached Zula by 19th March. On the 22nd the headquarters wing under Major Palliser started for the front and on 14th April it reached a place named Dildi, beyond Antalo. By that date however the campaign was already over. The only serious resistance offered by the Abyssinians was at Arogi on 10th April, and Magdala was entered almost without fighting on 13th April.

The disappointment of the regiment at being thus balked of the service for which they had so willingly endured the hardships of the voyage and of the march up-country may well be imagined. Their chagrin was not however allowed to interfere with the performance of their duties, which were exceedingly arduous in spite of the fact that the regiment was never engaged with the enemy.

While, as has been seen, the headquarters and one wing marched up to Antalo and beyond, the left wing under Captain Barnes was sent first to the port of Senafe and thence (leaving a troop behind) to Adigerat on the line of communications. This wing was employed entirely in escort and guard duties and in carrying mails between the places mentioned, and work of a similar nature also fell to the lot of the headquarters wing. Duties of this sort, and indeed field service of any kind in the climate of the Abyssinian littoral and in that barren and waterless country, were trying enough for the troops while for the horses they were deadly. The regiment left Calcutta with 457 horses and received 13 more in Abyssinia. Of these numbers only 191 returned to India, and several of those which were brought back died soon after landing.

Yet the stay of the regiment in the country extended to no more than two months. The headquarters began their return march to Zula on 27th April, the left wing started from Adigerat on 15th May, and the whole regiment was re-embarked by 21st May, and disembarked at Bombay by 16th June. The greater part of the baggage ponies were taken over by the Transport Train when the regiment embarked for India, only one pony for every six men being retained.

Without pecuniary assistance it would have been naturally impossible for a silladar regiment to support such losses as were imposed on the Tenth by the trials of the Abyssinian campaign, and soon after the return of the regiment to India the commandant was informed that compensation would be granted at the rate of Rs. 200 for each horse. The ponies taken over by the Transport Train were of course also paid for at a valuation. These credits did not however suffice to make up either to individuals or to the regimental funds for the expenses of the campaign, which in fact brought little satisfaction to the regiment except the permission to bear the name "Abyssinia" on the appointments. All ranks received a medal, Major Palliser was promoted to lieut.-colonel by brevet, and Risaldar Major Mirza Ata-ullah Khan was given the 2nd Class of the Order of British India.

A reminder of the expedition survived for many years in the persons of two Abyssinians who, as orphan boys, were found and adopted by the regiment, and of whom one was afterwards enlisted as a trumpeter and served until the last decade of the 19th century; the other was employed as an officer's servant and so remained until his death about 1890. A more lasting memento is a handsome and finely worked brass processional cross of the Abyssinian Christians, which is still in the officers' mess of Hodson's Horse.

Ten years of peaceful routine service now lay before the regiment. On arrival at Bombay in June, 1868, it was sent in the first place to Maligaon, on the western ghâts, there to await the end of the rains. On 2nd November, having meanwhile received orders to proceed to Sialkot, it left Maligaon and, moving by road and rail *viâ* Nagpore and Jabalpur to Delhi, marched thence by way of Saharanpur, Hushiarpur and Gurdaspur to Sialkot, where it finally arrived on 29th January, 1869. Thenceforward there are for several years but few events of sufficient interest to be recorded. Some ceremonial work in connection with a visit of the Viceroy (Lord Mayo) to Sialkot in April, 1870, was the sole incident to break the monotony of three years' service at that place. Then in November, 1871, the regiment once more moved towards the frontier and took up quarters at Nowshera, with detachments at the forts of Abazai and Mackeson, and, towards the end of 1873, with further detachments at Michni and Shabkadar. Thus it remained except for sundry camps of exercise until October, 1875, when it was ordered to Ambala.

Throughout all this time the Tenth earned consistently good reports from inspecting officers and expressions of satisfaction from the Commander-in-Chief at its efficiency, and in 1871 the commandant, Lieut.-Colonel Palliser, was made a Companion of the Bath. No other event requires notice here, nor were there any very remarkable changes in the administration or equipment. In 1871 twelve men of each troop received a subsidiary armament of carbines instead of pistols; in 1873 a new system of keeping the regimental accounts was introduced in accordance with orders of the Government of India, a change which was of considerable importance to the well-being of the silladar system; and in 1874 the title of the regiment was changed from "10th Bengal Cavalry (Lancers)" to the more reasonable form "10th Bengal Lancers."

Having marched out of Nowshera on 15th October, 1875, the regiment proceeded first of all to Delhi, where it took part in a camp of exercise, and then returned to Ambala, where it took up its quarters on 11th February, 1876. The close of this year and the beginning of 1877 were memorable in India for the great durbar, officially termed an "Imperial Assemblage," which was held at Delhi for the purpose of proclaiming the assumption by Queen Victoria of the title of Empress of India. The proposal for this change of title, though often attributed to Mr Disraeli, really came from the Queen herself and as is now known was accepted with reluctance by the Prime Minister. The official proclamation was made by the Viceroy, Lord Lytton, on 1st January, 1877. No one could have been found more inclined to enter fully into the spirit of the occasion, and although the Coronation Durbar of 1903 may have been more elaborate and gorgeous in its ceremonial, while that of 1911 gained increased lustre from the presence in person of the Sovereign and his Consort, yet the great Assemblage of 1877, the first of its sort in the history of British India, was remarkable alike for the purpose with which it was held and for the pomp and magnificence of its details.

The military force assembled to give dignity to this celebration amounted in all to about 17,000 men, including six regiments of Indian cavalry, of which the 10th Bengal Lancers was one. Moreover the regiment was more intimately associated with the ceremony through the person of Major Osmund Barnes, the second-in-command, who was appointed to be Chief Herald, and to read aloud the Imperial proclamation. Major Barnes was a man of great stature and of magnificient physique, and his selection was appropriate by reason both of his physical qualifications and of his record of service. Accompanying him was a troop of trumpeters selected from various regiments, among them being Trumpet-Major Hari Singh of the Tenth. In memory of the event the silver trumpet with its embroidered bannerol used by the Trumpet-Major was presented to the regiment and is still preserved in the officers' mess. More signal honour was bestowed on the Tenth a few months later when, on the occasion of the Queen's birthday, Her Majesty appointed H.R.H. the Duke of Cambridge, at that time Commander-in-Chief of the British Army, to be Honorary Colonel of the regiment. In the following year (India Office letter of 1st April, 1878) the additional title "The Duke of Cambridge's Own" was added to that of the Tenth Bengal Lancers.

The remainder of the year 1877 was passed by the regiment uneventfully at Ambala, but in April, 1878, it was ordered to raise an additional (4th) squadron which was to be attached to the 9th Bengal Cavalry, and to form part of a mixed division despatched from India to Malta. This service has been described more fully in the record of the Ninth. As has been shown there the move was entirely political. There was only a remote possibility of any active operations, and little advantage or satisfaction in any respect accrued to the regiments which took part in it. The squadron which was sent from the Tenth was commanded by Captain H. C. Greenaway, with whom was Lieut. A. Burlton-Bennet. It left Ambala by rail on 25th April, 1878, and rejoined the regiment at Nowshera on 8th December, having (on the outbreak of war with Afghanistan) been brought back to India under the command of Capt. S. D. Barrow, who had accompanied the expeditionary force to Malta as Brigade-

Major of cavalry. On its arrival at headquarters this squadron was broken up and absorbed in the other six troops of the regiment.

Some notes about the uniform and equipment of the 10th Bengal Lancers during the first twenty years of the separate history of the regiment may suitably be inserted here to supplement what has already been said on the subject.

It has been shown that very early and practical steps were taken by Major Palliser immediately after the Mutiny campaign to fit out the regiment with suitable uniform, saddlery and armament, the latter being from the first that of a lancer unit. One or two special points are worth noting in this connection. The lance was carried not in the usual leather bucket strapped to the stirrup but in a steel ring which was welded to the stirrup. For use with this ring the lance butt had necessarily to be of a special pattern. The steel socket holding the lance shaft terminated in a steel ball some two inches in diameter having beneath it a projecting butt end or tip three inches long and half an inch in thickness which, when the lance was held at the "carry" or slung, fitted loosely into the ring on the stirrup. The whole arrangement had the merit of being very neat in appearance, much more so than the rather clumsy-looking leather bucket, and the butt in ordinary use like an inverted, elongated cone, but on the other hand there was the objection that it was a feat of no little dexterity requiring long practice to drop the tip of the lance butt with certainty into the stirrup ring, especially if the rider's horse was moving at a trot. Nevertheless the pattern stood the test of use for over half a century.

It appears from the records of the regiment that the pattern of the blue alkalaks provided for all Indian ranks in 1859 continued in use until 1867, when, in preparation for the campaign in Abyssinia, the regiment was fitted out with dark blue serge frocks (or kurtas) with facings of scarlet braid round the neck, down the opening of the front, and on the cuffs, and with lancer piping along the seams. The opening which reached to the waist was fastened with four round brass buttons. This pattern, with but slight changes, continued in use until the Great War. The breeches, or pyjamas, in use at this time were of "Multani mutti" drill, but each man was also provided with dark blue cloth pantaloons.

The British officers' full dress uniform in the 10th Bengal Lancers was, until 1874, very much the same as that which has been described in the case of the 9th Bengal Cavalry (page 110)—namely, a blue tunic with gold lace, scarlet facings, and four quadruple loops of black cord hanging across the front. A pouch-belt was worn of gold lace with a narrow stripe of crimson in the centre. The pouch was of black japanned leather. Gold epaulets were worn on the shoulders, and the helmet was of French grey cloth. The undress patrol jacket was of dark blue cloth with five quadruple loops of black cord hanging loose across the front. The forage cap was of scarlet cloth (a pattern which had been worn since the formation of the regiment) with a gold band of 1¾-inch lace, and with two lines of gold tracing braid crossing the top of the cap at right angles, and a gold netted button over the crossing.

In 1874 considerable alterations in the uniform of the Bengal Cavalry were introduced by a general order of the Commander-in-Chief, and in 1875 these were embodied in a new edition of Bengal Cavalry standing orders. The changes were concerned for the most part with the dress of British officers, for whom (in the four regiments of Bengal Lancers) was now introduced a full dress uniform resembling that worn in the lancer regiments of the English army. At the same time the use of Indian dress by British officers when on duty with their regiments was authorized for the first time. The new uniforms were brought into use in the 10th Bengal Lancers at the beginning of 1876, and two years later, in April, 1878, details of the dress and equipment of the regiment were included in a book of regimental standing orders, which was prepared by Captain and Adjutant H. C. Greenaway. The lancer tunic as there described was blue with scarlet lapels (instead of the plastron generally worn in the English lancer regiments), scarlet cuffs, a scarlet piping along the seams, and double gold shoulder cords. The girdle was of gold lace with two crimson stripes. The pouch-belt was also of gold lace with a crimson centre stripe, the pouch of regimental pattern, having a silver flap. The pantaloons were dark blue with double gold stripes, between which was a welt of scarlet. The helmet was white with a white folded turban or puggri, the cap lines of gold cord. Black butcher boots with steel swan neck spurs, gold and scarlet swords slings, a gold acorn sword knot, and white gauntlets completed this dress.

The Indian full dress for British officers included a loose blouse or kurta of blue serge, with scarlet facing on the collar but none on the cuffs, which were trimmed with 1-inch gold lace. The seams, as in the tunic, were marked with a " light " of scarlet. There were shoulder chains of chain mail mounted on scarlet cloth. The kummerbund was crimson of Kashmir shawl pattern, with the embroidered ends hanging on the right side. The sword-belt, worn over the kummerbund, was gold lace with a centre stripe of crimson. Pouch and pouch-belt were as in the lancer dress. A blue lunghi with ends of gold thread, and a cap or kulla of gold was worn in this uniform. The breeches were white Melton, boots of Napoleon pattern, gauntlets, spurs and sword knot as before.

The undress patrol jacket and forage cap for British officers were unchanged.

The saddlery included a saddle of Hussar pattern with polished wood cantle bound with brass, a holster on the near side and a wallet of Crimean pattern on the off side, a bridle with plain brass bosses, a breast-plate and picketing chain, and a throat ornament consisting of a scarlet plume with plain brass ball-shaped holder.

Indian officers wore almost exactly the same uniform as the Indian dress of British officers, except that the cuffs of the kurta were more heavily laced with gold and that Sikh officers did not, of course, wear the gold kulla but a white "pug" instead.

The dress of the rank and file remained much the same as in the past, including the pyjamas of "Multani mutti" colour. The scarlet turbans which were formerly worn in full dress had been entirely discontinued and all ranks now wore none but the blue lunghi, with (for Muhammadans and Dogras) a scarlet kulla.

Space does not permit of further details of the uniform worn at different times and seasons. It should be mentioned however that the khaki drill uniform for the hot weather had now disappeared and that the whole regiment was dressed in white whenever the blue cloth or serge dress was not in use.

CHAPTER II.

THE AFGHAN WAR, 1878-80.

In 1878 events in India once more afforded to the Tenth an opportunity for active service. For some years past, in fact ever since the death of the Amir Dost Muhammad Khan in 1863, our relations with Afghanistan had been more or less unsatisfactory. Of late misunderstandings had grown into scarcely veiled hostility. At length a climax was reached in the summer of 1878, when news arrived that a mission had been received and welcomed at Kabul from the Czar of Russia, a ruler with whom Great Britain was at that moment not on cordial terms. It was one of our complaints against the Amir that he had repeatedly objected to receive a mission of this sort from India, and it was impossible therefore for the British Government to pass over without protest or remark his reception of representatives of another power. His action could not be interpreted otherwise than as an intentional defiance, in view especially of the aggressive policy which Russia was then pursuing in Central Asia. A despatch was at once forwarded to Kabul firmly demanding that a friendly mission of British officers should be received by the Amir on a footing similar to that accorded to the Russians. General Sir Neville Chamberlain, an old acquaintance of the Amir Sher Ali, was appointed to lead the mission, and after the lapse of a sufficient time since the despatch of the demand the party with a suitable escort started from Peshawar. On 21st September the mission encamped at Jamrud and on the same day Major Louis Cavagnari, the principal political officer, rode forward with an escort to Ali Masjid and demanded of the Afghan commander there permission to proceed to Kabul. The demand was met with a definite refusal. The mission thereupon withdrew to Peshawar and preparations were at once begun for war with Afghanistan. Opportunity was afforded to the Amir to apologise for the insult offered and to comply with the requests of the Government of India, but no reply or satisfaction having been received British columns crossed the Afghan border on 21st November, 1878.

The plan of operations consisted in the formation of three columns—namely, the Peshawar Valley Field Force under Lieut.-General Sir Sam Browne, the Kuram Valley Field Force under Major-General F. S. Roberts, and the Southern Afghanistan Field Force under Lieut.-General D. M. Stewart. The advance of the two first of these columns was designed to threaten Kabul, while the objective of the third was Kandahar. It is with the Peshawar Valley column that we are here concerned.

On 8th October, 1878, the Tenth received orders to prepare for field service. On the 23rd the regiment marched out of Ambala and on 4th December it reached Nowshera, where it joined the cavalry brigade of the 2nd Division, Peshawar Valley Field Force, commanded by Brigadier-General J. E. Michell, C.B., Royal Artillery, the commander of the Division being Lieut.-General F. F. Maude, V.C., C.B. The other regiments in the brigade were the 9th Lancers and the 13th Bengal Lancers. Meanwhile Colonel Palliser had been appointed to command the cavalry brigade with the Southern Afghanistan Field Force, and on his departure on 3rd November Major Barnes took up the command of the regiment.

The following months were both trying and

wearisome for the Tenth. While Sir Sam Browne's 1st Division had advanced into the Khaibar and taken Ali Masjid on 21st November, and the infantry of the 2nd Division moved forward to Jamrud early in December, the Tenth remained at Nowshera until 18th December, and even then moved only to Mardan, where the regiment remained (with detachments at Rustam Bakshali and Abazai, and a troop at Nowshera) until 9th March, 1879. But the duties which had meanwhile fallen to the lot of the cavalry with the more advanced forces were scarcely more interesting and no less trying than those of the reserve, as the Tenth were soon to discover. The fact was that after the fall of Ali Masjid no further resistance was offered by the Afghan forces, and the only fighting which took place during the ensuing months was between marauding tribesmen and punitive columns of our troops.

On 9th March the regiment received orders to advance into the Khaibar and by the 24th of that month it had taken over the posts in the pass as well as Jamrud. The headquarters under Major Barnes were at Dakka, one troop and a half at Lundi Kotal, 30 men at Ali Masjid and a troop at Jamrud. From this time onwards it was continually employed on escort and convoy duty and on daily patrols between Hari-Singh-ki-Burj and Basawal, tasks sufficiently onerous and responsible but for a space of six weeks without noticeable incident. Then on 22nd April an affair took place which to one troop at any rate afforded a welcome break in the prevailing monotonous routine. On the previous day, in consequence of information received at Dakka that the Mohmands on the left bank of the Kabul river were gathering with hostile intent against the line of communications in the neighbourhood of Kam Dakka, two companies of the Mhairwara Battalion under Capt. O'Moore Creagh had been despatched to that village, with the ostensible object of protecting the inhabitants. This force however was quite inadequate to check the advance of the hostile Mohmands. From dawn on the 22nd the enemy gathered in increasing numbers and pressed hard on Capt. Creagh's detachment, which had taken up a position with its left resting on the river. Until 3 P.M. the fight continued without slackening. By that time the ammunition of the Mhairwara Battalion was almost exhausted and the position was very critical. But meanwhile news had reached Major Barnes at Dakka of the state of affairs, and a relieving force consisting of parties of the 5th and 12th Foot and the 4th troop of the 10th Bengal Lancers, the whole under Capt. D. M. Strong, was despatched in all haste to Capt. Creagh's assistance. The distance to be traversed was about 5 miles and the track so bad in parts that only with difficulty could it be surmounted by cavalry. Capt. Strong pushed on with the infantry, which was better able to get over the ground, and arrived first at Kam Dakka. A few minutes later the troop of the 10th appeared under Lieutenant C. E. Pollock, when Capt. Strong immediately taking command delivered a vigorous charge against the crowds of tribesmen, with such effect that they were scattered in all directions, some throwing themselves into the river and the remainder taking to flight among the hills. Capt. Creagh's detachment was then withdrawn to Dakka under cover of the Tenth. In this affair only one sowar of the Tenth was wounded, but the chargers of Lieutenant Pollock and the Risaldar Major and eight other horses received wounds. Capt. Strong was very favourably mentioned in despatches for his conduct on this occasion and subsequently received a brevet-majority.

During the first phase of the operations in Afghanistan the Tenth had no other opportunity of active fighting. Thenceforward the regiment continued its monotonous duties of escort and patrol work. Meanwhile the Amir Sher Ali having died earlier in the year, his son Yakub Khan, who succeeded him, made terms with the British and a treaty was concluded with him at Gandamak on 26th May, 1879 a principal condition of which was the establishment of a British resident at Kabul. Thereupon the withdrawal of the field forces from northern Afghanistan was hastily begun and the Tenth, after moving for a few weeks to Landi Kotal, marched thence to Jamrud, where the headquarters and one squadron arrived on 9th July, detachments being still left at Landi Kotal and Ali Masjid. This retirement through the Khaibar at the height of the hot weather was a terrible experience. Not only was the heat intense both on the march and in the camps occupied by the regiment and its detachments, but to make matters worse cholera had broken out on the line of route, and in common with the whole force the Tenth suffered from the disease as well as from much sickness of other sorts. Among the British officers Major Greenaway, a cavalry leader of much promise, had died in January; Major Barnes, the officiating commandant, was obliged to take leave on account of ill-health in June, and in the same or the previous month four other officers were all given leave on medical certificate, of whom one, Lieutenant Burlton-Bennet, never rejoined and died in the following October.

At the beginning of September, 1879, when the regiment was thus depleted of British officers and still weakened by disease, the tragic news was received that the British residency at Kabul had been attacked and that Sir Louis Cavagnari, the resident, had after a gallant resistance been killed with all his escort. Vigorous steps were immediately taken to re-enter Afghanistan and to exact punishment for their treachery from the Amir and the people of Kabul. A strong force moved forward under Sir Frederick Roberts by way of the Kuram Valley, and such troops, including the Tenth, as had remained in and about the Khaibar after the peace of Gandamak were hastily reinforced and, advancing through the pass, were formed into a division of all arms under the command of Major-General R. O. Bright. These movements were seriously impeded not only by the sickly state of the troops at Peshawar and surrounding camps but even more by the great difficulties in the way of obtaining transport. All available carriage had been diverted to the Kuram for the use of Sir Frederick Roberts's division, and in the Khaibar so little transport remained that units had to be moved in detachments, the carts which carried their necessary supplies and baggage going forwards and backwards with each detachment. The prolonged delays, excessive fatigue for baggage guards, and generally increased wear and tear of such conditions can readily be imagined.

Thus it came about that it was not until 28th September that the Tenth marched from Jamrud, and on 30th September they reached Dakka, together with the 24th Punjab Infantry. Brigadier-General Charles Gough, who had been appointed to command the 1st Brigade of the Khaibar Field Force, arrived at Dakka on the same day.

Meanwhile Major Barnes had hastened back from leave on receiving news of the renewal of hostilities, but he was still quite unfit to withstand the hardships of the campaign, and within a week or two he was sent home for a year on medical certificate. In his place Major W. H. Macnaghten of the 13th Bengal Lancers was appointed to command the Tenth, Colonel Palliser being still employed with the South Afghanistan Field Force.

The general scheme of operations for the Khaibar Field Force was that the 1st Brigade should move forward through Gandamak to Jagdalak or beyond, and get into touch with Sir Frederick Roberts's force at Kabul, and that the 2nd and 3rd Brigades (Brigadier-Generals Arbuthnot and Doran) should secure the communications through the Khaibar. Major-General Bright with the headquarters of the division was at Jalalabad. On 17th October the distribution of the 1st Brigade was as follows :—

Fatehabad

(under Colonel Jenkins, The Guides).

Guides Cavalry (220 sabres).
No. 4 (Hazara) Mountain Battery.
2-9th Foot (300 rifles).
Guides Infantry (480 rifles).

Jalalabad

(Headquarters of 1st Brigade).

10th Bengal Lancers (2½ squadrons).
C-3 Roy. Arty. (4 guns).
2-9th Foot (wing).
24th Punjab Infantry (5 companies).
No. 6 Company, Sappers and Miners.

Ali Boghan.

24th Punjab Infantry (1 company).

Barikao.

10th Bengal Lancers (1 troop).
24th Punjab Infantry (2 companies).

A few days later the whole brigade moved further forward and on 24th October the Tenth was concentrated at Gandamak, where were also the brigadier and the main body of the brigade.

Here, while the brigade was halted for a few days, arrangements were made for a flying column to push forward under Brigadier-General Gough to join hands with a brigade from Kabul. This column, which carried seven days' supplies and the least possible amount of baggage, marched from Gandamak on the morning of 3rd November composed of the following units:—

	Strength.
Cavalry—	
Guides	200
10th Bengal Lancers . . .	100 (with headquarters)
Artillery—	
2 guns I.A. Royl. Horse Arty. .	38
Hazara Mountain Battery . .	200 (8 guns)
Infantry—	
9th Foot	425
Guides Infantry . . .	500
24th Punjab Infantry . .	500
Nos. 2 and 6 Companies Sappers and Miners	100
Total strength . .	1825

Jagdalak was reached at the end of the third day's march without any incident more important than the exchange of a few shots with hostile tribesmen, who kept at a respectful distance. The road, however, was very rough with frequent steep ascents and descents, which delayed the baggage camels so much that the rear-guard each day did not reach camp till after dark. At Jagdalak the force was divided into two columns, one to proceed through the Pari Dara defile, through which the Jagdalak stream flows, a road suitable for camels, the other to march by a higher road further to the west, which was easier for infantry. The Tenth formed the advance guard of the column which took the Pari Dara road, and the regiment was thus the first British unit to pass through that defile since 1842, when it was the scene of some of the most terrible incidents in the fatal retreat of Elphinstone's doomed brigade. Both columns accomplished the march without accident and arrived that day (6th November) at Kata Sang. Here at the summit of the *kotal* above Seh Baba, the Tenth met the 12th Bengal Cavalry forming the advance guard of the brigade from Kabul under Brigadier-General Herbert Macpherson.

At this point the advance of Gough's column ended, and communications having been thus opened with Kabul the return march began on the following day, the Tenth now forming the rear-guard of the force. Gandamak was reached on 10th November without incident.

Although the nights were very cold during this march yet the weather was bright and fine and the bracing air did much to improve the health of the troops and to dispel the sickness of the previous months.

Some weeks now passed without much movement, the only incident worth mentioning being the passage through Gandamak of the Amir Yakub Khan, who having surrendered to the British had been deposed for his share in the massacre of Sir Louis Cavagnari and his escort, and who was now being deported to India. The regiment was detailed to form the escort of the ex-Amir from Gandamak as far as Rozabad whence they returned forthwith to their former camp.

Meanwhile the situation around Kabul, where Sir Frederick Roberts had established himself after the battle of Chaharasia on 9th October, was becoming daily more threatening. Inflamed by the preaching of the *mullas* the Afghan tribesmen were gathering in rapidly increasing numbers, with the object of attacking the invaders, avenging the occupation of the capital, and if possible repeating the achievements of forty years before when the British had been expelled with disaster from the country. By the middle of December the aspect of affairs had become so serious that Sir Frederick Roberts (to whom the command of the whole of the troops in northern Afghanistan had now been intrusted) felt it necessary to call for the advance of the Guide Corps from Brigadier-General Charles Gough's posts about Jagdalak, and at the same time to warn Major-General Bright that he might possibly be obliged to summon to Kabul the whole of the 1st brigade. In order to relieve the Guides and to provide for the security of the communications in the event of Brigadier-General Gough being ordered to advance, efforts were made to move up more troops from India, and in the meantime 200 of the Tenth together with the 2nd Gurkhas were pushed on to Jagdalak. Here the brigadier arrived on 14th December and here on that date he learned that Sir Frederick Roberts had decided to concentrate his whole force in Sherpur, the old British cantonments. At the same time orders were received from the general that the brigade was to advance to Kabul forthwith.

For some days, however, it was impossible for the 1st Brigade to move forward. The posts on the Khaibar line of communications could not be left undefended, and no reinforcements had yet arrived to relieve Brigadier-General Gough's troops. It was not until 21st December that, sufficient additional force having reached the front, the infantry brigade (2–9th Foot, 2nd and 4th Gurkhas with the Hazara Mountain Battery) was able to advance from Jagdalak. It reached Kabul on 24th December, a day too late to take part in the complete defeat by Sir Frederick Roberts's troops of the Afghan attack on the British position.

Gough's brigade was accompanied as far as Seh Baba by the 10th Bengal Lancers, who furnished both advance and rear-guards, but from that place the Tenth returned with much disappointment to Jagdalak, and only a small detachment of 20 men under Lieutenant Drummond went on to Kabul with the infantry.

On the way back from Seh Baba the regiment was fired on, and for some distance a rear-guard fight was carried on, troops dismounting and retiring in turns; but the enemy avoided coming to close quarters and no casualties took place on our side. This was a prelude to a series of attacks on the posts on the line of communications, which had been expected for some days past. On 23rd December there was a vigorous effort against the post at Jagdalak Kotal, held by some 240 infantry and sappers under Major Thackeray, V.C., R.E., with 12 men of the Tenth and 2 guns of the Hazara Mountain Battery. The attack was beaten off after a brisk fight in which two sepoys were killed and Major Thackeray and one sepoy wounded. The hostile tribesmen, mostly Ghilzais, hung about the line of communications for some days more, threatening the movement of troops and supplies, and at length on 29th December a final attempt was made against Jagdalak camp by a large force of Ghilzais under Asmatullah Khan of Lughman, with whom were Muhammad Hasan Khan, late Governor of Jalalabad, and Faiz Muhammad Khan, formerly commandant at Ali Masjid, the same who had taken a leading part in opposing the advance of Sir Neville Chamberlain's mission the previous year. The garrison of the place, which was commanded by Colonel Francis Norman, 24th Punjab Infantry, consisted of 435 of that regiment and 216 of the Tenth Bengal Lancers under Major Macnaghten. The attack was delivered with considerable determination but was successfully repelled by the small garrison, and at 4 P.M. the arrival of three companies of the 51st Light Infantry and six companies of the 45th Rattray's Sikhs with four guns of 11-9 Royal Artillery completed the enemy's discomfiture. This defeat evidently disheartened the tribesmen for a time at least and no further fighting took place during the stay of the Tenth in the field force.

For with the arrival at the end of 1879 of considerable reinforcements for the line of communications, and on the dispersal of the hostile forces round Kabul consequent on their defeat by Sir Frederick Roberts on 23rd December, an opportunity arose for the relief of the Tenth and for their return to India. Ever since the previous March the regiment had been employed without intermission in the exhausting duties of convoys, guards and patrols, and that too in the peculiarly inhospitable and barren country of the Khaibar Pass, exposed to extremes of heat and cold, and to the ravages of disease with which the camping grounds were infected throughout the latter half of the year. It was indeed time that it should be given a respite from such conditions. The casualties among the British officers have already been mentioned. In addition to those enumerated Lieutenant Pollock, who rejoined from sick leave on the renewal of hostilities in September, 1879, was again invalided in the winter, and sent to England, where he died the following year. Among the Indian officers the losses were still heavier, five having died, among whom was Risaldar Man Singh Bahadur, a soldier of long and distinguished service and a great loss to the regiment.[1] The casualties among the rank and file amounted to 139, of whom 47 were killed or died of disease and 38 were wounded and invalided, the remainder being discharged. The loss in horses was also very severe, as was only to be expected in view of the nature of the work in a region where forage was almost unobtainable and water often very scarce. Exposure, privations and hard work had resulted in the death of 142 horses while many of those that survived were quite worn out and unfit for further service.

The headquarters of the regiment left Jagdalak on 5th January, 1880, and marched to Gandamak,

[1] Risaldar Man Singh was a Gurkha by race. He joined the British service in 1849, was appointed to Hodson's Horse during the Mutiny and won the Order of Merit in 1858 (vid. page 98). He was given a commission as jemadar in 1858, ressaidar in 1864 and risaldar in 1871. In 1876 he was awarded a parchment certificate of honorable service and in the same year he received the grant in perpetuity to himself and his heirs of a jaghir of 96 acres in the Gondah district, Oudh, as a reward for faithful and loyal service.

where they remained for another month. On 4th February they started on the return march to India, reached Peshawar on 12th February, and Mardan on the 25th. A month later the regiment moved once more to Peshawar and remained there till 31st August. Its final destination was Sialkot, but its arrival at that popular and fertile station, so much desired by all ranks, was still delayed by the uncertainty of affairs in Afghanistan and on the frontier. On the 1st September it was sent to Kohat, there to await events. On the same day however the complete defeat of the Afghan forces by Sir Frederick Roberts outside Kandahar made any further trouble highly improbable. The Tenth was kept at Kohat for two months more but without incident or adventure, and at length on 28th November, 1880, the regiment marched into Sialkot where it was able to recover from the trying experiences of the campaign in Afghanistan.

In 1881 authority was received for the words " Afghanistan 1878-80 " to be borne on the appointments of the regiment and all ranks received a special medal for the campaign. Colonel Palliser was made a K.C.B., Capt. D. M. Strong received a brevet majority and later a brevet lieut.-colonelcy, Capt. S. D. Barrow a brevet majority, Risaldar Man Singh the Order of British India (2nd Class) and Sowar Bhagwan Singh the Order of Merit (3rd Class).

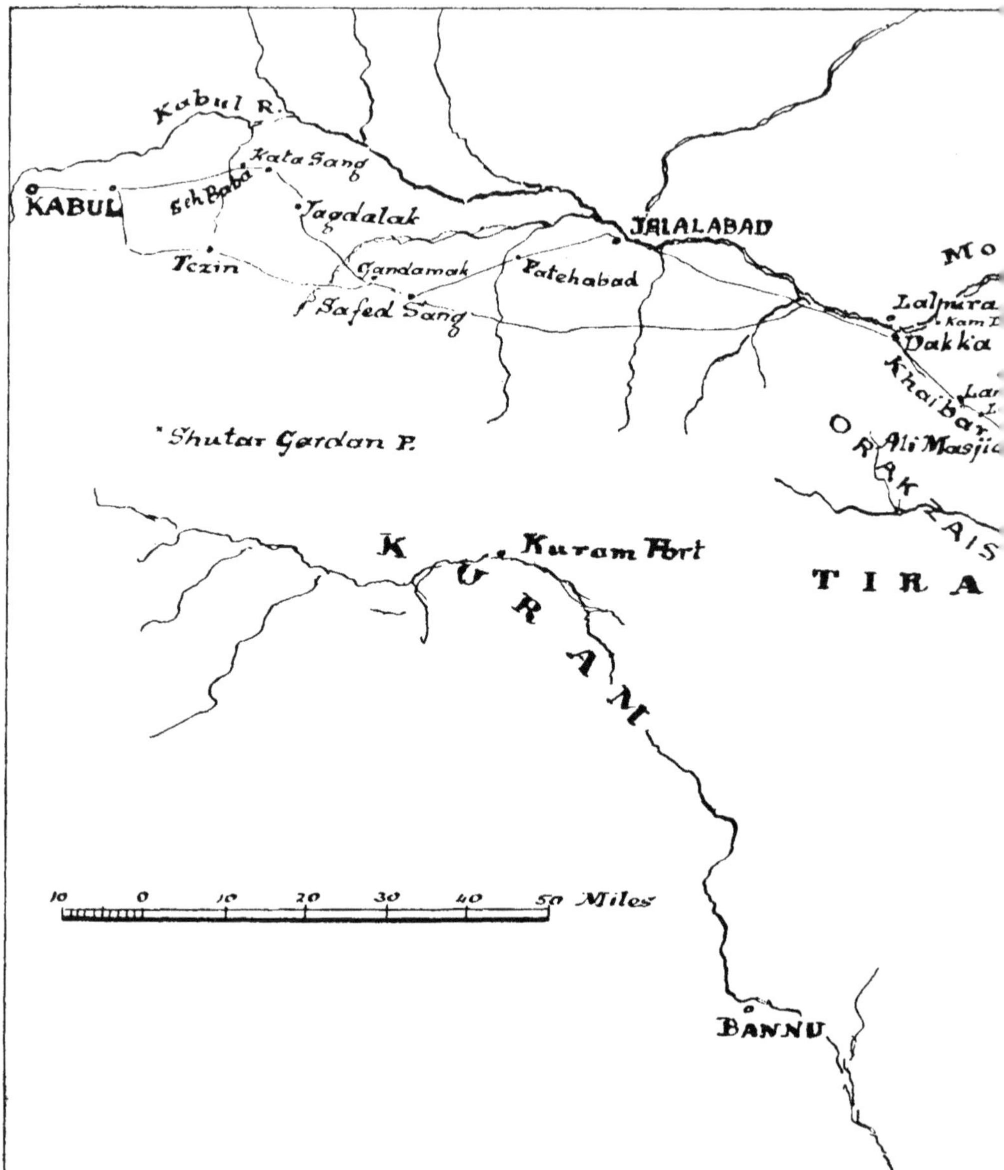

Kabul R.

KABUL

Seh Baba

Kata Sang

Jagdalak

JALALABAD

Tezin

Gandamak

Fatehabad

Safed Sang

Lalpura

Dakka

Khaibar

Shutar Gardan P.

ORAKZAIS

Kuram Fort

K U R A M

TIRA

10 0 10 20 30 40 50 Miles

BANNU

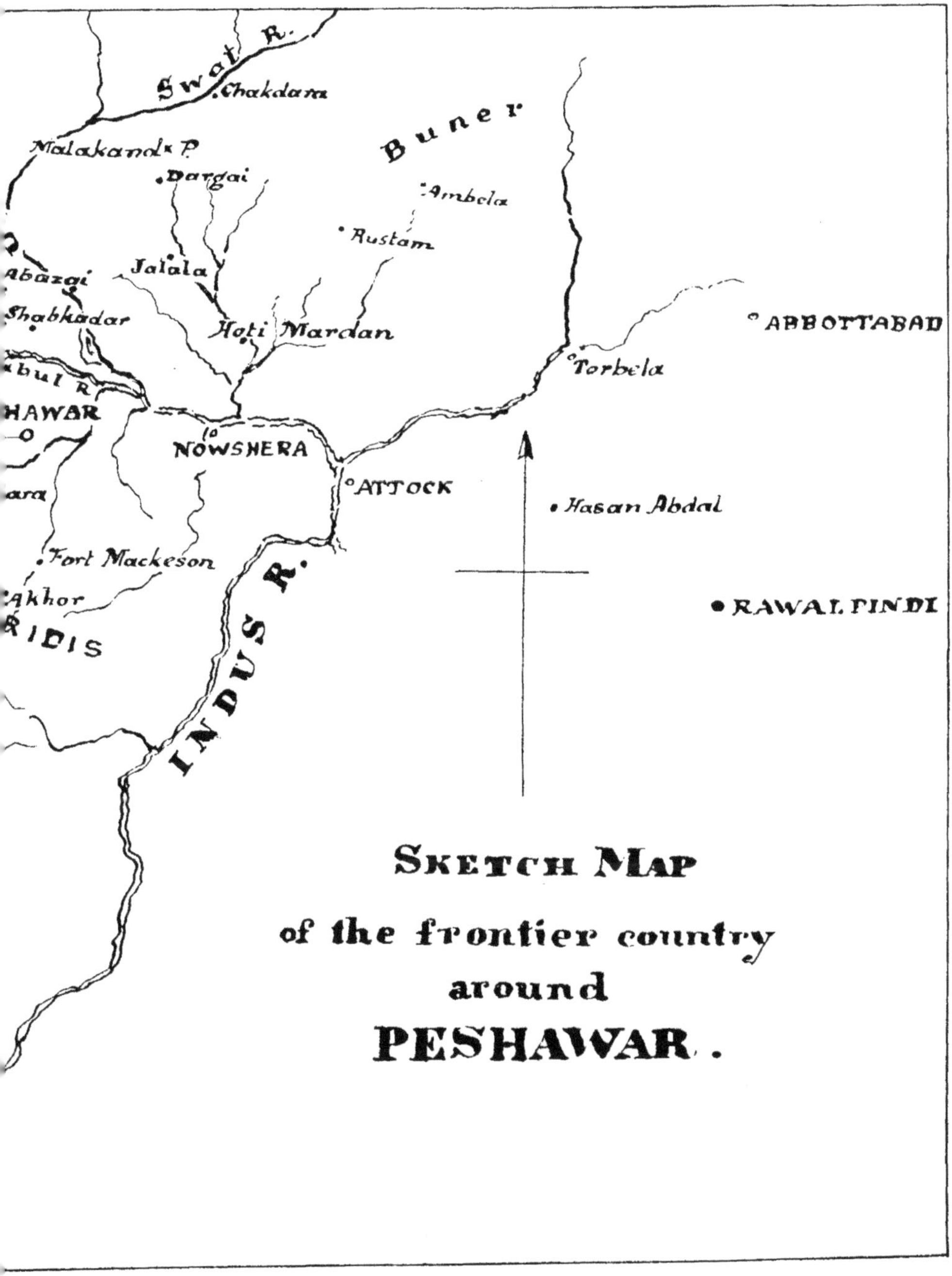

Swat R.
Chakdara
Buner
Malakand P.
Dargai
Ambela
Rustam
Jalala
Shabkadar
Hoti Mardan
ABBOTTABAD
Torbela
NOWSHERA
ATTOCK
Hasan Abdal
Fort Mackeson
RAWALPINDI
INDUS R.

SKETCH MAP

of the frontier country around

PESHAWAR.

CHAPTER III.

1881-1914.

There is little to record of the ensuing years. Until October 1883 the regiment remained at Sialkot. Thence it moved to Duki in Baluchistan where it arrived on 4th December of that year, detaching one squadron to Multan. The stay of some thirteen months in Baluchistan was almost equally without interest, the only incident being a march through the Zhob Valley by a mixed force under the command of Brigadier-General Sir O. V. Tanner, of which force the headquarters and one squadron of the 10th formed a part. Meanwhile the third squadron was still at Multan and other small detachments at Gandakindaf, Thulli, and Dabakot.

Duki was by no means an ideal station for cavalry, among its drawbacks being the fact that for a great part of the year green fodder for the horses was practically non-existent, a circumstance which necessitated the move of the regiment to Gumbaz in the Thal Chotiali Valley in December, 1884. It was therefore without regret that the Tenth marched out of Baluchistan in January, 1885, on the way to Multan, at which place they arrived at the end of the same month. Here too they spent four uneventful years, broken only by attendance at a large camp of exercise at Delhi towards the end of 1885, and hence in due course they moved once more on 3rd December, 1888, *en route* for Ambala.

Here we must pause to note some personal and

administrative changes of the previous eight years which are specially deserving of record.

Of these the first in date and importance was the disappearance from the rolls of the regiment of the name of Colonel C. H. Palliser, who after 22 years in command was transferred to the Brigade Staff of the army on the 5th of November, 1880. Colonel Palliser's appointment to command a brigade at the outbreak of the Afghan war has already been mentioned. At the end of the war he was appointed to command the Sialkot Brigade, and he was finally struck off the strength of the regiment on the date quoted above. The debt which the 10th Bengal Lancers owed to their first commandant has been recognised in a previous chapter. By those who served under him his memory was always held in honour and affection, and the tradition of his qualities as a commanding officer was long handed down as an example to his successors.

In the first of these, Colonel Osmund Barnes, who was appointed in Colonel Palliser's place, and who held the post for seven years, the regiment found another commander equally remarkable both for professional efficiency and for those personal qualities which are always important in regimental leaders, but especially so in officers of Indian regiments. Nor did the good fortune of the Tenth in this respect end here, for when Colonel Barnes retired in 1887 he was succeeded by Colonel Dawsonne Strong of whom it may be said that both in his relations with all ranks and in his administration and training of the regiment he could scarcely be surpassed as a commanding officer.

Mention should also be made here of the loss suffered by the regiment in the untimely death of Lieut.-Colonel Seymour Barrow in December, 1886. Lieut.-Colonel Barrow, a brilliant horseman and a soldier of exceptional qualities, was recognised by all as one of the most promising officers of the Indian army. He had held many important appointments

LIEUT. COLONEL AND RISALDAR MAJOR MIRZA ATA-ULLAH KHAN,
Sardar Bahadur, Order of Merit.

and had earned a brevet majority in Afghanistan, where he was seriously wounded at Patkao Shana, and a brevet lieutenant-colonelcy in Egypt in 1882. He was marked out for signal advancement when his career was cut short by disease to the great regret of all ranks in the Tenth.

Another notable personage who left the regiment during these years was Risaldar Major Mirza Ata-ullah Khan, Sardar Bahadur. We have seen in the first chapter how Ata-ullah Khan was one of those who, at the request of Robert Montgomery, raised a *risala* for Hodson in June, 1857, and in 1866 he was the first officer to hold the responsible post of risaldar major in the Tenth. In May, 1885, after twenty-eight years of loyal and valuable service in the regiment, he was appointed British Agent at Kabul, with the honorary rank of Lieutenant-Colonel, and left to take up this important position, which he filled for several years with credit and distinction. He was succeeded as Risaldar Major by his brother Mirza Abdulla Khan, who also had been one of those to join Hodson's Horse at Delhi in 1857.

Turning to administrative changes it is worth noting that in 1882 the strength of Indian cavalry regiments was raised to 550 of all ranks, while three years later the establishment was changed (in 1885) to four squadrons instead of three, the total strength of all ranks becoming 635. During the latter year two regiments, the 16th and 17th Bengal Cavalry, which had been broken up for reasons of economy in 1882, were raised again, and the Tenth was ordered to transfer to the new 16th the whole of its troop of Jats. The composition of the Tenth now became 1½ squadrons of Sikhs, 1 squadron of Punjabi Muhammadans, 1 squadron of Dogras, and ½ squadron of Pathans.

There were few important changes in armament and equipment during the years under notice. The Snider carbine had taken the place of the old Victoria carbine in 1878, just as the regiment was on the march to join the Peshawar Field Force (not a very suitable moment, one would have thought, for so vital a change !). In 1887 the pistol holsters, relics of a long-obsolete armament, were abolished, and the belts of the rank and file were altered to admit of the carbine being slung on the back when necessary. Finally in 1888 the mameluke bits, which have already been described, were modified so as to be used with curb chains.

Turning once more to the doings of the regiment we find again a period of almost a decade during which there is very little to chronicle except the daily record of service in cantonment. Leaving Multan, as has been stated, in December, 1888, the Tenth took part in instructional manœuvres at Mian Mir and then proceeded to Ambala, where it was stationed for nearly five uneventful but prosperous years. During this time, under the command of Colonel D. M. Strong, it reached a pitch of efficiency which is notable even in its unbroken record of merit, and the reports of cavalry inspectors and of the Commander-in-Chief himself are again and again remarkable for unstinted praise. In other respects too these years were fortunate. Both in 1890 and 1891 the team from the Tenth won the Bengal Cavalry Tentpegging Cup, and at a specially important Assault at Arms at Muridki in January, 1890, which was attended by H.R.H. the Duke of Clarence, the then eventual heir to the throne, the representatives of the regiment were first in the team performance of the lance exercise and in the competition of lance against sword.

In the same year the Martini-Henry carbine was issued to the regiment in place of the Snider, and the receipt of this excellent weapon was followed immediately by so marked an improvement in the musketry performances of the regiment that the Commander-in-Chief (Sir Frederick Roberts) went to the length of directing that his special commendation should be conveyed to the Adjutant, Lieutenant W. H. Fasken, whose efforts had been largely instrumental in bringing about such good results.

In the winter of 1891-92 the regiment attended a large concentration of troops at Alighar, followed by a ceremonial review by the Commander-in-Chief at Meerut, at the conclusion of which it returned to Ambala.

In October, 1893, its stay at that station came at length to an end and it moved to Jhelum where it arrived on 16th November. Here it was not so fortunate as in previous years in avoiding the dislocation of detachments. Throughout the whole of 1894 and until November, 1895, a squadron was detached at Kohat and another in the Kurram Valley, while in April, May and June, 1895, during the progress of the Chitral Relief Expedition the headquarters and two squadrons were moved temporarily first to Hoti Mardan and then to Nowshera. In June, 1895, however these squadrons with the headquarters returned to Jhelum where the regiment remained for another year until, on 1st August, 1897, it marched in relief to Nowshera.

Meanwhile in 1894 Colonel Strong's term as commanding officer came to an end and he was succeeded by Lieut.-Colonel E. J. F. Wood. In the Indian ranks a similar change had taken place in 1890 when Risaldar Major Mirza Abdulla Khan, Sardar Bahadur, retired on pension and was succeeded as Risaldar Major by Risaldar Khan Bahadur Khan.

With the arrival of the Tenth at Nowshera in August, 1897, there began one of those trying periods which every regiment, especially every cavalry regiment, has to experience at times, but which are none the less heart-breaking for individuals and a severe ordeal for the whole unit. The extensive operations on the North-West Frontier which marked the summer and autumn of 1897 had just begun, and it was naturally hoped that the regiment would have a share in them. In the event however its only part was (as in the case of the Chitral Expedition two years before) the exacting and thankless duties of occupying frontier posts vacated by more fortunate corps, of carrying out reconnaissance and patrol work, and generally of exercising a restraining influence over the tribesmen on one part of the frontier, while more active service was being done elsewhere.

The regiment did not pause at Nowshera, where it was eventually to take up its quarters, but moved on immediately to Hoti Mardan, whence the Guide Corps had proceeded on field service. Here the headquarters and two squadrons remained, while two squadrons were detached to join the so-called "Malakand Field Force," one being located at Jalala and Dargai and one at Rustam. The latter (C Squadron under Capt. W. L. Maxwell) in November and December, 1897, formed part of a mixed force which, under the command of Colonel R. J. F. Reid, carried out a march through the Utman Khel country. A month later, in January, 1898, the headquarters and two squadrons at Mardan moved forward to the Malakand, and the whole regiment was attached to a force which entered the Buner country. The tribesmen of that region were threatening to join the Afridis of Tirah against whom operations were then in progress, but the demonstration by the Malakand force had the desired effect of over-awing them and restraining their hostile desires. No opposition was offered and the British columns withdrew before the end of January, the headquarters and three squadrons of the Tenth remaining in the Swat Valley, while one squadron was sent back to Nowshera. Thither too the headquarters and another squadron returned early in June, but only to be ordered up to the Swat Valley once more in November, when they formed part of a movable column under Brigadier-General Reid and advanced as far as Chakdara and Landakai. Finally this squadron marched back again to Nowshera on 28th December,

1898, and there the headquarters and two squadrons remained until January, 1902. Meanwhile however one wing of the regiment remained detached and continued to form part of the Malakand Force until the end of 1901. All ranks received the medal issued for the frontier operations of 1897.

It was not to be expected that the regiment could for so long be broken up, and in such unfavourable conditions, without suffering to some extent, but it is a notable testimony to its general efficiency and to the quality of its officers that even the detached wing under Major F. A. Blyth was reported to be in a very satisfactory state in March, 1901, although it had then had several months of trying service in the Swat Valley.

In January, 1902, the regiment left Nowshera for Cawnpore, and a very short time in the more congenial surroundings of that station sufficed to restore the condition of the horses and the general well-being of the whole corps.

The stay of the Tenth at Cawnpore extended to four and a half years, from 28th March, 1902, till 1st November, 1906, but the time was uneventful and the few doings of interest can be recorded very shortly. In 1902 a party of nineteen non-commissioned officers and men under the command of Risaldar Gopal Singh went to England to attend the coronation of H.M. King Edward VII. Whilst in London Lance-Dafadar Chanda Singh and Sowar Gurditt Singh took part in the Military Tournament at Islington and were awarded gold medals for tent-pegging and feats of horsemanship. It may be added that they were the only representatives of the Indian Cavalry to be so rewarded. Risaldar Gopal Singh was admitted to the Order of British India in recognition of his selection to command the Coronation contingent.

At the end of the same year B Squadron, under Capt. W. E. Young, marched to Ambala and took part in the manœuvres around Delhi and in the subsequent Coronation Durbar at Delhi, returning to Cawnpore on 9th February, 1903.

It should also be recorded that in 1902 the regiment headed the list of Indian cavalry regiments in signalling, with a figure of merit of 645·30, and was specially commended for its efficiency in this respect.

In 1906 the bronze medal of the Royal Humane Society was awarded to Capt. A. D. Strong for his gallantry in attempting to save the life of a sowar who had fallen into a deep pool while watering his horse when on manœuvres in December, 1905. Capt. Strong dived several times into the pool in his endeavour to rescue the man but without success.

In the cold weather of each year there were instructional concentrations of troops to be attended, but in other respects the tour of service at Cawnpore was unbroken until at the end of 1906 the regiment returned once more to the Punjab and was quartered at Jullundur where it remained for almost six years.

Meanwhile, in the previous decade there had been several changes in personnel. Lieut.-Colonel Wood completed his tenure of command in 1901 and was succeeded by Lieut.-Colonel F. A. Blyth, who in turn gave place in 1907 to Colonel M. Cowper. Among the Indian officers Risaldar Major Khan Bahadur Khan, who in 1895 was appointed Indian aide-de-camp to the Lieut.-General in the Punjab command, was succeeded as Risaldar Major by Risaldar Sultan Muhammad Khan. The latter retired in November, 1903. Risaldar Sher Baz Khan then became Risaldar Major, and he again two years later was succeeded by Risaldar Gopal Singh who held this important position with conspicuous success until his retirement in 1910.

Some brief notes must here be inserted of changes made in the various uniforms subsequent to 1878, the orders about which were embodied in a new and very complete edition of the regimental standing orders prepared by Lieutenant and Adjutant W. H.

Fasken in October, 1890. This invaluable guide to the interior economy of the regiment was brought up to date, but very little altered, by Lieutenant and Adjutant W. N. Evans in 1896 and again by Lieutenant and Adjutant J. E. Moir in 1904. The most noticeable and far-reaching change was the introduction of khaki uniform at the time of the campaign in Afghanistan, and as in subsequent years the use of this dress became more and more general for all except ceremonial purposes other changes from time to time were all in the direction of simplification and of the abolition of unessentials.

Thus in 1882 ankle boots and black puttees were introduced for all ranks in "field day order," a great relief from the clumsy and unpractical Napoleon boots. The latter were entirely discontinued for British officers about 1890, their place being taken by calf knee boots, but the use of Napoleons survived for other ranks in drill and review order for another four or five years. By 1896 however the rank and file wore none but ankle boots and puttees in every variety of mounted dress.

At this period the full dress of British officers, when not on duty with the regiment, remained much the same as that laid down ten years before, nor did it vary materially afterwards. For mounted duties with the regiment the full dress shawl pattern kummerbund was discontinued about 1895, and a plain scarlet Kashmir kummerbund, which for some years had been in use for field dress, was thenceforward worn on all occasions, whether with blue or khaki uniform. The red facings on the neck of the blue kurta were abolished in 1882. At the time of the Afghanistan campaign a very serviceable regimental pattern of frog sword belt of brown leather, with broad cross belt, was introduced for all mounted uses except full dress, and in 1906 the undress pouch, belt, and sword slings which had been worn for unmounted duties for the previous twenty years were abolished, and the frog sword belt was thereafter used in their place. This latter change carried with it the discontinuance of the steel sword scabbard, except for full dress. Ever since the brown leather frog was introduced the sword had on all mounted parades (except full dress) been carried in a sensible wood scabbard covered with black leather, and the use of this now became general. In 1888 a blue serge fatigue coat was introduced for wear in camp, and some ten years later this took the place of the cloth patrol jacket, which was abolished.

The uniform of non-commissioned officers and men, except as mentioned above, remained almost unchanged as it had been since 1867, the most noticeable alteration being the use of white cotton pyjamas instead of the garments of Multani mutti colour, which were discontinued about 1878. Finally in or about 1900 the use of khaki in field day and marching order was extended so as to include the lunghi, kulla, kummerbund and puttees, and from this date in those orders of dress all ranks were dressed entirely in khaki.

The armament of the regiment with Martini-Henry carbines in 1890 has already been mentioned. In 1902 Lee-Metford carbines were substituted for the Martini-Henrys and a year later the Lee-Metford short rifle was issued in place of carbines. A minor change was the reduction of the length of the lance from 10 ft. to 8 ft. 6 in. in 1902. At the same time an order was issued discontinuing the use of the sword by lancers, but subsequently second thoughts prevailed on this subject and in 1903 the order in question was cancelled.

A very notable change in the designation of the regiment was made in 1901 when the name "Hodson's Horse" was for the first time officially added to its title, while a further change in 1903 abolished the erroneous and misleading name "Bengal" and made the regiment the "10th Duke of Cambridge's Own Lancers (Hodson's Horse)."

. . . .

This record has now been brought down to the year 1907 and the "Jubilee" of Hodson's Horse, and (as in the case of the Ninth which has already been described) the Tenth set about arrangements to celebrate the occasion. These involved no small amount of careful consideration. For obvious reasons it was impossible to hold the celebrations at the exact date when the regiment was formed, for this fell in the middle of the hot weather, and it was finally decided to set apart the week from 15th to 21st December for the purpose. The question of finance was a difficult one, for the plan in view was to hold a reunion to which would be invited as far as possible every man who had ever served in the regiment, and to entertain them all for the period of the "Jubilee." This would of course necessitate a large outlay. The money was eventually made available by a grant from the "miscellaneous fund" and by personal subscriptions by all ranks. Assistance was given by the railway authorities who issued tickets at special cheap rates, and this enabled the regiment to pay the fares of all pensioners. For the accommodation of the guests a number of tents were borrowed from other corps at Jullundur, but this arrangement proved unnecessary, for the men serving in the regiment were quite pleased to lodge the whole of the visitors in their houses.

The result was a complete success. Upwards of 350 pensioners of all ranks arrived at Jullundur (including thirteen mutiny veterans) not a few of them accompanied by a son or a grandson. The ceremonies included a reception at the officers' mess of all Indian officers and all mutiny veterans, a sports meeting, a race meeting, and a regimental parade, at which the veterans stood at the saluting base and took the salute, the regiment carrying out all the ceremonial evolutions of marching past as laid down in the Cavalry Training. Most important of all there was a great Durbar at which Colonel Cowper presided, surrounded by all the officers, British and Indian, while the other ranks, past and present, crowded round as closely as possible. Colonel Cowper delivered a short speech of welcome to which retired-Risaldar Akbar Ali Shah made a reply in excellent and well-chosen language, followed in similar strain by Ressaidar Kashi Naud, Rai Bahadur. Then followed the business of the Durbar—to wit, the presentation of a great number of petitions, most of which concerned the grant of land in the canal colonies. All of these petitions were carefully considered and duly noted and no less than thirty letters were subsequently written by Colonel Cowper for the petitioners in support of their claims. The satisfaction afforded both to past and present members of the regiment by this sympathetic attention to their desires will be readily understood, and the tightening of the bonds of good fellowship, affection and *esprit de corps* which certainly resulted from this part alone of the Jubilee celebrations amply fulfilled all the hopes which had been formed with regard to the occasion.

When the week came to an end the guests of the regiment departed with regret, having enjoyed their entertainment and thoroughly appreciated the genuine heartiness of the welcome extended to them.

Among many pleasant and some amusing incidents one may be recorded here. One of those who preferred petitions at the Durbar was a certain mutiny veteran, a Muhammadan, who represented that through the kindness of the Colonel he had received a piece of yellow paper by means of which he had been able to get a railway ticket to Jullundur and back for a single fare. "But," he went on, "it is now my intention to go on a pilgrimage to Holy Mecca. May your honour therefore be pleased to give me another such piece of paper so that the cost of my journey may be halved." "Assuredly," replied Colonel Cowper, "I will give you a certificate if you so wish, but you must bear in mind that if you achieve your 'Haj' at half price you must expect to receive only half the benefit." The petitioner after this did not press his request.

After this notable anniversary four years elapsed without any event that requires record. From time to time the regiment moved out to camps of exercise but otherwise the time was passed at Jullundur. The reports by inspecting officers during this period were uniformly favourable, culminating in one of special excellence in 1911. At the end of that year Colonel Cowper's tenure of command was completed and he was succeeded by Lieut.-Colonel W. L. Maxwell, while in the previous year the Risaldar Major, Gopal Singh Bahadur, retired on pension and his place was taken by Risaldar Major Sardar Khan.

The stay of the regiment at Jullundur had now extended to more than the normal period at one station and orders were received for a move to Dera Ismail Khan; but in consequence of the scarcity prevailing in northern India that year these orders were cancelled and the move was eventually deferred until the autumn of 1912. Meanwhile in November 1911 the regiment marched to Delhi where, as divisional cavalry with the 3rd Division, it took part in the great concentration of troops on the occasion of the Imperial Coronation Durbar, returning thereafter to Jullundur.

At length, after nearly six years at that station, the regiment marched on 27th September, 1912, for Loralai, where it arrived on 18th November. The heat in the Punjab when the march began was excessive, but by the time that the highlands of Baluchistan were reached the winter in those regions had set in and the cold encountered was as extreme as the previous heat, and all the more trying by reason of the sudden change. The regiment went into cantonments at Loralai, with small detachments at Gumbaz, near Duki, Murgha, Maratangi, and Musa Khel, and here it was still serving when in August, 1914, the great European war burst on the world.

The events of the succeeding years must be left to another chapter, this record of the preceding period of peaceful service being concluded with a mention of some changes in the personnel and in the domestic affairs of the regiment.

On the 7th March, 1914, all ranks were both shocked and grieved by the sudden death of the commanding officer, Lieut.-Colonel W. L. Maxwell. He had been in failing health for some time past, and he was about to proceed on leave to England pending retirement, but the fatal termination of his illness was quite unexpected. Lieut.-Colonel W. E. Young was appointed to the vacant command.

In July, 1914, Risaldar Major Sardar Khan, Bahadur, retired on pension after 32 years' service and Risaldar Bijai Singh was appointed Risaldar Major in his place.

In 1912 two changes of considerable moment were introduced in the interior economy of the Tenth. The first of these was the abolition of the regimental bunnias. Improved communications and increased facilities for obtaining supplies of all sorts had long rendered the employment of these agents unnecessary. When the regiment marched to Delhi at the end of 1911 and was for two months absent from Jullundur the regimental bunnias were left behind as an experiment. The result of the experience so obtained was to convince all ranks not only that they were unnecessary but even that the absence of their intervention was a pecuniary advantage to all concerned, and from the date of the regiment leaving Jullundur for Loralai their services were accordingly dispensed with.

The second innovation in 1912 was the establishment of a regimental co-operative savings bank, in order to inculcate habits of thrift and independence of money-lenders, and incidentally to assist in the liquidation of the debts owed to the regimental bunnias. There was at first a good deal of opposition to this scheme especially on the part of some of the Mussulmans in the regiment. Nevertheless the plan was ultimately successful, and by the end of the year the bank had 260 members, holding an aggregate of 1506 shares of 10 rupees each. The bank was affiliated to the Punjab Co-operative Village Banks and arrangements were made for its accounts and proceedings to be annually inspected and audited by qualified authorities.

CHAPTER IV.

THE GREAT WAR.

MESOPOTAMIA, 1916-1918.

FROM the date of the despatch of the expeditionary force from India to France in 1914 all regiments which were not fortunate enough to be included in the divisions for field service were inevitably called upon again and again to supply officers, both British and Indian, non-commissioned officers and men to make good casualties in the field. These demands increased in number as the months went by and as operations developed in Gallipoli and in Mesopotamia. The Tenth, stationed in such a backwater as Loralai, where the likelihood of the regiment itself being required to take the field was very remote, was peculiarly liable to requisitions of this sort, with which its officers naturally enough were keenly anxious to comply. Thus it came about that between the autumn of 1914 and the summer of 1916 the regiment was denuded of nearly half its British officers and a large number of experienced Indian officers, also of many of the best of its non-commissioned officers and of some three hundred and fifty of its trained men. All of these served with credit in the various theatres of war and many with marked distinction, a fact which made it all the more regrettable that, with such fine material in its ranks, the regiment should have no opportunity of earning in the field that credit and renown which individually its members showed themselves so well able to acquire. The services of these individuals and the honours and rewards which they won are enumerated elsewhere.[1] Meanwhile those who remained with the regiment had an arduous time in training not only recruits and young horses to fill the vacancies caused by the despatch of drafts, but also considerable numbers in excess of the normal strength. In this manner, in addition to the ordinary routine of regimental training and station duties, the months were very fully occupied until on 13th August, 1916, the 10th Lancers received orders to proceed to Mesopotamia to join the Indian expeditionary force (which was known officially as force "D") and to relieve the 16th Cavalry.

It is scarcely necessary to recapitulate the events in Mesopotamia since the beginning of operations there in November, 1914, the occupation of the Basra vilayet in 1915 followed by the successful advance to Kut el Amara, the check at Ctesiphon on 22nd-24th November, the retirement of the British force and subsequent investment of Kut, the unsuccessful attempts to relieve that place, and finally its surrender to the Turks on the 29th April, 1916.

The fall of Kut was followed by a period of several months during which the British Governments in India and at home and the military commanders on the spot devoted strenuous efforts to retrieve the position lost through previous mistakes. The reorganisation of the force in Mesopotamia, the improvement of communications and transport by water and road, the provision of every description of supplies—food for troops, forage for animals, medicines, and sanitary stores of all sorts, in short everything which is essential for the existence of an army in the field, and which had been woefully deficient in the past—these, and the myriad tasks which they involved, were the almost superhuman labours undertaken by Lieut.-General Sir Percy Lake and his subordinates

[1] See page 331.

in the summer of 1916. And if the work was a heavy and difficult one in any circumstances, the conditions of climate and of the country in which it had to be done rendered the undertaking many times more arduous. In a region where even the normal summer temperature is probably the highest in the world the heat this year was abnormally great, and it had to be endured without the relief of any mitigating shade, while its terrors were increased and life rendered almost unendurable by the horrible plague of flies which made food scarcely eatable and rest or sleep more than ever difficult to secure. The misery of the trenches in Flanders was great, but it is questionable whether any other British forces during the great war endured for so long without relief such horrible conditions as those which the troops in Mesopotamia suffered during the summer months which followed the fall of Kut. Moreover these trials fell upon men who were already wasted by disease and privations, and who above all were disheartened by failure. The campaign was discredited and ill-omened, and the best that those taking part in it could hope was to wipe out by some measure of success the memory of what had gone before.

To be launched into the midst of such a service as this offered but a cheerless prospect. Not the smallest spark of glamour remained about the warfare on the banks of the Tigris. Very different were the feelings of those who in 1916 embarked at Karachi for Basra, from the high hopes and enthusiasm of the expeditionary force which set forth for France two years before. It is true that all were confident of ultimate success, but the hardships to be endured and the difficulties to be overcome had been so cruelly emphasised that only a very high standard of *moral* could enable the troops now leaving India to face the future without misgiving. Nor should the fact be forgotten that, as has been shown above, the Tenth had been seriously depleted of experienced officers, non-commissioned officers and men since the outbreak of war. Its ranks included an entirely undue proportion of young officers of little experience, and of men who were scarcely more than recruits. It is therefore all the greater credit to the regiment that, as will be shown in the following pages, it so gallantly stood the test of very trying service, and fully maintained its reputation throughout the four arduous years that lay before it.

The regiment was so far fortunate that the worst of the summer heat was over before it reached Basra, although some weeks of very high temperatures still remained. Moreover considerable progress had been made in the provision of the most necessary supplies, though in this respect as well as in the matter of transport much remained to be done. The energy and determination of Sir Percy Lake in the opening months of the summer, seconded by equal activity on the part of Lieut.-General F. S. Maude when he succeeded to the chief command on 28th August, had effected a material change in the condition of the expeditionary force, and had improved the prospect of successful operations in the future. For the present however all question of forward movement was in abeyance, and the mission of the force was confined, by the instructions of the British Government, to the protection of the oil-fields and pipe-lines in the vicinity of the Karun river, to the maintenance of control over the Basra vilayet, and to the denial of hostile access to the Persian Gulf and to southern Persia.

Such was the general position when the Tenth was warned for service. It was ordered to move by wings, the first wing to proceed to Mesopotamia and to take over the horses, saddlery and equipment of a wing of the 16th Cavalry, which was thereupon to return to India and proceed to Loralai. The second wing was not to start from Loralai until the arrival of this advanced wing of the 16th, to which it was to hand over its horses, &c., and then proceed in its turn to Mesopotamia to relieve the second wing of the 16th.

The advance wing (B and C Squadrons) under Major H. G. Young, with a strength of five British and six Indian officers and two hundred and twenty of other ranks, left Karachi on 12th September in the s.s. *Chakdara*, and arriving at Basra on 17th September, went on by river steamer to Amara, where the horses of a wing of the 16th Cavalry were taken over according to orders.

The headquarters and second wing (A and D Squadrons) under Lieut.-Colonel W. E. Young, and composed of nine British and nine Indian officers and two hundred and sixty-four of other ranks, left Karachi on 22nd October, reached Basra on the 28th, and left there by road for Amara on 4th November, having in charge three hundred horses which were handed over to the Remount Depot.

The regiment formed part of the "Tigris Defences" (under Brigadier-General H. H. Austin) and was strung out along the line of communications between Amara and Sheikh Saad throughout the ensuing months. The headquarters were located at Sheikh Saad, A Squadron was at Ali Gharbi, B was for some weeks at Sheikh Saad, C was for six weeks split up between Amara and Ali Gharbi and then moved up to Sheikh Saad, D was at Mudelil with one troop at Amara. Little of interest occurred during these first weeks in the country, the regiment for the most part being employed only on routine duties, parades with the movable columns, a certain amount of intelligence work and the like. The effects of the climate began quickly to be felt and B Squadron at Sheikh Saad suffered especially in this respect, Major H. G. Young, who was in command of the squadron, and the two other British officers being all attacked with jaundice, which was very prevalent among the troops all through the year. In the middle of December this squadron moved up to Es Sinn and after being employed there for a fortnight with the G.H.Q. of the expeditionary force it was detailed on 1st January, 1917, to form part of a composite regiment, the other squadrons being one from the Herts Yeomanry and two from the 32nd Lancers. This regiment was commanded by Lieut.-Colonel H. G. Young, with Captain R. T. Lawrence of the 10th as adjutant. B Squadron was now under the command of Major T. E. Hulbert, 3rd Skinner's Horse (attached), with Lieut. B. H. Wiles, I.A.R.O., as squadron officer. The composite regiment was to act as corps cavalry with the III Corps (Lieut.-General W. R. Marshall) in the operations which General Maude had now initiated against the Turkish positions at and about Kut el Amara.

The general condition of the Mesopotamia force having been by this time vastly improved General Maude began a forward movement on 16th December. This advance met with such success that by the first week of January the British lines had been considerably advanced and now closely threatened the Turkish position east of Kut, which was spread out along several miles of river frontage.

The weather, which had been fine at the opening of the operations, broke up in the latter part of December, and thick mists and heavy rain, resulting in the country becoming generally water-logged, interfered seriously with the progress of the attack. Nevertheless the General did not pause in his plan of advance. By the middle of January the Turks had been almost completely dislodged from their positions on the right bank of the Tigris below Kut, and General Maude thereupon turned his attention to clearing them similarly from the salient on both banks of the Hai river, immediately opposite and above the town. This task was allotted to the III Corps. The attack began on 25th January and by the 5th February the Turks had been driven from the position astride of the Hai and had fallen back to a line farther westwards but still on the right bank of the Tigris. After only four days' delay for reorganisation the attack was renewed. On 15th February, in the midst of a heavy downpour of rain, the III Corps delivered a final assault on the position with complete success, and on the following morning the whole right bank was clear of the enemy to a distance of upwards of twelve miles west of Kut. The threat to the Turkish line of communications was now obvious and this fact coupled with the sense of defeat materially lowered the enemy *moral*, while that of the British troops was correspondingly restored. Nevertheless the formidable obstacle offered by the Tigris in winter flood inevitably caused a check to the advance. That this delay was not prolonged was due to the admirable forethought of the high command and the excellent work of all concerned. On 23rd February the passage of the river was effected by the III Corps in three places, the opposition of the enemy was broken down, and the next day the whole Turkish army was in full retreat.

In these operations B Squadron of the Tenth had such share as falls to the lot of corps cavalry, and on the succeeding days, when, first at Imam Mahdi and then on 26th February at the Nahr al Kalek bend of the Tigris, the Turkish rear-guard attempted to hold up the British advance, the composite regiment to which the squadron was attached took a rather more prominent part, in conjunction with the 14th Division, in dislodging the enemy.

The Turks were now a beaten force and the British, full of enthusiasm after their recent successes, were (as General Maude wrote of them) at the top of their form. There seems no doubt that in the course of this advance the Army Commander looked for the accomplishment of more decisive results by the cavalry, and comparisons have been drawn between the performance of the mounted arm in Mesopotamia and that of General Allenby's Desert Corps a year and a half later in Palestine and Syria. But in any such comparison the most vital point of all is the manner in which the mounted arm was used by the two Army Commanders. General Allenby told his cavalry commander what he wanted done and left him free to do it. General Maude tied the leader of his cavalry tightly, with stringent orders to communicate with him every hour! Damascus would not have been taken in thirteen days by methods such as these. The contrast between the action of a cavalry general and that of an infantryman is instructive. However these incidental reflections do not directly concern the present narrative. B Squadron of the Tenth advanced with the corps cavalry regiment as far as Aziziya, which was entered on 1st March; it was then sent back to Kut, thus missing the final fighting at the passage of the Diyala, and the occupation of Baghdad. The squadron moved up to Aziziya again at the end of April and remained there on line of communications duty for upwards of seven months during which all the country within twenty miles was reconnoitred.

To return now to the headquarters of the regiment and Squadrons A, C and D, which as has been seen were posted at the beginning of the year, A Squadron at Ali Gharbi, headquarters and C at Sheikh Saad, and D at Mudelil with one troop at Amara. There is little to relate about the two first of these squadrons during 1917. Both continued to be employed on the line of communications, each forming part of the mobile columns for defence of the several posts, and being also engaged on escorts and in reconnaissance duties, completing as far as possible the many unsurveyed places on the existing maps. In the course of these duties C Squadron under Captain S. D. N. Cahusac was more than once engaged with hostile Arab raiders especially in January, when a quantity of arms and ammunition was captured, and again on 14th March, when the squadron had a smart and successful skirmish.

In March A Squadron moved from Ali Gharbi

to Bughaila, and in December the headquarters of the regiment with both A and C Squadrons moved up to Baghdad, being joined on the way at Aziziya by B Squadron.

Meanwhile D Squadron under Major A. W. M. Kemmis, after spending some six months of the year between Mudelil and Amara, proceeded by river to Baghdad in June and was attached to the I Corps. Thence on 4th July the squadron marched to the Euphrates to join the 7th Infantry Brigade at Falluja, one troop however being detached to join the 24th Punjabis at the Hindiya barrage lower down the river.

Since the beginning of March the Mesopotamia field force had captured Baghdad and had established control of the Diyala as far as Jabal Hamrin and of the Tigris as far as Samarra. Falluja on the Euphrates was also secured before the end of March, but Ramadi, the next place of some importance up that river, was still held by the Turks in an entrenched position, with a force of about one hundred and fifty cavalry, a thousand infantry and six guns. This force Sir Stanley Maude now (at the beginning of July) determined to expel if possible by a sudden surprise attack. Operations in the great heat of the summer season were sure to be very trying, but special measures were taken to minimise the effect of this, and it was believed that the undertaking could be carried through rapidly and without excessive exposure of the troops.

Unfortunately for all concerned on the very day before the operations were to begin a wave of intense and, even in that fiery climate, abnormal heat swept over the country. The temperature in Baghdad reached 122° and in tents and dug-outs it was much higher. The heat of the sandy ground was so great that it blistered the feet through the soles of thick boots. It was declared by the people of the country to be the hottest weather within the memory of man. In such conditions the column for the attack on Ramadi concentrated at Falluja on 7th July. It consisted of the 7th Infantry Brigade, fourteen guns R.F.A., four armoured cars and D Squadron 10th Lancers, less one troop, under Major Kemmis, together with a weak squadron of the 32nd Lancers which was also placed under Major Kemmis's orders. The whole column was commanded by Lieut.-Colonel Haldane.

Dhibban, a stony hill about ten miles from Falluja, was occupied by an advanced detachment of the 2/7th Gurkhas on the night of the 8th-9th July, and the cavalry, with a company of the Connaught Rangers and some artillery, moved forward to the same place on the morning of the 9th to cover the advance of the main column, which concentrated here early on the following day. At 5.30 P.M. on the 10th the cavalry and armoured cars made a further advance of 10 miles to Madhij and secured the defile there between the neighbouring low hills and the river. The main column followed later in the evening and after a rest at Madhij advanced again at 1 A.M. on 11th July. At 3.15 A.M. the leading troops were fired on by a piquet at the point of Murshaid ridge about two miles from the Turkish lines, but the position was quickly carried.

The infantry now continued the advance towards Ramadi, but the cavalry was ordered to move to the right front towards the Euphrates and to work slowly forward under cover of the palm-groves towards the mouth of the Habbaniya canal, ready to pursue the enemy as soon as he should be dislodged. The squadrons moved off at about 4.45 A.M. and reaching the river bank some two miles east of the canal mouth advanced slowly westwards, with several long halts to allow the infantry attack to develop. At length at 9 A.M. a place was reached where the palm-groves and gardens ended. A thousand yards further on the enemy held the line of the canal and opened fire on the cavalry, while fire was also opened at the same time by Arabs on the further bank of the river. Notwithstanding the time that had now elapsed there were no signs of the Turks being moved from their position, and a reconnaissance to the left showed that although the British attack had succeeded in capturing the first line of trenches it had made no further progress. Major Kemmis therefore decided to retire to a point a mile further back where there was better cover, and there to await developments. This withdrawal drew fire from Arabs in the neighbouring villages, who had previously shown no signs of hostility, as well as from others across the river and from the Turks, but little damage was done and the hostile fire was quickly silenced by the squadron Hotchkiss guns.

Meanwhile things had not been going well with the infantry attack. The heavy sandy going had delayed it in the first place, and when at 6.30 A.M. the enemy had been driven from his forward position on the Habbaniya canal it was soon found that his main defences were much stronger than had been anticipated and the ground in front of them very open. To make matters worse a furious dust-storm now set in, which rendered further advance for the time impossible. By 10 A.M. the heat was so intense that the commander decided that the attack could not be pressed; it was however equally impossible to withdraw the forward troops before dusk, and there was no course open but to remain stationary through the day, every possible effort being made meanwhile to keep the exhausted troops supplied with water.

The cavalry remained unmolested until 1 P.M. when, it being evident that the infantry attack had come to a standstill, and it being impossible to communicate with the headquarters of the column owing to the blinding sand-storm, Major Kemmis determined to withdraw to the shelter of the Murshaid ridge. This withdrawal was again the signal for a hot fire to be opened by Arabs concealed in the neighbouring gardens and villages; but the retirement was carried out quietly and steadily, the half squadrons dismounting alternately and providing covering fire, and the only casualties were one sowar wounded and Major Kemmis's horse shot. The gardens under the Mushaid ridge were reached at 2.30 P.M., and here the squadron off-saddled and watered.

At 7 P.M., on receipt of a report that the position at Ramadi had been abandoned, the commander of the column sent the cavalry and armoured cars forward to reconnoitre, but it was soon found that the rumour was without foundation.

During the night and in the early hours of the 12th July the infantry was withdrawn to the palm-groves about Murshaid point, where the force rested during that day. At 2 A.M. on the 13th the return march to Madhij started. As soon as it was daylight Arabs swarmed out of all the neighbouring villages and hung on to the rear of the column throughout the day, during which the cavalry was employed as a flank-guard on the hills south of the road and was not engaged, the Arabs keeping to the low ground near the river. Madhij was reached at 7 A.M. The same evening the march was continued to Dhibban, the cavalry again furnishing a flank-guard in the hills and therefore not taking any part in the desultory fighting with Arabs by whom the infantry was again annoyed.

Although this undertaking against Ramadi was unsuccessful, the failure was due to a great extent to the climatic conditions, which were abnormal. Certainly no blame for the lack of success could be imputed to the troops, who in extraordinarily trying conditions displayed the utmost courage and endurance. It was the first occasion on which D Squadron of the Tenth had been in action and the men showed perfect coolness and steadiness throughout.

The attack on Ramadi was not renewed until 28th

September, when the whole of the 15th Division and the 6th Cavalry brigade were employed. The position was occupied, almost the whole of its garrison taken prisoners, and a great quantity of railway material, engineering stores, arms, ammunition and guns was captured. On this occasion D Squadron, which was again included in the force, was left at Dhibban to keep open communication with Falluja and took no part in the attack. The squadron remained on the Euphrates until the end of November when it returned to Baghdad. There the headquarters and other three squadrons arrived in December, but just as there seemed a possibility of the regiment coming together again D Squadron was once more despatched on detached duty, this time with the 52nd infantry brigade, which left Baghdad on 16th December for the Hilla area on the middle Euphrates.

During the past year many changes had taken place among the officers of the Tenth. Lieut.-Colonel W. E. Young, who had brought the regiment from India, was invalided in May and did not again return. Lieut.-Colonel H. G. Young, who took his place, left three months later on being appointed to the command of the 22nd Cavalry. Major A. W. M. Kemmis then assumed the command of the Tenth and continued in the appointment until Lieut.-Colonel Ricketts arrived from India in December.

The 1st of January, 1918, found the headquarters and three squadrons at Baghdad and D Squadron with the 52nd brigade in the neighbourhood of Hilla. Thither it was shortly afterwards followed by B Squadron and No. 2 machine-gun subsection. Headquarters and the remainder of the regiment under Lieut.-Colonel Ricketts left Baghdad early in January and marched to Falluja and thence to Sheikh Habib, where they remained until 10th February carrying out reconnaissances of the surrounding country. They then marched to Ramadi, where from 15th February to 10th March they were employed with one squadron of the Herts Yeomanry and the 42nd and 50th infantry brigades in reconnaissance work and frequent skirmishes with Turkish cavalry, and finally in the operations which resulted in the occupation of Hit and Sahiliya and the retirement of the Turks to their main position at Khan Baghdadi.

The question now arose how far the defeated Turkish forces should be driven northwards in order to render the country on the Euphrates from Hit downwards secure from further trouble. At a conference between the Commander-in-Chief, Lieut.-General Sir W. R. Marshall, and Major-General Sir H. T. Brooking, commanding the 15th Division, it was decided that operations should be undertaken on 24th March with the primary object of expelling the enemy from his position at Khan Baghdadi, of driving him as far north as possible, and subsequently of occupying a suitable covering position. Major-General Brooking's force, which concentrated about Sahiliya for this purpose, was composed of the 11th cavalry brigade under Brigadier-General Cassels, the 10th Lancers (less two squadrons), one squadron of the Herts Yeomanry, forty-eight guns and howitzers, the 12th, 42nd and 50th infantry brigades, No. 275 M.G. company and three L.A.M. batteries. All the preliminary arrangements were made as secretly as possible and precautions were taken to prevent the enemy from anticipating any serious attack on the Euphrates front. The forward movement was to begin on 25th March, when a mixed column of all arms under Brigadier-General Andrew (described officially as Andrew's Brigade Group) was to advance from Sahiliya at 9 P.M. supported by a similar "group" under Brigadier-General Lucas, with a third "group" in reserve. Simultaneously a wide turning movement west of Khan Baghdadi was to be made by the cavalry brigade, with the object of threatening the enemy's communications and if possible of cutting off his retreat.

Andrew's group, consisting of the 10th Lancers,

215th Brigade R.F.A., the 50th Infantry Brigade and the 48th Pioneers moved off as arranged at 9 P.M. on the 25th March and at 1 A.M. on the 26th reached a point on the Aleppo road 10 miles from Sahiliya. As soon as it was light enough a reconnaissance was made by three artillery officers escorted by A Squadron of the Tenth under Lieutenant O. B. P. Russell, and a position having been found whence the guns could enfilade the Turkish trenches two batteries were sent under escort of A and C Squadrons to open fire. Meanwhile the preliminary attack of the infantry began at 2 A.M. and was further pressed at 10.15, supported half an hour later by Lucas's group. During this time the cavalry turning movement had progressed favourably and unobserved by the Turks, and it continued through the day until by 4 P.M. the cavalry brigade was astride of the Aleppo road behind Khan Baghdadi. Up to that hour the enemy was apparently unaware of the threat to his retreat, and he clung to his position until nearly nightfall, obstinately resisting the British infantry attack. When it appeared that he was beginning to withdraw from his advanced trenches A Squadron of the 10th with No. 1 M.G. section was ordered to operate on the left of the advance, and pushed forward accordingly ahead of the infantry, engaging the retiring enemy with dismounted fire until held up by machine guns from the main Turkish position. At 8.20 P.M. the attack was broken off but it was resumed at 3 A.M. on the 27th March. It was then found that the Turks had retired in the night. The pursuit was at once taken up, the 10th Lancers escorting the guns, until at the Wadi Hauran, which was reached at 7 A.M., it was found that practically the whole of the Turkish force had surrendered to Brigadier-General Cassels.

In view of the completeness of this success Major-General Brooking determined to push on at once to the towns of Haditha and Ana, and if possible to capture the Turkish garrisons there and the considerable stores of ammunition, &c., which Ana was known to contain. A brigade group under Brigadier-General J. M'K. Hogg was ordered to carry out this operation, the troops composing it being the 10th Lancers and Herts Yeomanry, a half battalion each of the 1/5th Queen's and 2/39th Garhwalis, two M.G. sections in Ford vans, No. 8 L.A.M. battery and 1072 battery R.F.A. with double-horse teams. This force advanced rapidly as far as Ana, where 5254 prisoners were taken, as well as twelve guns, forty-seven machine guns, and a large quantity of stores.

The news of these entirely successful operations came at a moment of tense anxiety on the western front and were all the more welcome on that account. The troops and their commanders received cordial congratulations from the Commander-in-Chief at Simla, from the Secretary of State, and from the War Cabinet, and His Majesty the King wired to Lieut.-General Marshall—

"In the midst of the great struggle in western Europe I wish to assure you that I follow with constant interest the splendid progress made by the gallant troops under your command. I congratulate you and all ranks on the success of your latest achievements."

The entire destruction of the Turkish force at Khan Baghdadi, and the subsequent advance to Ana, practically brought the war to an end in this area. The 10th Lancers were withdrawn to Sahiliya, where they went into standing camp for the summer undisturbed by any incident of importance. On two occasions however the routine monotony was broken by minor operations in which the squadrons at Sahiliya took a prominent part. The first of these was a raid on 5th-6th June, which was carried out by C Squadron under Captain Cahusac, against a settlement of disaffected Arabs believed to be harbouring patrols of mounted Turkish soldiery. The raiders were successful in capturing a Turkish cavalry officer and eight men. On the 22nd-23rd June another raid was carried out by a mixed force under Lieut.-Colonel Ricketts, including the 10th Lancers, under Major Hodson, four armoured cars, a section of field artillery and two companies of the 1st Oxfordshire Lt. Infy. The object of the operation was to round up a squadron of Turkish cavalry which was known to be hanging about in the neighbourhood of Khan Baghdadi. The intended movement was however betrayed to the Turks, who managed to withdraw in time to avoid capture, but their advanced troop was surrounded by the Tenth and taken prisoners, numbering one officer and sixteen of other ranks.

While these events were in progress on the upper Euphrates B and D Squadrons of the Tenth under Major Kemmis were employed with the 18th Division in the Hilla-Najaf area, a duty involving a good deal of reconnaissance work in this part of the country which had not before been visited by our troops, and also including various minor operations against hostile Arabs. Each squadron carried out raids on Arab villages during the early weeks of the year, and on one occasion the two executed a combined raid. In January, when B Squadron was reconnoitring the country near Najaf, fire was treacherously opened on it from the walls of the town and two sowars were killed. This hostile act was followed a few weeks later by a more flagrant outrage, when on 19th March Major Marshall, the political officer at Najaf, was murdered in the streets of the city. Such gross violence as this, the proof of a deep-seated conspiracy, demanded immediate punishment. But Najaf is one of the cities most venerated by Shiah Mussulmans and contains three of their most holy shrines. The British authorities were therefore unwilling to expose it to bombardment, and instead of this a close blockade of the place was ordered, which was to be maintained until the ringleaders in the recent outrages should be handed over for punishment.

In the first three weeks of this blockade the wing of the Tenth, supplemented by a squadron of the Patiala Lancers, the whole under the command of Major Kemmis, was continuously employed. The work was very trying and exhausting. It entailed constant outpost duty with consequent exposure, and the strain of incessant watchfulness, and in addition short rations for the men and inadequate forage and water for the horses. Day and night the piquets and even the camp got no rest from the annoyance of snipers from the walls of the town and neighbouring enclosures, and the fact that the hostile fire was ill-directed and caused few casualties merely reduced and did not remove the resultant tension. The blockading line was pushed gradually closer to the town, Captain R. H. W. Welsh (16th Cavalry), Lieutenant J. C. Platts (17th Cavalry) and Lieutenant B. H. Wiles in the squadrons under Major Kemmis distinguishing themselves in their efforts to make the operations effective. At length on 11th April the wing of the regiment was relieved from this exacting duty, although it was not until 30th April that the blockade of Najaf was brought to an end on the proscribed persons being handed over. For another seven weeks the squadrons remained in the Hilla area, constantly employed on reconnoitring and escort duties, in the course of which the horses suffered greatly from over-work and insufficient food. It was a welcome relief when orders were received to join the headquarters at Sahiliya, and there the regiment was once more united at the end of June after being split up into detachments for nearly two years.

From Sahiliya on 25th July Lieut.-Colonel R. L. Ricketts, who had been commanding the 10th since the previous December, left the regiment to take up an appointment on the staff of the Northern Persia force, and Major A. W. M. Kemmis, D.S.O., assumed the command which (with the exception of two months at the beginning of 1919 when Lieut.-

Colonel P. R. Chambers, D.S.O., temporarily rejoined from staff employ) he continued to hold for the rest of the time that the regiment remained in Mesopotamia.

In October the force at Sahiliya was withdrawn to Hit in order to simplify the work of supply, the latter place being the highest navigable point for river steamers. Here the regiment remained quietly until the end of the year, the time being passed in steady training work, and here in November the momentous intelligence was received of the armistice with Turkey, followed immediately afterwards by the cessation of hostilities in Europe.

CHAPTER V.

ON THE UPPER EUPHRATES, 1918-1920.

THE GATHERING STORM.

"I think there has seldom been such a series of hopeless tangles as the West has made about the East since the armistice."—GERTRUDE BELL, 1920.

THE period of two years which now began was destined ultimately to bring for the regiment experiences more trying and losses more severe than any that they had previously suffered. Moreover these losses were incurred not in fair fighting nor in legitimate war, but in a guerilla contest of the most harassing sort and against an enemy peculiarly treacherous and savage. At the time of the armistice the prestige of England in the middle east was higher than ever before. The failures of 1915 and the humiliation of Kut had been amply retrieved by the brilliant successes of Baghdad, Shergat, Khan Bhagdadi and Damascus. But within two years the position thus recovered had again been lost, not as before by military failures but by a succession of circumstances for which the army had no responsibility. The course of international policy with regard to the Asiatic territories occupied by the Allies in the war was the primary cause of the trouble. Doctrinaires at Versailles set to work to deal out mandates and to construct kingdoms. But the primitive savages who were the victims of their experiments cared for none of these things. Moreover in Mesopotamia the steps taken to establish a settled government were not within the comprehension of the population. The Turkish power had been overthrown, and the people looked for a firm rule in its place; but they found only an officialism unsuited to their needs and an unwonted burden of taxes the object of which they could not understand. It was not surprising that they responded readily to the promptings of Damascus and Moscow, and the Arab war of 1919-20 was the result.

Under the arrangements for garrisoning the country on the withdrawal of the 15th Division, to which the Tenth had for several months past been attached, the regiment was allotted to the 17th Division at Baghdad and was detailed, with no other assistance than that of the 6th L.A.M. battery, to hold the whole area on both banks of the Euphrates from Falluja to Deir-ez-Zor, a distance of no less than three hundred miles. According to this plan the headquarters and three squadrons were to be located at Ramadi and one squadron at Hit. In pursuance of these orders the regiment, leaving C Squadron at Hit, marched from that place on 10th February, 1919, and reached Ramadi on the following day. Here a good standing camp was made on the banks of the river at the bridge of boats below the town, and here almost the whole year passed uneventfully. Several long reconnaissances of the country were carried out by British officers with a few Indian ranks and with Hotchkiss guns on Ford vans, and the regiment was put through a very thorough course of individual and collective training. On one or two occasions the squadron at Hit went out to assist the political officers in protecting local tribes from raiding parties of other tribesmen in the southern desert, but it was on no occasion engaged with raiders. In other respects no events of interest occurred. Leave was opened to India and to England for all ranks, officers of the Indian army reserve were demobilised, many men were sent to India for discharge and drafts arrived in their places, and there seemed no reason why the regiment should be involved in any further active operations before its relief and return to India.

But all this while hostile agents were at work fomenting trouble and inciting the Arab tribes to rebellion. As has been seen the course of political events since the end of the war had created enmity and unrest both in the new Arab kingdom of Damascus and in a large section of the population of Iraq, while emissaries from the Soviet government of Russia fostered the growing discontent. Moreover the withdrawal of large numbers of our troops from the country, and the dispersal over a wide area of those who remained, gave rise to the impression among the local populations that these latter might be attacked with impunity. At length the malcontents determined to put the matter to the test, and on 11th December, 1919, a raiding party, with the goodwill if not with the actual assistance of Damascus, attacked and seized the small isolated post of Deir-ez-Zor in the extreme north of our area of occupation on the Euphrates, and captured the solitary political officer and his escort of two armoured cars and their crews.

On the news of this outrage reaching Ramadi Major Kemmis with B and D Squadrons of the Tenth at once marched for Khan Baghdadi (55 miles) where they arrived on 14th December. Here they were joined on the 19th by a composite squadron made up of A and C, the squadrons being very weak owing to the absence of men on leave in India. At Khan Baghdadi the regiment was ordered to establish a suitable position which should serve as a rallying-point for any forces that might fall back from Ana, and to cover the concentration of troops moving forward from places lower down the river. Meanwhile a column of armoured cars and infantry in Ford vans under Colonel H. G. Young pushed on to Ana and thence to Abu Kemal, whence the political officer of the district with his small escort had withdrawn after the hostile occupation of Deir-ez-Zor, 70 miles further north. No further advance was made, nor was Colonel Young permitted to reoccupy Deir-ez-Zor or to attempt to rescue the British captives; on the contrary after a delay of a few days terms were arranged with the Arabs, Deir-ez-Zor being left in their hands on the condition that the prisoners should be released. Thereupon Colonel Young's column was broken up, and the 10th fell back to its former positions at Ramadi and Hit, the political officer being left at Abu Kemal with an increased escort and a good supply of rations and ammunition.

The result of such a policy was inevitable. The *ballon d'essai* of the Arabs had succeeded beyond—one would imagine—their utmost expectations. Having driven the British out of Deir-ez-Zor they were not likely to abstain from following up their advantage. But their next move was more serious because the Damascus Government was now definitely hostile. On 11th January, 1920, the Arabs attacked Abu Kemal, Damascus putting forward at the same time a claim to all the country down to and including that place.

Once more it became the lot of the Tenth to march up the Euphrates. Major Kemmis was given the local rank of colonel and was placed in command of a column of all arms which was to concentrate as rapidly as possible in the disturbed area. The regiment (less C Squadron which was left as a garrison at Ramadi) marched on 12th January under the command of Captain Platts to the relief of Abu Kemal, 190 miles distant, where it arrived on the 22nd. Here it was found that not only was Deir-ez-Zor in the hands of emissaries of the Arab government of Damascus but also Meyaddin, some thirty miles further south, while officers and agents of the Arabs were active in all the country down to and about Abu Kemal, inciting the local tribes to rebellion and organising raids on convoys and on any weak detachments of British troops. The garrison of Abu Kemal consisted of a section of armoured cars and a wing of the 126th Baluchis, of which the other wing was at Ana, about half-way to Ramadi. Two more battalions of infantry and a field battery were on their way to the district. The 10th Lancers who had now arrived were about a hundred below war establishment owing to the absence of men on leave and to the non-arrival of drafts from India.

The duties which had now to be performed by this force were peculiarly difficult and exacting. For the Tenth they involved daily escorts with convoys moving up and down the road or foraging expeditions to obtain grain and bhoosa from neighbouring villages. In the execution of either of these duties the isolated detachments of the regiment were liable at any moment to be suddenly attacked by raiding bands of Arabs in superior numbers. Moreover they moved always in a country of which the inhabitants, though supposedly neutral, were ready to seize any opportunity of joining our assailants; thus a village or area peaceful one day would be openly hostile the next, and if a raiding party attacked a convoy every Arab within sound of the firing would hasten to join in the attack for the sake of any loot which might be obtainable. The country consisted of open rolling downs interspersed with rocky hills, precipitous cliffs and deep, dry wadis. In parts near the river there were flat cultivated areas with patches of dense tamarisk and poplar jungle. At frequent intervals there were places where considerable numbers of men could lie concealed, so that the nature of the ground, as much as the character of the enemies likely to be encountered, required the greatest precaution and constant vigilance. The service was rendered more arduous by the intense cold which lasted till the latter part of February, with snow and hard frosts, during which the horses suffered severely from having only one blanket at night. A final difficulty was the supply of fodder. From the second half of February onwards only half rations of bhoosa were available, and by the end of the month the horses were in very poor condition. Owing to the nature of the work to be done it was often difficult even to water the horses properly. It frequently happened that there was no opportunity for watering them between the very early morning before daylight, when they would drink little or nothing because of the cold, and after dark in the evening, when they would be utterly exhausted. The distances covered in these protracted and trying services were considerable. Thus between 12th January and 29th February the average distance marched by A, B and D Squadrons, including the march from Ramadi, was about 400 miles, and during March, by which time C Squadron had rejoined from Ramadi, all squadrons did an average of 380 miles. Continuous hard work of this sort for so long, on insufficient food and in very trying climatic conditions, reduced the horses so much that by the end of March over a hundred were unfit for work and many of the rest could hardly raise a trot.

Meanwhile in the course of the escort and other duties during these two months squadrons of the regiment were involved in several skirmishes and engagements with Arab raiders in some of which serious casualties were incurred. Up to the end of February these affairs were of minor importance, the most considerable of them being on the 16th of that month when the regiment, less one squadron, with one section R.F.A. was sent to take action against some villages, in reprisal for a recent serious attack on a convoy in which a British officer and several men of the 126th Baluchis had been killed. Having marched six miles down the Ana road the regiment and guns then worked back up the river bank, burning all the villages which had been concerned in the attack on the convoy. No opposition was offered until the regiment came into the bend of the river just below the town of Abu Kemal, when fairly heavy fire was opened on the troops from the opposite (left) bank of the river. The fire was returned but it was of course impossible to get at the enemy, the river intervening. The casualties in the Tenth were one man killed and one mortally wounded and fourteen horses wounded.

On the 26th February C Squadron under Lieutenant Waters, with men from leave and a draft of fresh men from India, arrived from Ramadi, bringing the regiment practically up to full strength. On the same date B Squadron under Lieutenant A. T. Oates was sent fifteen miles down the river to Al Qaim with orders to remain there and to furnish escorts and protection to convoys as required. Four days later, on 1st March, Lieutenant Oates marched from Al Qaim with his squadron, less one troop, and two machine-guns in Ford vans to meet a convoy which was expected from Nahiya. The convoy not having appeared when half the distance to Nahiya had been covered Lieutenant Oates sent the Ford vans on to look for it while he followed with the squadron. The country here was very broken and cut up by nullahs. Just east of Wadi Safra the advanced guard was suddenly fired on. Lieutenant Oates galloped forward with two troops to occupy a commanding ridge but fell mortally wounded, and the small force found itself confronted by some two hundred or two hundred and fifty Arabs advancing from the low hills in front and from the river. Meanwhile the Ford vans returning brought the news that the looked-for convoy was not coming. The squadron and Fords thereupon retired to Al Qaim, the former handled with great coolness and skill by Risaldar Laurasib Khan. The Arabs followed up the retiring force for a distance of four miles, endeavouring by moving along the river bank to cut it off before it could get clear of the broken, hilly country. In this however they were unsuccessful, Risaldar Laurasib Khan withdrawing his men with very small loss to open ground where the enemy did not venture to follow. Besides the death of Lieutenant Oates, a valuable and promising young officer, the casualties in this affair were two men wounded (one of them mortally), one horse killed and two wounded, of which one had to be destroyed. The Military Cross was awarded to Risaldar Laurasib Khan for his conduct on this occasion.

Immediately on receipt of news of the attack on B Squadron and the death of Lieutenant Oates, Lieut.-Colonel Kemmis (who had now resumed the command of the regiment) with two squadrons of the Tenth and a section R.F.A. left Abu Kemal, and reaching Al Qaim at 1.45 A.M. on the 2nd March pushed on to the scene of the attack near the Wadi Safra, which was reached about 7 A.M. Not a sign of any enemy was to be found and the regiment could only return to Al Qaim, bringing in the convoy which had been expected on the previous day. Camp was reached at 8 P.M., the squadrons and guns having covered 50 miles in the previous 23 hours.

Only three days later still more serious losses were incurred in another affair similar to that described above. On 4th March C Squadron under Lieutenant C. E. Waters, a young officer who, as has been mentioned, had just joined the regiment from Ramadi, proceeded on escort duty to Salahie, twenty-five miles up the Euphrates from Abu Kemal, on the road to Deir-ez-Zor. The next day, 5th March, he left Salahie with the squadron escorting a convoy back to Abu Kemal. About six to seven miles south of Salahie the road passes below some cliffs whence it is open to attack, and Lieutenant Waters rode forward with three troops in order to secure this position, leaving his fourth troop with the convoy. As he approached the cliffs his advanced guard was fired on, but Lieutenant Waters probably did not realise at first either the strength of the enemy or the extreme difficulty of the ground, which here as on the Wadi Safra is much broken up into low hillocks and intersected with water-courses. He continued to move forward to support his advanced party[1] with the object of driving the Arabs off the heights, at the same time sending his two Hotchkiss guns on to a low hill on the right front to cover his advance. Now however more Arabs appeared on his left, and to hold them in check he sent part of his remaining force in that direction, while he himself joined the Hotchkiss guns. One of these had jammed and Lieutenant Waters sent it down to the road below. At the same time or earlier, having realised the considerable strength of the attacking force, he sent word to the convoy to retire while he endeavoured to withdraw his own men. But meanwhile the Arabs had been working up the nullahs and broken ground until they reached within a short distance of where Lieutenant Waters and a small party were left with the remaining Hotchkiss gun. The troop on the left was beginning to withdraw when the Indian officer in command at that point suddenly saw about a hundred Arabs charge down on the Hotchkiss gun, and the whole party with it was overwhelmed.

The squadron brought the convoy back to Salahie, having suffered a loss of Lieutenant Waters and eight men killed and three men taken prisoners.

The daring shown by the Arabs in this attack was not allowed to pass unpunished, but the stiff fight which resulted on the 7th March was a good example of the difficulties besetting the small bodies of cavalry when called upon to secure the safety of slow-moving and unwieldy convoys against an active and wily foe in peculiarly unfavourable country.

When news was received at Abu Kemal of the second attack by the Arabs and the death of

[1] The advanced party occupied the point marked A on the sketch on p. 309, but were driven off it before Waters could reach them.

Lieut. Waters, Brigadier-General Coningham, who had arrived there with the headquarters of the 51st Infantry Brigade, at once sent the 10th Lancers, less two squadrons, under Lieut.-Colonel Kemmis to the scene of the fighting, in the hope of punishing the raiders and also to bring in the convoy from Salahie. Marching from Abu Kemal at 3 A.M. on 7th March with A and D Squadrons, Lieut.-Colonel Kemmis made his way before daylight to a point[1] on the edge of the low plateau which here falls in precipitous escarpments to the Euphrates valley a few miles from Salahie. The road ordinarily used runs in the plain below these cliffs and it was here that the Arabs attacked Lieut. Waters two days before. But another track ascends to the high ground at a little distance from the town and Lieut.-Colonel Kemmis now signalled that the convoy, with a section of R.F.A., should take this upper road.

The position which he had taken up commanded all the plain below, and as the morning mists cleared away a large body of Arabs was seen moving across to the cliffs with the intention of again intercepting the convoy, which was now starting from Salahie. It soon became evident that the enemy was making for a height[2] at some distance south-east of that occupied by the regiment. They did not yet appear to be aware of any larger force than a piquet on the hills above them and Lieut.-Colonel Kemmis sent off A Squadron, hoping to seize the enemy's objective before they should arrive there or realise the presence of the cavalry. Unfortunately however, before the squadron could reach the point, the Arabs appeared on the crest and immediately opened fire.

By this time the field guns from Salahie had joined Lieut.-Colonel Kemmis and came into action[3] against the Arabs, preventing them from making much progress against A Squadron which was now

[1] Marked A on adjoining sketch.
[2] Marked B on sketch.
[3] Position of guns about F on sketch.

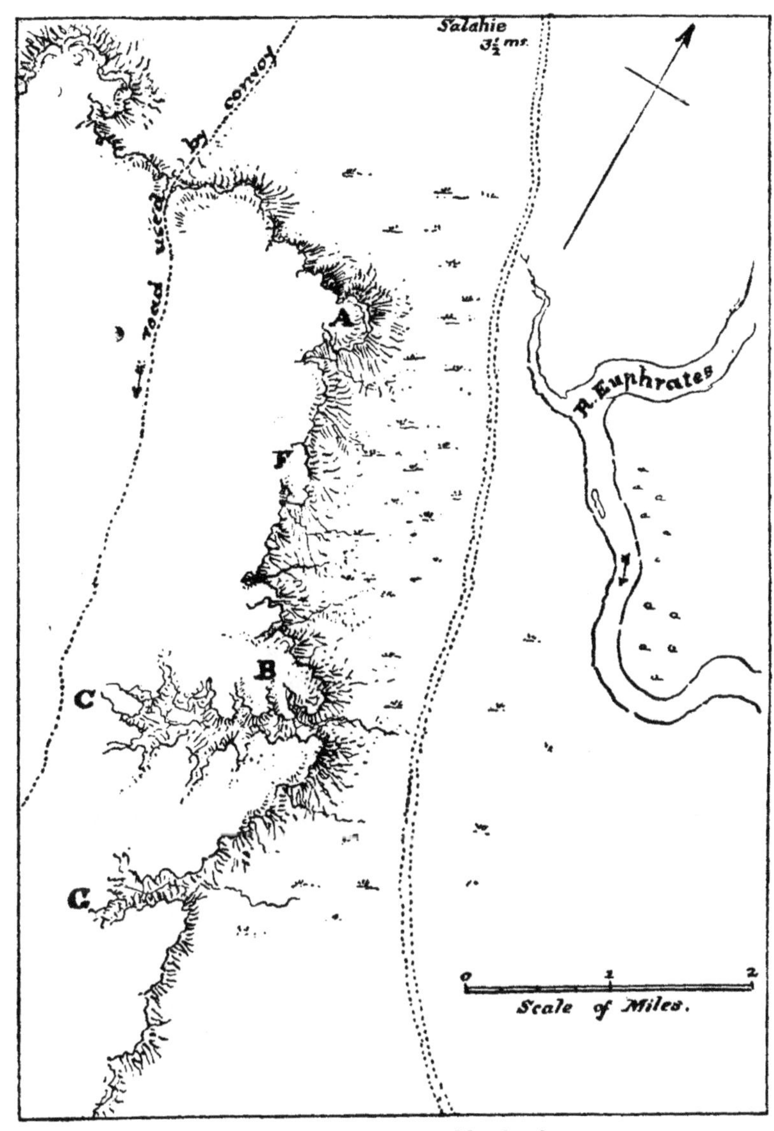

Sketch of the action on March 7th.

dismounted. Notwithstanding this artillery support however, the broken nature of the ground, with many abrupt inequalities and nullahs, allowed the Arabs gradually to get round more to the south-east and thus to threaten the right[1] of A Squadron. To create a diversion and relieve this pressure Lieut.-Colonel Kemmis now sent two troops of D Squadron under Captain J. C. Platts (Adjutant) and Captain O. B. P. Russell to make a mounted attack on the right of the Arab position about B on the sketch. Covered by Hotchkiss gun fire and the two field guns this attack was delivered in a most determined and gallant manner along the crest of the cliffs, and it was entirely successful in clearing the Arabs out of the nearest nullahs and driving them into the plain below. But as the main force of the enemy was approached the ground became quite impossible for mounted action. Captain Platts thereupon dismounted his men and continued the attack on foot, driving the Arabs out of the nullahs and off the high ground with considerable loss. The enemy resisted however in spite of their casualties and, seeing how small was the number of their assailants, clung to the ground with tenacity. The fighting was at close quarters and even at times hand to hand. Captain Platts took a lance from one of his men and had already killed two Arabs with it when, as he was pursuing another of the enemy, he was shot dead at only five yards' range. Captain Russell who took command when Captain Platts was killed continued to press the enemy with great determination, himself shooting five of them with a rifle at ranges under fifty yards.

In the meantime the convoy had come up to Lieut.-Colonel Kemmis and was ordered by him to move southwards to more open ground further away from the cliffs. Captain Russell having heavily punished the Arabs in front of him withdrew without difficulty and furnished an advance guard, while

[1] A squadron was in action about C. The Arabs endeavoured to attack from the nullah G.

A Squadron also fell back covered by fire from the guns. As soon as the difficult country cut up with nullahs and bluffs was left behind the enemy did not attempt to follow and the column proceeded without further incident, except a little sniping, and reached Abu Kemal the same evening.

The casualties in this action were Captain Platts and six men killed, Risaldars Bishn Singh and Rai Singh and nine men wounded. The losses inflicted on the enemy were certainly considerable, about fifty dead being counted in front of D Squadron alone. All ranks in the regiment behaved with coolness and courage, and Captain O. B. P. Russell was awarded the Military Cross for his gallantry.

During the remainder of March the regiment was employed on several occasions, together with the 158th Battery R.F.A., the 85th Carnatics and 126th Baluchis, in carrying out operations on both banks of the Euphrates, and punishing the tribesmen for their complicity in the recent attacks on convoys and escorts. A number of villages between Abu Kemal and Salahie were burnt, but no serious fighting took place. Their losses on 7th March had made the Arabs unwilling to come to close quarters, and they contented themselves with comparatively harmless firing on our troops from the safety of the opposite bank of the river. Escort duty was also carried out at intervals but without any repetition of the attacks previously encountered.

By the end of the month the horses, as has already been mentioned, were utterly worn out with constant heavy work and exposure and insufficient food, and it was a fortunate respite for the regiment that April passed without any operations of importance so that men and animals were able to get some much-needed rest.

Meanwhile the local political officers had demanded submission and compensation from the hostile tribes about Abu Kemal for the recent outrages, and they, persuaded by the punitive operations in March, at length gave way and undertook to pay the fines required of them. In the course of April the money was forthcoming, but the good effect of this temporary firmness was quickly dissipated by a repetition early in May of the policy which had left Deir-ez-Zor in the hands of the Arabs. In response to renewed demands by the Damascus Government, the whole country down to and including Abu Kemal was surrendered, and most of the 51st Brigade and attached troops were withdrawn to Ana. As if however to ask for further trouble an isolated post was still left at Al Qaim, fifty miles from Ana and only fifteen from the now hostile position at Abu Kemal. Convoys had of course to be escorted between Al Qaim and Ana, and it was equally a matter of course that these convoys were exposed to constant attacks from raiding tribesmen. On the very day before the withdrawal from Abu Kemal a determined attack was made on a convoy between that place and Al Qaim, and two men of the Tenth were wounded, as well as four killed and four wounded in the 80th Carnatic Infantry. Even on 8th May, the day on which Abu Kemal was handed over to the Damascus Government, the retiring column was fired on and threatened so persistently that the field guns were brought into action, a circumstance which clearly showed in what light the Arabs regarded our withdrawal. In their eyes we had been driven by them out of the country and it was merely in the natural sequence of events that they should harass our retirement.

During the remainder of the month squadrons of the regiment (of which A and C were at Al Qaim with a movable column under Lieut.-Colonel Kemmis and B and D at Nahiya) were on several occasions engaged with parties of hostile Arabs while escorting convoys, the most considerable affair being on 11th May when a squadron from Al Qaim under Lieutenant Crichton (5th Cavalry) surprised and attacked a party of Arabs, who were lying in wait for an expected convoy. Six or more Arabs were killed and some rifles and ammunition were seized in this encounter, in which Lieutenant Crichton himself was wounded.

At length at the end of May it seemed that the Tenth had reached the termination of their arduous and thankless service in Mesopotamia. On the 24th of that month the 5th Cavalry arrived to relieve them and the regiment thereupon assembled at Ana preparatory to marching down to Baghdad *en route* for India. It was indeed time that they should have some respite from the strain which they had endured. The horses had put on a little condition during April but the renewed hard work of the past three or four weeks had quickly reduced them again. They had been terribly weakened by the trying conditions of the previous months and were quite unequal to long marches in the hot weather which had now set in. A board was assembled at Ana to examine them and it reported that seventy per cent were unfit for further cavalry work.

On 26th May the regiment was inspected by Brigadier-General F. E. Coningham, C.M.G., D.S.O., commanding the 51st Infantry Brigade, when there were present on parade Lieut.-Colonel A. W. M. Kemmis, D.S.O., Captain O. B. P. Russell, M.C., Lieutenant D. E. Abercrombie, Lieutenant L. Steedman (25th Cavalry), Lieutenant W. M. Tatham (16th Cavalry), Captain Tha, I.M.S., nine Indian officers and 399 non-commissioned officers and men. The following order of the day was published by the Brigadier, whose reputation as a commander as well as his intimate knowledge of the work done by the Tenth on the upper Euphrates render his testimony peculiarly valuable :—

HEADQUARTERS 51ST IFY. BRIGADE.
ANAH.
26th May 1920.

ORDER OF THE DAY.

" On the eve of the departure of the 10th Lancers to India I wish to place on record my great appreciation of the excellent work done by the Regiment, throughout the operations on the Upper Euphrates.

The Regiment has had a great deal of very hard marching and a considerable amount of severe fighting under very difficult conditions, but whatever it has been called upon to do, it has invariably carried it out well and cheerfully.

The actions of the 1st and 7th March were both masterpieces of tactical handling, and the fighting record of the Regiment throughout has been one of which any Regiment might well be proud.

I congratulate all ranks on not only having maintained but for having added to the high reputation of the Indian cavalry in general and the 10th Lancers in particular.

Sd. F. E. Coningham, Brig. General
Commanding 51st Infantry Brigade."

The regiment was now nominally on the way to Baghdad but notwithstanding this fact, and in spite of the deplorable condition of the horses, the amount of escort work which had to be done in the existing unsettled state of the country, and the paucity of mounted troops for the work, necessitated the continued employment of squadrons of the Tenth again and again, backwards and forwards, between posts on the road between Ana and Ramadi. Meanwhile the heat had become very severe, and although officers and men were quite fit the horses suffered much, especially from lack of water, which was only obtainable in the morning and evening. During the month of June the headquarters and C Squadron marched over two hundred miles, while the other squadrons covered upwards of three hundred miles, thus completing some thirteen hundred miles since leaving Ramadi on 12th January.

At last on 30th June the regiment was concentrated at Karada, just outside Baghdad. 1st to 5th July were spent in Baghdad handing over the horses and saddlery and such equipment as was not to be taken back to India, and on 6th July the Tenth embarked on P. steamer 51 for Basra, fondly believing that their service in Mesopotamia was at an end.

On the eve of their departure the following farewell orders were published by the G.O.C. 17th (Indian) Division and by Lieut.-General Sir Aylmer Haldane, K.C.B., D.S.O., Commanding-in-Chief the Mesopotamia Expeditionary Force :—

Farewell Order by Major-General G. A. Leslie, C.B., C.M.G., Commanding 17th (Indian) Division, Mesopotamia.

Baghdad.
29th June 1920.

"On the departure of the 10th Lancers to India the Divisional Commander wishes to express his appreciation and gratitude for the fine services rendered by the Regiment during the last three and a half years of active service in the Great War.

The 10th Lancers arrived in Mesopotamia in November 1916 and since July 1917 they have been employed continuously on the Euphrates.

Their services included the first attack on Ramadi, carried out on the hottest day of the hottest year on record, and the operations at Nejaf.

As Divisional Cavalry, 15th (Indian) Division, they took part in the complete victory over the Turks at Khan Baghdadi, and as Divisional Cavalry, 17th (Indian) Division, they have for the last six months been engaged in the arduous and successful operations against the Arabs on the Upper Euphrates, 285 miles from Baghdad, in the course of which the Regiment covered 1320 miles.

Throughout their service in Mesopotamia they have maintained the highest standard of efficiency and during the late operations, under very trying conditions of cold, heat, long marches, shortage of water and of rations, they have shown themselves imbued with the highest qualities of alertness and of the true fighting spirit."

General Headquarters,
Mesopotamia Expeditionary Force,
Baghdad.
26*th June* 1920.

"*7th Cavalry Brigade.*

The following is to be conveyed to the 10th Lancers on their way through Baghdad:

"The C.-in-C. desires, before the departure of the 10th Lancers, to express his high appreciation of the work done, and hardship endured by the regiment under command of Lieut.-Col. Kemmis on the Euphrates.

Since the commencement of the disturbances in the Deir-ez-Zor Area in December, this Regiment has been continuously employed in operations and convoy duty in all weathers and trying climatic conditions, and owing to the difficulty of carrying out reliefs in the circumstances which have obtained their return to India has been unavoidably delayed.

The C.-in-C. wishes all ranks to know how thoroughly he has appreciated their devotion, endurance and determination, and he regrets that he has been unable to convey personally his congratulations in their success and his thanks to the 10th Lancers for their services on the Euphrates."

CHAPTER VI.

THE ARAB REBELLION, 1920.

Early on the morning of 10th July, 1920, the 10th Lancers reached Basra on their way to India. But unfortunately for them, during the ten days that had elapsed since they left Baghdad events had occurred which frustrated their hopes of relief and necessitated the indefinite postponement of all arrangements for their departure. The storm of which the first warning had been supplied in the attack on Deir-ez-Zor seven months before, and which had gathered strength in the constant hostilities on the upper Euphrates during the spring, had now spread over the length and breadth of the land. It burst with sudden violence on 30th June when the local tribesmen attacked the garrison of a small post at Rumaitha, on the Hilla branch of the middle Euphrates, and besieged the place for three weeks until a considerable force had to be employed for its relief. At the same time the railway lines were cut in many places; in a very few days communication between Baghdad and Basra by rail was completely severed, and the whole country was in revolt.

In this emergency, and faced with the duty of restoring and maintaining order over an enormous tract of country with entirely inadequate forces (the mobile force at the disposal of the Commander-in-Chief on 1st July is stated by him to have been no more than 500 British and 3000 Indian troops [1]), it was impossible for Sir Aylmer Haldane to release any troops that were still under his control. The 10th Lancers, therefore, instead of embarking for India, were ordered to proceed to No. 1 Indian Base Depot at Basra, where they were placed under the orders of the G.O.C. Lines of Communication. As has been seen the regiment was now without horses, but from this time forward for several months it was employed on all sorts of duties, none of which would normally be performed by cavalry soldiers. In spite, however, of the unusual and often uncongenial nature of the work which they were now called upon to do, and although it came at the end of six months of prolonged and strenuous exertions, above all in spite of the disappointment caused by the further postponement of their return home after four years of foreign service, all ranks in the Tenth responded to this fresh call with the same cheerfulness and loyalty which they had shown throughout the previous campaign. The regiment was very short of British officers, but Captains Cahusac and Bidie rejoined in the course of the summer. A certain number of men whose presence at their homes was urgently needed were despatched to India as soon as possible, but the rest of the regiment settled down for a further period of several months in Mesopotamia.

The first call made on the Tenth was for a non-commissioned officer and nine men to man the machine guns on the inland-water-transport gunboat F 11. This party, taking five Vickers guns, embarked on 11th July and proceeded to Nasiriya.

The next day a British offiçer (Second-Lieutenant Rikh) with fifty men of C Squadron was detailed to reinforce the infantry guard at the base magazine at Basra, Second-Lieutenant Rikh taking over command of the magazine defences.

[1] 'The Insurrection in Mesopotamia,' by Lieut.-Gen. Sir Aylmer Haldane, page 72.

On 14th July Captain O. B. P. Russell, M.C., with sixty men of C Squadron and four Hotchkiss guns took over No. 3 armoured train and left for Ur junction with orders to work between that place and Khidhr.

A few days later Lieutenant Steedman (25th Cavalry) with some non-commissioned officers was detailed to instruct the personnel of four companies of a Labour Corps, whom it was proposed to employ as guards over Turkish prisoners of war.

On 22nd-23rd July Lieutenant Abercrombie with Jemadars Sohan Singh and Ishar Singh and sixty rifles from A Squadron, with six machine guns, embarked on three "T" class two-decked screw steamers for patrol work in the "Narrows" between Basra and Amara, and to protect the river transport in that vulnerable area.

The foregoing catalogue shows the sort and variety of jobs to which parties of the regiment were put in the first month of the emergency, and similar duties were continued throughout the summer. The guard at the magazine was increased to a hundred and fifty men early in August, the Tenth now finding the complete guard there. In the course of this month the detachments on two of the patrol ships in the Narrows were relieved by the Royal Navy and rejoined the regiment. The gun crew on "F 11" was engaged in a sharp encounter with Arabs on 15th August when, in company with the steamer *Grayfly*, a gallant but unsuccessful attempt was made to refloat a third defence vessel, the *Greenfly*, which had run aground about five miles above Khidhr. All three vessels were under heavy fire and were riddled with bullets, and every man of the machine-gun detachment was wounded. It therefore became necessary to relieve the party, all of whom were reported to have behaved very well in very trying circumstances. Another detachment was supplied for duty on F 11 as well as similar parties for F 2 and F 7.

On 14th August Lieut.-Colonel Kemmis with the

headquarters of the regiment, C Squadron and one hundred and fifty details from the base depot at Basra proceeded by rail to Ur junction, with orders to take command there and to strengthen the defences of this important railway centre which was threatened by further risings in the neighbourhood. This task was undertaken at once, a scheme of defence being planned on the evening of the 14th and arrangements made for the distribution of the work, which was pushed on during the following days until 22nd August. On that date the 2/117th Mahrattas arrived and relieved the Tenth, who returned to Basra the following day.

At the end of this month orders were received for two squadrons to be mounted on horses which were to be handed over by the 37th Lancers, and in the early part of September B and part of C Squadron were engaged in getting these and other horses received from the remount department into condition and in fitting saddlery.

Meanwhile at the end of July D Squadron (Dogras), one hundred strong, had proceeded to relieve the sixty men of C Squadron who for the previous fortnight had been working No. 3 armoured train under the command of Captain Russell above Ur junction. Captain Russell himself was not relieved but took command of D Squadron and now took over No. 1 armoured train with headquarters at Samawa.

The state of affairs at that place was bad, and it was soon to become much worse. Situated on the Euphrates just below the junction of the Hindiya and Hilla branches of the river, and only some twenty miles from Rumaitha which had been the scene of the first outbreak in June, Samawa was in the very centre of the insurgent area. In spite of hostilities on several occasions and constant tampering with the railway line, communications by rail and river were kept open during July with Khidhr and Nasiriya (Ur junction, 55 miles). On 13th August however Khidhr was evacuated and in the withdrawal an armoured train was derailed and abandoned, thus completely interrupting all railway communication with Samawa. Nor was communication by river in any better condition. A convoy carrying provisions, ammunition and some small reinforcements did indeed get through as late as 28th August, but it was subjected to a galling fire from close range all the way up and a steamer and two barges were lost before Samawa was reached.

At the date when Captain Russell and D Squadron took over No. 1 armoured train the garrison of the place consisted of two and a half companies of the 114th Mahrattas, one hundred men of the 125th Rifles, and some details from other units, the whole under the command of Major A. S. Hay, 31st Lancers. The garrison also had the support of the protected defence vessel *Greenfly*, which however unfortunately ran aground between Samawa and Khidhr (as has already been mentioned) on 10th August, and was not refloated. With the reinforcements which reached Major Hay on 28th August (forty-five men of the 2/123rd Rifles) he had under his command the equivalent of about a battalion in all.

But his difficulties were greatly increased by the manner in which, on his arrival at the end of June, he found the small garrison split up into no less than four different posts. After taking over command he made various improvements in the defences, which subsequently enabled the main post to stand a siege of six weeks, but the outlying posts were not at first abandoned nor the garrison concentrated.

The main camp was situated on the river bank about a mile up-stream (N.W.) from the walled town of Samawa. Some two thousand yards west of the main camp is Barbuti bridge, where the railway line crosses the Euphrates, and two thousand yards south-east of the main camp is the railway station, only about a couple of hundred yards from the wall of the town. There was also a supply camp on the river bank some four hundred yards west of the main camp. From the date when railway communication between Samawa and Ur junction was interrupted No. 1 armoured train was kept at Samawa railway station, and was occupied by seventy men of the 10th Lancers with three Hotchkiss guns; there was also a 13 pdr. gun on the train. The railway station, a mud-walled enclosure south of the line, was manned by thirty of the 2/125th Rifles under Second-Lieutenant Fleming. The whole post was under the command of Captain Russell. The remaining thirty men of D Squadron, together with a small infantry detachment, held Barbuti bridge. The rest of the garrison, some six hundred strong, was concentrated in the main camp.

On 28th August, the same night on which the river convoy and reinforcement reached Samawa, a serious attack was made on the railway station post, which for several previous nights had been annoyed with intermittent sniping. The post was peculiarly ill-sited for purposes of defence. It was, as has been described, not more than two hundred yards from the wall of the town, while on the other side some brick-kilns at a distance of three hundred yards gave cover to assailants. Nor was there free communication between the post and the main camp, from which it was hidden by palm-groves near the town. Most serious of all, the water supply, after the place was invested, consisted only of such amount as could be stored in tanks in the station yard, and these when the firing became heavy were pierced with bullets and leaked badly. The armoured train stood on an embankment slightly above the station and between it and the town. It was composed of two bullet-proof trucks at each end, with carriages and unarmoured trucks in the centre.

The attack on the 28th was beaten off but it was renewed on the 29th when three determined attempts were made by the insurgents to overwhelm the little garrison. Each of these was repulsed with heavy loss although on the last occasion the enemy succeeded in getting close up to the train. The Arabs then closely invested the post keeping the station and train continuously under fire. These days must have been a time of acute suffering for the occupants of the train. The terrible hot weather of Mesopotamia was at its height and the heat in the steel trucks must have been almost insupportable. It was impossible even to get a little air through the loopholes because, owing to the closeness of the enemy's fire, these could not be kept open for any length of time during the day. Captain Russell lived with his men in the train all this time, directing the defence, and only going out to the piquet in the station for a little fresh air at night. Meanwhile as soon as the leakage of the tanks was noticed Captain Russell informed Major Hay that he would not be able to hold the station post for more than four days. In consequence of this report a plan for the evacuation of the post was elaborated. It was arranged that a covering party of two hundred rifles of the 114th Mahrattas should move out of the main camp on the 3rd September and proceed to the rifle butts, which were close to the railway line about half-way between the station and Barbuti bridge. The detachment of the 10th Lancers at the bridge was also to move out along the line, to guard against the risk of envelopment. Meanwhile the detachment of the 125th Rifles under Second-Lieutenant Fleming was to evacuate the station, followed by the armoured train as rear-guard. The signal for the movement to begin was to be given on the arrival of two aeroplanes from Baghdad, which were to bomb the enemy and help to cover the withdrawal.

The foregoing plan was put into execution on the morning of the 3rd September as arranged. Two aeroplanes appeared from Baghdad, but their co-operation was brief and of little value if any. The covering force of the 114th Mahrattas moved out from the main camp and at 7 A.M. the evacuation of the station began, the detachment of the 125th Rifles leading the way with orders to withdraw a short distance, and then await the movement of the train. If the latter was moving all right the withdrawal was to continue. The train started from the station at about 7.15, a heavy fire being kept up from the trucks to cover the retirement of the infantry. After going a hundred yards or so it stopped and Captain Russell sent an orderly to tell the infantry that they could go on. The train then moved on another hundred yards and came to a sudden stop, apparently owing to the engine going off the rails at a point where the line had been tampered with. Captain Russell, who was in the armoured truck immediately behind the engine, jumped down on to the line and, with Captain Pigeon, I.M.S., who was also on the train, ran back along the train shouting to the men to get out and run for it. At the same time the Arabs, seeing the train at a stand-still, swarmed out of the neighbouring gardens two or three hundred yards away and rushed towards the train to the number of nearly three thousand. When Captain Russell reached the last truck the Arabs were too close for it to be possible to escape, so he and Captain Pigeon climbed in and shut the door, telling the occupants that they must fight it out and keep the Hotchkiss gun going, which would be heard in the main camp and help would be sent to them.

Some of the men who were in the forward trucks succeeded in overtaking the retiring infantry. Large numbers of the enemy were following up this retirement, and from the position then reached it was not seen that fighting was still going on at the rear of the train. On the contrary the officer in command of the covering party fully believed that all except those who had come in had been already overwhelmed and killed. He accordingly fell back on the camp and on arrival there reported to Major Hay that Captains Russell and Pigeon and about thirty men of the 10th Lancers were missing.

But in the meantime a desperate struggle was in progress in the two rear trucks of the armoured train, which were still held by Captain Russell and the remnants of his men. All through the morning the defence was continued with such good effect that the Arabs, though in overwhelming numbers, dared not get to close quarters. About 2 P.M., the numbers of the garrison being considerably reduced, Captain Russell withdrew all the survivors, about twelve in number, into the last truck and continued the fight, all the time encouraging his men by word and example, and exhibiting the most resolute bravery and coolness. But now the assailants finding it beyond their powers to capture the defenders' stronghold by direct attack went about to burn out the occupants by setting alight to the sleepers below. With this object they crawled under the empty trucks and so to the occupied one. Captain Russell had foreseen the possibility of this method of approach and had kept open the communicating door between the two last waggons, and several Arabs were shot by him from above as they endeavoured to pass underneath. At length however, at about 4 P.M., as he stood at the open door, he was wounded in the side by a rifle shot, and although he assured a sowar who went to his assistance that it was only a flesh wound yet he had to leave his post of observation and the communicating door was shut. The end was now near. The ammunition of the defenders was almost exhausted. Some of the Arabs were setting a light to the sleepers below the waggon. Others getting bolder climbed up and began to tear off the roof. Captain Russell, still determined to the last and still encouraging his men, told them to jump from the train and sell their lives as dearly as possible. Most of them did so and were immediately overpowered and killed. One or two remained in the truck with their wounded commander. In a moment the savage assailants swarmed in and Captain Russell was cut down. One of the sowars who remained with him in the truck was taken alive and, with two others captured earlier in the day, was subsequently released.

This tragic story is related above as told by the surviving sowar, Rallah Singh, who was with Captain Russell to the last, and as substantiated by the other survivors. When it was referred to Major Hay and to the officers who took part in the withdrawal of the infantry from the station post they were disposed to throw doubt on its credibility, being naturally loth to believe that the fight round the train could have been continued for so long, unknown to the garrison of the main camp less than a mile and a half away. But a very careful and searching investigation made by order of Sir Aylmer Haldane succeeded in securing evidence from Arab witnesses which confirmed in every essential particular the sowar's story. They related the details of the attack on the train and how the fight was prolonged till between 4 and 5 in the afternoon, and all spoke with enthusiasm of the courage and determination of the British commander, whom they called "Abu sil Sillah" (the Father of the chains) on account of the steel chains which he wore on the shoulders of his jacket. They particularly referred also to the bravery of another Englishman, who may have been Captain Pigeon, or possibly the corporal who was in charge of the 13 pr. gun. These facts are related in Sir Aylmer Haldane's valuable and interesting record, 'The Insurrection in Mesopotamia,' wherein too the author shows how it came about, by reason of the peculiar atmospheric conditions in Mesopotamia, and the position of the scene of the fighting in relation to the main camp, that no sound or sign of the struggle reached the troops at the latter place through the long, sweltering hours of that terrible day.

Some months later when incontestable evidence

had been collected of Captain Russell's bravery and devotion a recommendation was put forward for the grant of a posthumous Victoria Cross on his account, but on the score of the time that had elapsed it was negatived. None the less the story of the defence of the armoured train at Samawa and of the steadfastness and determination with which this young officer, scarcely more than a youth, faced the fearful ordeal which he was called upon to endure, cheering and encouraging to the end his staunch and gallant men, and meeting death undismayed, is one which will bear comparison with the records of any roll of honour. He fought and died worthy of his regiment and of his country, while of his Dogra soldiers to whom he set so fine an example there can be no higher praise than to say that they maintained the great traditions of their race.

The news of the disaster to the armoured train and of the deaths of Captain Russell and of a third of the men of D Squadron who were with him at Samawa came as a great shock to the Tenth. After four years of foreign service in peculiarly trying conditions, and especially after the harassing experiences of the past six months, during which the regiment had suffered serious casualties, including the loss of three very gallant and promising young officers, this last blow created a very painful impression. And these feelings of distress were intensified by the nature of the service in which the losses were endured. Men will face death with equanimity in a good cause and a fair fight, but to be butchered by overwhelming numbers of blood-thirsty and remorseless savages is a fate the prospect of which might well shake the resolution of the most veteran troops.

In spite however of the natural feelings of horror at the Samawa disaster, the regiment stuck manfully to the multifarious duties for which it was detailed, buoyed up with the hope that it would have a share in avenging the recent losses before the time came for its long-deferred relief. This opportunity came at the end of September when the headquarters and two squadrons made up from A, B, and C, and mounted on horses received from the 37th Lancers, the whole under Lieut.-Colonel Kemmis, D.S.O., were detailed to form part of a column of all arms which assembled at Nasiriya under the command of Brigadier-General F. E. Coningham, with orders to march to the relief of the beleaguered garrison at Samawa camp.

The other units with this column were the following:—

10th (How.) Battery R.F.A.
13th Battery R.F.A.
69th Company 2nd Sappers and Miners.
8th Battalion Machine-gun Corps.
1st Battalion King's Own Yorkshire Light Infantry.
3/5th Gurkha Rifles.
3/8th Gurkha Rifles.
1/11th Gurkha Rifles.
3/23rd Sikh Pioneers.

The regiment proceeded to Nasiriya by rail, and on 1st October the march to Samawa began. Progress was slow because, as the column advanced, it was necessary to repair the damaged railway line, upon which the force was entirely dependent for all supplies and on some marches even for water. At first no opposition was encountered, the tribes within three marches of Nasiriya not having joined the insurgents. From the 4th to the 14th October however, on which latter date the column entered Samawa, the regiment was in touch with the enemy every day. At Khidhr on 6th October and outside Samawa on the 13th the enemy offered determined resistance in entrenched positions, from which they were driven with heavy loss. On both occasions the regiment was employed in protecting the left flank of the infantry advance from a threatened attack by strong bodies of Arabs. Our casualties on each day were slight.

On the 14th October the column marched into Samawa and relieved the garrison, which had been besieged in the camp since the middle of August. Among the garrison were one Indian officer and sixty-five non-commissioned officers and men of D Squadron, out of the total, one hundred strong, who had formed the crew of the armoured train. Here also were ten non-commissioned officers and men who had manned the machine guns on the steamer F 2. All were reported to have done excellent work, the machine gunners on F 2 being particularly commended. On the same day the bones of the men killed in the armoured train were collected and buried.

Here the regiment remained for thirteen days, being employed on several occasions with punitive columns against hostile villages in the neighbourhood, and here on 16th October was received the supremely welcome intelligence that the regiment was to be relieved by the 37th Lancers in the immediate future and was thereupon to proceed to India. On 27th October the Tenth had a final encounter with the Arabs in the course of a punitive expedition against two neighbouring villages. During the retirement B Squadron, when covering the infantry withdrawal, was heavily attacked by the enemy in great force on both sides. One man was wounded and the squadron had considerable difficulty in getting him away, but their steadiness prevented the Arabs from getting too close and the latter part of the retirement was not molested. In this affair Ressaidar Hayat Khan and Dafadar Ghulam Baquir Khan particularly distinguished themselves by their coolness and gallantry. The casualties in the squadron were one man killed and three wounded, two horses killed and four wounded.

On the evening of the same day the headquarters and two squadrons of the 37th Lancers arrived at

Samawa to relieve the regiment and take over the horses. On the 28th the Tenth entrained for Basra, arriving there on the 29th. On their arrival at the wharf at Basra a large draft from the depot was met, just disembarking. These men joined the regiment forthwith and the whole embarked and sailed for India on 30th October.

. . . .

The 10th Lancers had been on foreign service for four years and three months during which time the following casualties had occurred in the regiment:—

	Killed.	Wounded.	Died of disease.
British officers	4	1	1
Indian officers	1	3	1
Other ranks	52	33	17

The personal rewards for gallant and distinguished service in the field are detailed in Appendix IV.

The battle honours awarded to the regiment for the Great War were "Mesopotamia 1916-18," and "Khan Bhagdadi." It is noticeable that no such award was made for service in the Arab rebellion, although the losses suffered then were more numerous, and the service certainly more arduous than during the period of hostilities against Turkey.

Below is given a list of all British officers who served with the 10th Lancers in Mesopotamia; those marked with an asterisk served in the Arab rebellion.

Lieut.-Colonel W. E. Young.
Lieut.-Colonel H. G. Young, D.S.O.
Brevet Colonel R. L. Ricketts.
Major J. E. Moir (died, 26th Jan. 1917).
Lieut.-Colonel P. R. Chambers, D.S.O.
*Brevet Lieut.-Colonel A. W. M. Kemmis, D.S.O.
Major J. E. Hulbert (3rd Skinner's Horse).
Major V. C. P. Hodson.

Major D. O. W. Lamb, O.B.E.
Major M. G. P. Willoughby, M.C.
Major R. P. J. Michell (37th Lancers).
*Major S. D. Cahusac, M.C.
Captain R. T. Lawrence, M.C.
Captain R. H. W. Welsh (16th Cavalry).
*Captain A. G. C. Bidie.
Captain A. R. Whistler.
*Captain J. C. Platts (17th Cavalry, killed in action).
*Captain O. B. P. Russell, M.C. (killed in action).
*Lieutenant A. T. Oates (killed in action).
Lieutenant B. H. Wiles (I.A.R.O.)
Lieutenant A. Mellish (I.A.R.O.)
Lieutenant E. Wilks (I.A.R.O.)
Lieutenant A. N. F. Moore (I.A.R.O.)
Lieutenant E. F. Rawlins.
*Lieutenant D. E. Abercrombie.
*Lieutenant E. F. Waters (killed in action).
*Lieutenant W. J. Ekin.
Lieutenant A. M. Dore (22nd Cavalry).
*Lieutenant W. M. Tatham (16th Cavalry).
*Lieutenant J. C. Crichton (5th Cavalry).
*Lieutenant L. Steedman (25th Cavalry).
*Second-Lieutenant A. Rikh.
*Second-Lieutenant W. G. Barker.
*Captain R. R. H. O. Tha, I.M.S.

Note.—It has been shown in Chapter IV that during the first two years of the war, when the 10th Lancers remained at Loralai, many British and Indian officers and men were detached from the regiment and were employed in the field in various capacities. The extra-regimental services of some of these British officers and of others (which are not mentioned in the foregoing narrative) are outlined below :—

Lieut.-Colonel R. L. Ricketts, see Appendix III.

Major A. D. Strong; special service with Jodhpur Lancers in France, Sept. 1914 to March 1918. (Severely wounded; mentioned in despatches three times; D.S.O. and Brev. Lieut.-Colonel). Raised the 151st Sikh Infantry and commanded it in Mesopotamia, March to June 1918, in the Afghan campaign, 1919 (despatches), and in Palestine, 1920. Commanded the 2/41st Dogras in Waziristan, 1921.

Major J. Peters; squadron commander 34th Poona Horse in France, November 1914 to May 1918.

Major P. R. Chambers; see Appendix III.

Captain K. O. Goldie; Asst. Mil. Sec. 3rd Indian Army Corps, Mesopotamia, Oct. 1916 to Apr. 1918; D.A.A.G. Mesopotamia, Apr. 1918 to Feby. 1919 (despatches twice; O.B.E.)

Captain A. W. M. Foster; attached to 9th Hodson's Horse, France, Oct. 1914; A.P.M. 2nd Ind. Cav. Div. Jany. to Jul. 1915; G.S.O. III. 2nd Ind. Cav. Div. Jul. 1915 to Jan. 1916. Mesopotamia, Bde. Maj. Cav. Bde. Jan. 1916 to Mar. 1917. G.S.O. II. Mar. 1917 to May 1918 (despatches three times).

Captain D. O. W. Lamb; D.A.A. and Q.M.G. Waziristan, June to Sept. 1917. Mesopotamia, D.A.A.G. Gen. H.Q. Dec. 1917 to Nov. 1918. Afghanistan, D.I.G.C. Jamrud, May to June 1919 (despatches twice; O.B.E.)

Lieutenant G. de la P. Beresford; accompanied the 9th Hodson's Horse to France and served with that regiment throughout the war (despatches; M.C.)

Lieutenant E. C. Braddyll; attached to the 3rd Signal Squadron, 3rd Cav. Div. B.E. Force, Sept 1914, and accompanied that unit to France. Killed in France, 5th September 1915, while serving with the Royal Flying Corps.

Lieutenant M. G. P. Willoughby; Mesopotamia, attd. to 16th Cav. Mar. 1915 to July 1916 (Bde. Machine Gun officer, 6th Cav. Bde. Jan. to June 1916); attd. to 12th Cav. July to Aug. 1916 (despatches three times; M.C., brev. major).

Lieutenant C. F. L. Stevens; accompanied the 9th Hodson's Horse to France and served with that regiment throughout the war until invalided from Syria in October 1918 (despatches; M.C.)

CHAPTER VII.

THE DEPOT, 1916-1920.

A CHAPTER must here be devoted to the doings of the regimental depot during the absence of the Tenth on field service. When the regiment was mobilised in August, 1916, Captain M. G. P. Willoughby, M.C., who had lately returned from Mesopotamia on a month's leave after being on field service since March, 1915, was appointed to command the depot. The other officers were Second-Lieutenant N. Walker, I.A.R.O., and Second-Lieutenant O. B. P. Russell, just joined from the Quetta Cadet College.

The Indian officers were Risaldar Mahan Singh, Jemadar Musahib Khan, Jemadar Sohan Singh and pensioned Jemadar Dayal Singh.

Non-commissioned officers were difficult to find, the regiment having been drained of its best men ever since the war began. Dafadar Harnam Singh, A Squadron, did good work and was much the best of those available.

The first orders received were that the depot should move to Allahabad, but this destination was fortunately changed and it was ordered to Multan. Arriving there after no small amount of trouble in the task of transporting the whole of the impedimenta, public and private, by road and rail from Loralai, the regiment took over part of the cavalry lines. These had been unoccupied for two years and were in a tumble-down condition. Much of the woodwork had been looted, the mud walls were collapsing, and it was necessary to expend a good deal of labour on them before they were habitable.

Work was started at once on training instructors and recruits and the care of field accounts. A draft was asked for by the headquarters very soon and twenty men were sent. The next reinforcement was furnished by the Guides Cavalry, but thereafter all drafts for the regiment were sent by the depot. It took a little time to get an efficient system of training going, the lack of good non-commissioned officers being severely felt, but fortunately a number of good and active pensioners offered their services which were gladly accepted. Pensioned dafadar Lachman Singh (D Squadron), a most efficient quartermaster dafadar, resumed his old duties, and the regiment had good reason to thank him as well as dafadars Lehna Singh and Chanda Singh (A Squadron) and Teja Singh (D Squadron) and many others for their invaluable services as instructors. It was also a great help when a big batch of Indian officers and non-commissioned officers, who had been attached to other units, came back to the regiment. Some of these remained at the depot and were invaluable, especially Jemadar Amar Singh, who as Wordi Major distinguished himself by his tact, efficiency and tireless energy. He was afterwards given well-earned promotion into the 42nd Cavalry.

Many British officers of varying qualities and various antecedents passed through the depot. Second-Lieutenant Russell soon went out to the Tenth in Mesopotamia and it has been told how he won the Military Cross in the Arab war, and how by his defence of the armoured train at Samawa, when he and most of those with him perished, he added a brilliant chapter to the history of the regiment. " By far the best," wrote Captain Willoughby with regard to the officers at the depot, " was young Oates, a youth of rare promise, efficient at all skill at arms, fine horseman and keen sportsman. He picked up the language like magic, and had the gift of galvanising others with his own vitality. What a loss to the regiment when he was killed in the Arab rebellion!" Others who did good work were Captain C. W. Chater and Captain D. Lambart (who served at the depot for upwards of two years), Lieutenant H. G. Sheldon, Lieutenant E. M. Egan, and Lieutenant C. G. Lowe, as well as those whose names occur in the following narrative.

With regard to horses, the depot started work with the few that had been surplus to the requirements of the 16th Cavalry when the latter relieved the regiment at Loralai, mostly good workers but old. They were supplemented later by a batch of really good ponies bought at the Amritsar fair, and again by others sent by the remount officers, some good but most of them indifferent. However no better could be got and so the best was made of them.

In addition to the normal work of a regimental depot the commanding officer had to look after the horse-breeding farm, which had lately been started near Okara in the Montgomery district. A good beginning had been made by the enthusiastic efforts of Lieut.-Colonel W. E. Young, but the venture in those early stages required watchful attention for which there was little opportunity in the midst of a great war. Captain Willoughby found an invaluable assistant in Risaldar Sant Singh, subsequently succeeded by pensioned Risaldar Ganda Singh and Risaldar Nur Khan, and the concern had already shown itself a financial success when in 1921, in pursuance of the general policy with regard to silladar regiments, the whole estate was handed back to the Government.

In 1917 it became evident that under the existing system the clerical work of a depot in the course of a prolonged war was far more than could be dealt with by the ordinary depot staff, and generous allowances were authorised to enable the work to be properly carried on. In this and in other respects the war showed the silladar system to be entirely unsuitable for modern conditions and its eventual abolition became inevitable.

After the difficulties at the start had been surmounted work went on steadily at Multan until the beginning of 1918. In the course of the previous year a sudden interest was excited by the unexpected arrival at the depot of Jemadars Matiullah Khan and Jahandad Khan and eight of other ranks, who had been sent to Persia in 1913 as consular guard at Shiraz. When the war broke out the British Consulate was seized and the Consul (Sir Frederic O'Connor) and his escort were made prisoners by the local gendarmerie, who were commanded by Swedish officers with strong German sympathies. Every effort was made by persuasion and threats to induce the Indian officers and men of the escort to throw off their allegiance to the British and to join the *jehad* which was preached against us. These attempts were entirely unsuccessful. The loyalty of the escort was unwavering throughout in spite of both temptations and hardships. They with the Consul were marched to the fort at Borazjun and remained in captivity till August, 1916, when they were at length set at liberty and finally made their way back to India. Sir Frederick O'Connor afterwards wrote an emphatic testimony to their loyalty and steadiness, which reflected the greatest credit on the regiment. "I can vouch for the fact," he said, "that these men did good and loyal service under very trying conditions, much more trying in many ways than are the conditions of actual warfare, and that their conduct and bearing throughout were unexceptionable."

In the spring of 1918 trouble among the frontier tribes afforded the officers and men of the depot at Multan an experience outside the ordinary course of depot work and one which was of value in the training of the young recruits. Following on several instances of ill behaviour by the Marris, at the end of February the neighbouring Khetrani tribe beleaguered the district police officer with a small escort in the post of Bar Khan, severed communications with district headquarters at Dera Ghazi Khan, and on 2nd March sacked and burnt Fort Munro. There was only a handful of police to protect Dera Ghazi, and on receiving the news of these outrages the Commissioner telegraphed to Multan for troops. Major Willoughby, who happened to get early news of the facts, at once proposed to Brigadier-General Miles, commanding at Multan, to mobilise three hundred men from the depot and march forthwith to Dera Ghazi Khan. Permission to mobilise was given and orders to march were issued six hours later. This brief interval was not much in which to fit out three hundred men and horses. There was no mobilisation gear of any kind, very little saddlery or equipment, most of the men were quite raw recruits, and many of them had not even got uniform. Fortunately Major Willoughby had had two days' warning of possible trouble and had promptly ordered that all horses were to be shod (they usually worked unshod at Multan). By dint of very hard work the farriers succeeded in getting all the horses well shod all round by the time they were wanted to march out. At 12.15 P.M. on 3rd March, six and a half hours after the order to mobilise had been given, one hundred and fifty-five men with three hundred horses left the parade ground at Multan for Ghazi Ghat, a truly surprising performance in all the circumstances of the case, and one which was only accomplished through the vigour and energy of Major Willoughby supported by all ranks, and especially by Risaldar Muhammad Ashruf Shah, Jemadars Amar Singh, Jahandad Khan, and Saidan Shah. Lieutenant A. T. Oates went in command of the mounted men, accompanied by Lieutenant J. G. Drummond, I.A.R.O. In order to save the horses as much as possible Major Willoughby took the remainder of the men by train and reached Ghazi Ghat at 8 P.M., the mounted party arriving at 9.15 after a march of forty-three miles.

Next morning, 4th March, the new unit, known as the 2/10th Lancers, marched for Dera Ghazi Khan, an advance party going on in front under Lieutenant H. J. Morel in *tum-tums*. At Dera Ghazi it was organised into two squadrons at field service strength—viz., one squadron of Hindus and one of Mussulmans. There was also a surplus of men over and above these two squadrons.

Six weeks or so were now spent for the most part in hard marching through the Khetrani and Marri country. Much reconnaissance work was done, and together with the 55th Coke's Rifles the regiment took part in several punitive expeditions against outlying places. The country was very rough and the marches, often twenty or thirty miles and on one day as much as fifty miles, were very trying for the horses. Moreover the weather during the first three weeks was most inclement, with much rain and wind and bitterly cold. Nevertheless when the force broke up and the 2/10th Lancers returned to Multan on 26th April the men were in much better health than when they started. The young soldiers in the regiment had shown great keenness during the expedition and put up with considerable discomforts without a murmur. Brigadier-General Miles sent a most complimentary message on the 2/10th Lancers leaving the force, congratulating Major Willoughby and the regiment on their performance, and acknowledging the good work which they had done. Major Willoughby was requested to send in two names of those who had done specially good work, and he accordingly submitted the names of Jemadar Amar Singh and Lance Dafadar Ganda Singh (D Squadron).

On the way back from the Marri expedition Major Willoughby received orders to raise forthwith a squadron to be sent to a new regiment (the 42nd) then about to be raised. With the approval of the regimental headquarters a composite squadron was formed, half Sikhs and half Muhammadans. Risaldar Muhammad Ali, Jemadar Nawab Khan and Jemadar Mal Singh (III) came back from Mesopotamia to be troop leaders and Jemadar Amar Singh joined it from the depot with the rank of Ressaidar. Every endeavour was made to turn out a thoroughly good squadron both in respect of personnel, horses and equipment. In due course the squadron (commanded temporarily by Lieutenant Drummond) proceeded to join the 42nd Cavalry at Baleli near Quetta. Subsequently when this regiment was broken up after seeing service in the Afghan War and in Persia, Ressaidar Amar Singh and a few of the other ranks returned to the 10th Lancers.

In August, 1918, Major V. C. P. Hodson relieved Major Willoughby in command. In the following year some extra work was thrown on the depot in assisting the civil authorities of Multan to ensure the preservation of the peace during a period of general unrest in the Punjab. In the same summer a composite squadron, composed of one troop from each squadron at the depot, formed part of a movable column which was sent across the Indus to protect the country on the right bank of the river from small bands of raiders, who seized the opportunity of the hostilities with Afghanistan to come down on marauding expeditions from the Suliman range. This force constructed a camp at Mangrotha and remained there until November when the squadron of the 10th rejoined the depot.

Nothing further need be recorded of the events at Multan until the return of the Tenth from active service. The regiment landed at Karachi on 10th November, 1920, and proceeded thence by rail to Multan where it arrived on 15th November.

. . . .

It has already been narrated in the record of the

9th Hodson's Horse how the close of the great war was very shortly followed by considerable changes in the strength and organisation of the Indian cavalry, including the amalgamation of regiments and the abolition of the silladar system. The decisions on these points had been promulgated before the return of the Tenth to India. Each of them required a good deal of preparation and the early months of 1921 were fully occupied with these matters.

The reduction of strength necessitated by the amalgamation of two regiments into one was a measure which involved the sad and invidious task of selecting those of all ranks who must be called upon to leave the service. Some of course were ready to retire, especially after the experiences of the past four years. Others who had been only recently enlisted could be sent back to civil life without serious regrets. But when all these had been taken into account it still remained necessary to dispense with the services of many whose departure meant a severance of old ties, painful alike to those who went and to those who remained.

The prospect too of the disappearance of the 10th Lancers as a separate entity could not but be a melancholy one to all who had been identified with the regiment through some of the best years of their lives. It was fully recognised—as has already been said of the Ninth—that both regiments of Hodson's Horse were more fortunate in this respect than some other units. They could find comfort in the fact that in being amalgamated once more they simply reverted to the form in which the corps was originally raised. But in spite of this the sadness and regret at the contemplated changes were in the case of each regiment alike natural and inevitable.

In the midst of these preparations the Tenth moved by rail on 21st April, 1921, under the command of Captain S. D. N. Cahusac, M.C., from Multan to Lahore. Under the new organisation now about to be introduced this was to be the permanent centre of the cavalry group to which Hodson's Horse was in future to belong. There the regiment arrived on the following day and there on 15th July its actual conversion from the silladar to the regular system took place, all clothing, equipment, &c., being taken over by the state and compensation and the value of their assamis being paid to the men.

A fortnight later, on 3rd September, as has already been described, the 9th Hodson's Horse arrived from Ambala and the two regiments were forthwith merged in one another. For the ensuing six months the Army Lists showed two commandants of the composite regiment, Lieut.-Colonel C. H. Rowcroft of the 9th and Colonel R. L. Ricketts of the 10th; but the latter officer was absent on the staff and Lieut.-Colonel Rowcroft alone exercised the command until March, 1922, when his tenure expired as well as that of Colonel Ricketts. They were succeeded by Colonel P. R. Chambers, D.S.O., late of the 10th Lancers, the first to hold the undivided command of the "4th Duke of Cambridge's Own Hodson's Horse." *

Note.—A Squadron and L/C (Sikhs) are now represented by A Squadron of the present regiment.

B Squadron (Punjabi Muhammadans) is contained in the present B Squadron.

R/C Squadron (Pathans) has disappeared.

D Squadron (Dogras) is contained in the present C Squadron.

* In 1927 the designation of the regiment was again changed and it became: "Hodson's Horse (4th Duke of Cambridge's Own Lancers)."

EPILOGUE.

THE records of the Ninth and Tenth are closed but those of Hodson's Horse continue unbroken. Just as sixty-five years earlier the crowd of horsemen drawn from the chaotic and warring elements of the Punjab was welded into a corporate body through the genius of its founder, and imbued with his spirit of loyal and devoted service, and just as the twin regiments which sprang from the parent stock were inspired by the same ideals, so in its reunion the new corps does not fall from the high traditions of its forerunners, and aims at no less than to serve in peace and war with equal devotion and equal loyalty as did the Sikhs and Mussulmans, the Dogras and Pathans who followed Hodson at Delhi and mourned him at Lucknow.

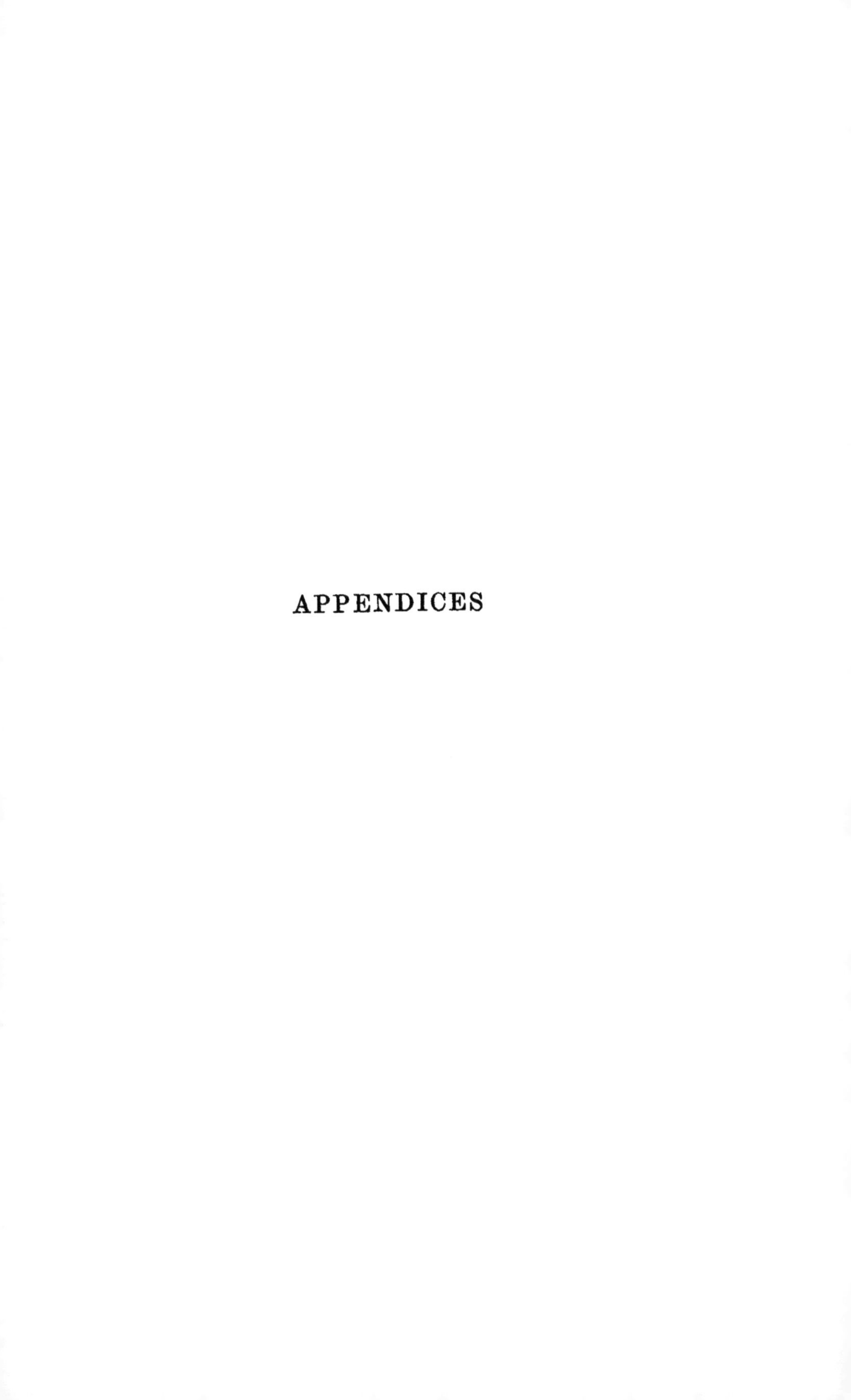

APPENDICES

APPENDIX I.

HODSON.

A MEMOIR AND AN APPRECIATION.

William Stephen Raikes Hodson was born at Maisemore Court near Gloucester on the 19th March, 1821. He was the third son of the Revd. George Hodson who was afterwards Archdeacon of Stafford and Canon of Lichfield. After being educated at home until the end of his sixteenth year Hodson was sent to Rugby early in 1837. There Dr Arnold was his headmaster and Tom Hughes one of his most notable contemporaries. Tall, slight, fair, well-proportioned and of active build, he was specially remarkable at school as a runner. He was also a fair scholar, and was of sufficiently outstanding character for him to be selected by Arnold to be head of a house where the discipline had become somewhat slack, a distinction which was justified by the results.

From Rugby Hodson went to Trinity College, Cambridge, in 1840 and took his B.A. degree in 1844. But all his tastes and inclinations were in the direction of an active career and, notwithstanding that he had passed the age at which cadets were usually appointed to the H.E.I.C.'s service, he managed by taking first of all a commission in the Guernsey militia to secure a nomination to the Bengal Native Infantry, and he left England in 1845, landing in Calcutta on 20th August.

He was posted to the 2nd Bengal Grenadiers and joined them just in time to take part in the Sutlej Campaign. The following year he was transferred to the 1st European Bengal Fusiliers, with which regiment however he did but little service, for he had already made the acquaintance of Colonel Henry Lawrence, with whom he formed a fast and enduring friendship. Through Lawrence's influence Hodson was appointed as second in command to the Guide Corps, then just raised by Henry Lumsden, and with the Guides he did remarkable service in the Jullundur district during the troublous times of the second Sikh War, receiving the special thanks of the Governor-General for his work. He was also present at the battle of Gujrat and was mentioned in Lord Gough's despatch.

After this Hodson served for three years as an assistant commissioner, first at Lahore and later at Amritsar, until in September, 1852, he was appointed to officiate for Lumsden as commandant of the Guides. Meanwhile on 5th January, 1852, he had married at Calcutta Susan, widow of John Mitford, Esq., of Exbury, Hants, and daughter of Captain C. Henry, R.N.

Hodson led the Guides with distinction in the Black Mountain expedition of 1852-3 and against the Afridis in the Bori Valley in 1854, and during this time he also had civil charge of the Yusafzai district. Then came a period of trouble and eclipse. It began with charges of misuse of his civil powers. A murderous attack was made upon Lieutenant Godby, adjutant of the Guides. Hodson believed that a certain sardar named Khadar Khan was concerned in the outrage and he had him arrested. Nothing however could be proved against the man and Major Herbert Edwardes, commissioner of Peshawar, took the matter up with some warmth, reporting Hodson's action to Lord Dalhousie. The latter agreed with Edwardes, removed Hodson from the Guides and ordered his return to his regiment.

Simultaneously charges of misappropriating the funds of the Guides and of other similar malpractices were made against him by one of his British officers. To crown his misfortunes his little daughter, born in 1853, died in the summer of 1854.

A court of enquiry on the Guides' accounts sat at Peshawar in December, 1854, and reported that they were "most unsatisfactory," but certainly the proceedings of the court were no less so. While evidence and complaints against Hodson were openly solicited from all quarters he was not even heard in his own defence. He vehemently protested his innocence and appealed to the Chief Commissioner (John Lawrence) for a court-martial. Lawrence thereupon deputed Major Reynell Taylor, an officer of the highest reputation, to go into the whole matter. Taylor after exhaustive study of the accounts wrote to Lawrence declaring that he was completely satisfied and urging that, as Hodson had been publicly condemned, the court of enquiry should reassemble and give equal publicity to his exoneration. This letter was forwarded to the Government of India but Lord Dalhousie would not agree to the course proposed. "Major Taylor," he wrote, "is entirely unprejudiced and competent to judge. He being satisfied with the correctness of the accounts I would give Lieutenant Hodson an acquittance and close this harassing and painful business."

This was all very well but it did not publicly reinstate Hodson. Taylor submitted a detailed report of his examination of the accounts and Hodson again pleaded that the court should reassemble to consider this report. But his efforts were unavailing, and he was still pressing for satisfaction in this respect when the mutiny broke out in May, 1857. What followed thereon is told in the first part of this book. After the siege of Delhi and the events which followed the capture of the city Hodson went on leave to see his wife at Ambala, but his stay there was cut short after only a fortnight. He left

Ambala on 14th November, 1857, and did not again return before his death in March, 1858.

Hodson has been virulently abused and extravagantly praised. Neither the one extreme nor the other can be justified. Yet it is not difficult to read his character from the story of his life and from the opinions of his contemporaries, many of which are on record. Eager, quick-witted, intelligent, ambitious, responsive to sympathy but impatient of control and intolerant of dulness or inefficiency, with immense confidence in himself and in the rightness of his own judgment, his faults which were not concealed even in his school-days were intensified during his early years in India. His rapid advancement made him more and more sure of himself, more than ever inclined to be guided by his own judgment. His success came too soon and was the cause of his undoing. He has been called cruel. This he certainly was not, but relentless in the punishment of wrong and unhesitating in its execution. The charges against him of dishonesty are based generally on the flimsiest grounds. None of them would have survived but for the fact that the trouble over the accounts of the Guides happened to be published just at the time when, for quite other reasons, he was removed from the command of that Corps. The accusations brought against him in this respect were, as shown above, completely refuted. He was careless in money matters and, in colloquial phrase, casual. The drudgery of account-keeping irked him and he evaded it as much as he could. But of more culpability than this there is no proof whatever.

Of his bravery there has never been any question. He knew no fear and was audacious even to rashness. Admired by all but the most prejudiced, beloved by many, it is incredible that the chosen and intimate friend of such men as Henry Lawrence and Robert Napier could have resembled even remotely the character depicted by his detractors. Truth prevails in the end over the most persistent malignity and history recognises Hodson as a public servant of great ability, high talents and inexhaustible zeal, as a brilliant leader who earned the devotion of his men, and as a very gallant soldier whose name stands high in the honourable roll of the Indian army.

APPENDIX II.

CHRONOLOGICAL LISTS.

(*a*) Honorary Colonels.

10th Bengal Lancers, 1878-1904.

Field-Marshal H.R.H. George William Frederick Charles, Duke of Cambridge, K.G., K.T., K.P., etc. etc., Commander-in-Chief.

9th Hodson's Horse.

Major-General F. W. P. Angelo 1912-

10th D.C.O. Lancers (Hodson's Horse).

Major-General M. Cowper, C.B., C.I.E. 1916-

(*b*) Commandants.

W. S. R. Hodson 1857-1858
H. D. Daly 1858-1859

1st Regt. (Ninth)		2nd Regt. (Tenth)	
W. C. Grant .	May-Sept. 1859	C. H. Palliser . .	1859-1880
J. P. Caulfield . .	1859-1863	O. Barnes . . .	1880-1887
C. H. Mecham . .	1863-1864	D. M. Strong . .	1887-1894
H. L. Campbell . .	1864-1881	E. J. F. Wood . .	1894-1901
T. Dayrell . . .	1881-1882	F. A. Blyth . . .	1901-1907
T. J. Watson . .	1882-1885	M. Cowper . . .	1907-1911
A. P. Palmer . .	1885-1888	W. L. Maxwell . .	1911-1914
D. H. Robertson .	1888-1893	W. E. Young . .	1914-1918
E. E. Money . .	1893-1894	R. L. Ricketts . .	1918-1922
G. L. Garstin . .	1894-1901		
F. W. P. Angelo .	1901-1907		
A. G. Peyton . .	1904-1911		
R. B. Low . . .	1911-1916		
G. A. H. Beatty .	1916-1920		
C. H. Rowcroft . .	1920-1922		

P. R. Chambers 1922-1924

(c) Risaldar-Majors.

1st Regt. (Ninth)	
Man Singh . . .	1866-1877
Manowar Khan . .	1877-1880
Harditt Singh . .	1880-1889
Hira Singh . . .	1889-1892
Mirza Ghulam Ahmad Khan . .	1892-1894
Muhammad Ali Beg	1894-1907
Ram Singh . . .	1907-1916
Mir Jafar Khan . .	1915-1916
Malik Khan Muhammad	1916-1920
Dost Muhammad .	1920-1921
Nur Ahmad Khan .	1921

2nd Regt. (Tenth)	
Mirza Ata-ullah Khan	1866-1885
Mirza Abdulla Khan	1885-1890
Khan Bahadur Khan	1890-1895
Sultan Muhammad Khan	1895-1903
Sher Baz Khan . .	1903-1905
Gopal Singh . .	1905-1910
Sardar Khan . .	1910-1914
Bijai Singh . . .	1914-1915
Rijhu	1915-1916
Uttam Singh . .	1916-1918
Mahan Singh . .	1918-1920
Sant Singh . . .	1920-1921
Labh Singh . .	1921-1927

APPENDIX III.

BIOGRAPHICAL NOTES ON COMMANDING OFFICERS AND SOME OTHERS.

ANGELO, Frederick William Pakenham (1859-). Major-General.

Born at Simla 15 Sept. 1859; eldest son of Major John Angelo, 15th Bengal Cav. 2nd Lieut. 63rd Manchester Regt. 30 Jan. 1878. Transferred to Bengal Staff Corps and joined 9th Bengal Cav. 5 May 1880. Sudan campaign, Suakin, 1885, as Adjt. of the Regt.; actions of Hashin and Tamai (Medal with clasp, and Khedive's Star). Chitral campaign 1895; relief of Chitral (Medal with clasp). China expedition 1900; D.A.A.G. of Cavalry Brigade (Medal with clasp—Despatches—Brevet Lt.-Col.). Comdt. 9th H.H. 20 Dec. 1901 till 1 Dec. 1907. Appointed to command Faizabad Brigade 1 Dec. 1907; transferred to Risalpur Cavalry Brigade 1 May 1908. Maj.-Gen. Feb. 1911. Retired May 1911. Appointed Colonel 9th H.H. 10 Dec. 1911.

BARNES, Osmond (1834-). Colonel.

Born in London 23 Dec. 1834; son of John Barnes of Chorley Wood House near Rickmansworth. Entered Bombay Army 4 Mar. 1855; Ensign 13th Bombay N.I. 29 Aug. 1855. Served in Indian Mutiny campaign 1857-9; operations in Central India; siege of Kotah (Medal with clasp). Appointed to the 10th Bengal Cavalry (Lancers) 28 Sept. 1864, and served with the regiment till his retirement. Abyssinia campaign 1868 (Medal). Herald at Lord Lytton's Delhi Durbar and proclaimed Queen Victoria "Empress of India" on 1 Jan. 1877. Second Afghan War 1878-9; relief of Kam Dakka (Medal). Zhob Valley expedition 1884. Commanded 10th Bengal Lancers from 29 Dec. 1880 to 29 Dec. 1887. Colonel 4 Mar. 1885. C.B. 1893.

BARROW, Seymour Duncan (1845-1886). Brevet Lieutenant-Colonel.

Born at Norfolk Island, Pacific Ocean, 15 Aug. 1845. Son of Samuel Barrow. Cornet, Bengal Army, 4 Jan. 1862. 19th Hussars, 30 July 1862. 13th Bengal Cav. 1869. Appointed to 10th Bengal Lancers, 22 Oct. 1869. Bde. Maj. of Cav. Malta Expedy. Force, 1878. Afghan War 1878-80, with 10th B.L. till Feb. 1880; Orderly Officer to Brig.-Gen. Palliser, July 1880; action of Patkao Shana, severely wounded. (Medal and clasp; brevet majority). Egypt Expeditionary Force, 1882; Bde. Maj. 2nd Cav. Bde.; battles of Tel-el-Kebir and Kassassin. (Medal and clasps, bronze star; brevet lieut.-colonel.) Commanded (offg.) 10th B.L. in Zhob Valley expedition, Nov. 1884. Died at sea on the way home 12 December, 1886.

BEATTY, Guy Archibald Hastings (1870-). Major-General.

Born 22 June, 1870; third son of Surg.-Gen. T. B. Beatty, H.E.I.C.S., of Derry, Ireland. Lieut. the Royal Irish Regt., 21 Dec. 1889. Appointed to 9th Bengal Lancers, 27 May, 1892. N.W. Frontier (Tochi expedn.) 1897, as Provost Marshal (Medal and clasp). China, 1900 (Despatches; medal). Great war; with 9th Hodson's Horse in France 13 Nov. 1914-1917; commanding the regiment June 1916-Dec. 1917; Givenchy, 1914; Somme, 1916, including Bazentin and Flers-Courcelette; Cambrai, 1917. (Despatches twice, 2 medals and star, D.S.O., and bar, C.M.G.). Commdg. Lucknow Cav. Brig., Egypt, 1918. G.O.C. British Forces in Transcaucasia Oct. 1918-Apr. 1919 (Despatches, medal and clasp). Afghan Campaign, 1919, comdg. 6th Inf. Brig. (Chora operations) (Despatches, medal and clasp; brev. colonel). G.O.C. Rawal Pindi divisional area, 1920. Mesopotamia, 1921, comdg. 75 Inf. Brig. (Despatches, medal and clasp, C.S.I.).

Colonel Comdt. 1st Ind. Cav. Brigade 1921-25.

Appointed Mil. Adviser in Chief, Indian States forces, 22 June, 1927.

C.B. 1 Jan. 1923. A.D.C. to the King 14 Feb. 1924. Coronation Durbar medal and golden star of Bokhara, 1st Class.

Major-General, November, 1926.

BLYTH, Frederick Augustus (1857-1919). Bt. Colonel.

Born at Castlemaine, Victoria, 20 Nov. 1857; son of Maj.-Gen. Frederick Samuel Blyth C.B. late comdg. 40th Foot. Sub.-Lieut. 40th Foot 11 Feb. 1875; transferred to 73rd Foot 18 Dec. 1875. Wing Officer 40th Bengal N.I. 20 Jan. 1881. Squadron Officer 10th Bengal Lancers 30 Jan. 1882 and served with the regiment till his retirement. Zhob Valley expedition 1884. Adjt. 1 Jan. 1885 to 10 Feb. 1886. N.W.F. 1897-8, Buner expedition, offg. comdt. (Despatches, medal with clasp). Commanded 10th Lancers 14 Oct. 1901 to 31 May, 1907. Bt. Col. 13 June, 1904. Died 26 Feb. 1919.

CAMPBELL, Herbert Lowe (1826-1881). Colonel.

Born 16 Aug. 1826; 3rd son of Rev. Charles Campbell of Weasenham, Norfolk. Entered the Bengal Army 21 Ap. 1844; Ensign 52nd B.N.I. 13 Feby. 1845. Second Sikh War 1848-9; siege of Multan; action of Surajkhund; battle of Gujrat (Medal with two clasps). Appointed Adjt. 9th Irregular Cav. 29 July 1856. Served at siege of Delhi 1857 (Medal with clasp). Appointed comdt. 9th Bengal Cavalry 20 Sept. 1864 and commanded the regt. till his death. Colonel 21 Apr. 1875. Died 14 Aug. 1881.

CAULFEILD, John Palmer (1820-1863). Brev. Major.

Born in India 7 Aug. 1820; eldest son of Lt.-Gen. James Caulfeild, C.B., Bengal Army, M.P. for Abingdon. Entered Bengal Army 12 Dec. 1837; Ensign 57th N.I. 1 Sept. 1838. Second Sikh war 1848-9; with Bdr. Wheeler's force in the Jullundur Doab (Medal). Indian Mutiny campaign, doing duty with 1st Bengal Fus.; action of Najafgarh; assault of Delhi (wounded); actions of Karnaul, Gangiri, Patiali and Mainpuri (Medal with clasp, Brev. Major). Appointed Comdt. 1st Regt. Hodson's Horse 8 Sept. 1859. Died at Benares 14 Apr. 1863.

CHAMBERS, Philip Roper (1881-). Colonel.

Born, 1 June, 1881; son of Colonel C. J. O. Chambers, Ind. Army. 2nd Lt. Ind. Army, 17 Jan. 1900. Appointed to 10th Bengal Lancers 17 Oct. 1901. Great War. G.S.O.3 England, 1914-15; G.S.O.2 Egypt, Gallipoli and Macedonia. 1915-16; France, Apr.-May, 1916; G.S.O.2 Mesopotamia and Persia, 1917-19 (Despatches 3 times; wounded; 2 medals and star; brevets of major and lieut.-colonel; D.S.O.) Afghanistan, 1919. Commandant, 4th D.C.O. Hodson's Horse, Mar. 1922-1 Feb. 1924. G.S.O.1. Dy. Dir. Staff Duties, India, 1924. Retired 25 Dec. 1925. Colonel 1 Jan. 1923.

COWPER, Maitland (1859-). Major-General.

Born at Ramsgate, Kent, 24 Dec. 1859; youngest son of Capt. C. C. G. Cowper, H.E.I.C.S. Entered the Army

1880 as 2nd Lieut. 100th Foot (Royal Canadians), subsequently 1st Bn. The Leinster Regt. Transferred to the 3rd Bengal Cav. 1884. Appointed to 10th Bengal (D.C.O.) Lancers 1885, and served with the regiment for 27 years. Commandant 1907-11. Asst. Inspecting Officer, Imperial Service Cav., Punjab and Kashmir, 1891-5. Chitral Relief Force 1895 (Medal with clasp). D.A.Q.M.G., Army H.Q., 1900-5; A.A.G. Rawal Pindi Div. 1905-7; G.S.O.1, Burma Div., 1912; A.A. and Q.M.G. Meerut Div. 1912-3; Comdg. Allahabad Bde. 1913-5; Poona Divisional Area 1915. Served in Mesopotamia 1915-6, as A.Q.M.G., I.E. Force D. (Despatches 3 times; C.B.; Order of the White Eagle, Servia, 2nd cl.) C.I.E. 1912; C.B. 1916. Maj.-Gen. 1915; retired 1919. Other services: Investigation of Horse Breeding in Punjab and N.W.P. 1900 (received thanks of Govt. of India); Officer in charge of C.-in-C.'s camps at the Coronation Durbars, Delhi, 1903 and 1911, and H.R.H. the Prince of Wales's camps 1905.

Appointed Colonel, 10th Lancers (H.H.), 22 Feby. 1916.

DALY, Sir Henry Dermot (1821-1895). General.

Born I. of Wight 25 Oct. 1821; younger son of Lt.-Col. Francis Dermot Daly, 4th Light Dgns. Entered Bombay Army 1 Sept. 1840; Ensign 1st Bombay European Regt. Second Sikh war 1848-9; siege and surrender of Multan; battle of Gujerat; pursuit of Sikh army (Medal with two clasps). Raised 1st Punjab Cav. 1849, and commanded it on service on N.W. frontier 1850-2 (Medal with clasp). Commanded Corps of Guides at siege of Delhi (twice wounded, and horse kld.). Appointed Commandant of Hodson's Horse 12 Mar. 1858, and commanded the Regt. and subsequently the Brigade of Hodson's Horse throughout the campaign in Oudh 1858-9 (Medal with clasps). Commanded Central Indian Horse 1861-71. Agent to Governor-General for Central India, 1871-81. General 1 Dec. 1888. C.B. 21 Jan. 1858; K.C.B. 29 May, 1875; C.I.E. 1880; G.C.B. 25 May, 1889. Died at Ryde, I.W., 21 July, 1895.

DAYRELL, Thomas (1838-1890). Lieut.-Colonel.

Born 6 May 1838; son of Rev. Thomas Dayrell, Rector of Long Marston, Yorks. Entered Bengal Army 6 Jan. 1857; Ensign 58th B.N.I. Served at the siege and assault of Delhi, and was severely wounded at the assault, 14 Sept. 1857 (Medal with clasp). Appointed Adjt. 1st Regt. H.H. 13 May, 1859. Comdt. 9th B.C. 14 Aug. 1881 till 1 Oct. 1882. Retired as Lt.-Col. 1 Oct. 1882. Died 12 Jan. 1890.

GARSTIN, George Lindsay (1851-). Colonel.

Born at Cherrapunji 10 June 1851; 3rd son of the Rev. Anthony Garstin, Chaplain Indian Establishment. Ensign 33rd (Duke of Wellington's) Regt. 3 Sept. 1870; transferred to 63rd Regt. Oct. 1871. Transferred to Bengal Staff Corps; to the 9th Bengal Cav. 3 Aug. 1877. Second Afghan war 1880 (Medal). Sudan campaign 1885; Suakin; actions of Hashin and Tamai (Medal with clasp, and Khedive's Star). Comdt. 9th Bengal Lcrs. 20 Dec. 1894 till 19 Dec. 1901. Chitral campaign 1895; relief of Chitral (Medal with clasp). Tirah expedition 1897-8: operations in the Bara Valley 7-14 Dec. 1897 (Clasps). Colonel 3rd Sept. 1900. Retired 6 Nov. 1908.

GOUGH, Sir Charles John Stanley (1832-1912). General.

Born at Chittagong 28 Jan. 1832; 2nd son of George Gough, of Rathronan, co. Tipperary, B.C.S. Entered Bengal Army 20 Mar. 1848; Cornet 8th Bengal Lt. Cav. Second Sikh war 1848-9; action of Ramnagar; passage of the Chenab; battles of Sadulapur, Chilianwala and Gujerat. (Medal with two clasps.) Indian mutiny 1857-8; with Corps of Guides at siege and capture of Delhi; with Hodson's Horse at Gangiri, Patiali, Mainpuri and Shamshabad (wounded), action of Mianganj, siege and capture of Lucknow. (Medal with two clasps; brev majority; V.C. for gallantry on four occasions, on the first of which, at Kharkaodah, on 15 Aug. 1857, he saved the life of his brother who was wounded.) Transferred, 1858, to 19th Hussars. Appointed to command 5th Bengal Cav. 1864. Bhutan expedn. 1864-5. (Medal with clasp.) Afghan War 1878-80; Ali Masjid; relief of Sherpur; commanded a brigade. (K.C.B., medal with two clasps.) Commanded Hyderabad Contingent, 1881. 1st Class district in Bengal 1886-90. General 1 Apr. 1894; K.C.B. 22 Feb. 1881; G.C.B. 25 May, 1895. Published in 1897 (in collaboration with Mr Arthur Innes) 'The Sikhs and the Sikh Wars.' Died 6 Sept. 1912.

GOUGH, Sir Hugh Henry (1833-1909). General.

Born at Calcutta 14 Nov. 1833; 3rd son of George Gough of Rathronan, co. Tipperary, B.C.S. Entered Bengal Army 4 Sept. 1853; Cornet 3rd Bengal Lt. Cav. Appointed adjt. of Hodson's Horse in July 1857 and served with the regiment throughout the Mutiny campaign until Mar. 1858, repeatedly distinguishing himself by his gallantry; twice wounded; two horses killed under him. (Medal with three clasps, several times mentioned in despatches, V.C.

for gallantry on several occasions, especially at Alambagh 12 Nov. 1857). Invalided to England after the capture of Lucknow. Afterwards appointed to 19th Hussars. Appointed to command 12th Bengal Cav. 1867; Abyssinia campaign, 1868 (Medal). Afghan War, 1878-80 (wounded); commanded cavalry of Kuram F.F.; commanded cavalry brigade at Sherpur and in Kabul to Kandahar march. (K.C.B., medal with four clasps and bronze star). Commanded Lahore div. 1887-92. K.C.B. 22 Feb. 1881. G.C.B. 26 May, 1896. General 1894. Keeper of the Crown Jewels 1898-1904. Lt.-Gov. of Channel Is. 1904. Author of 'Old Memories,' 1897. Died 12 May, 1909.

GRANT, Walter Colquhoun (1822-1861). Captain.

Born 1822; son of Colonel Colquhoun Grant, Chief of the Intelligence Dept. in the Peninsular war. Cornet 2nd Dgns. 26 Feb. 1841. Served with "Eastern Army" in the Crimea from Nov. 1854 as Capt. Comdt. of the Mounted Staff Corps, with local rank of Major (Medal with clasp for Sebastopol). Cornet 2nd D.G. 30 Nov. 1855; Capt. do., without purchase, 31 May, 1859. Comdt. 1st Regt. Hodson's Horse 13 May till 8 Sept. 1859. Died at Saugor 27 Aug. 1861.

HAVELOCK-ALLAN, Sir Henry Marshman, 1st Bart. (1830-1897). Lieut.-General.

Born at Chinsura 6 Aug. 1830; eldest son of Maj.-Gen. Sir Henry Havelock, K.C.B., V.C. Ensign 39th Foot 31 Mar. 1846. Served in Persian war 1857 (Medal). Indian Mutiny 1857-9; awarded V.C. for gallantry at Cawnpore 16 July, 1857; relief and defence of Lucknow (twice wounded); siege and capture of Lucknow; relief of Azimgarh; with Hodson's Horse in final operations against Oudh rebels 1858-9 (Medal with two clasps). Appointed 2nd-in-Comd. 3rd Regt. H.H. 9 Sept. 1858; exchanged to 1st Regt. 24 Dec. 1858; vacated his appointment Mar. 1859 and reverted to the 18th Royal Irish. Maori war 1863-4, as D.A.Q.M.G. (Medal). A.Q.M.G. in Canada 1867-9; A.A.G. in Ireland 1869-72. Lt.-Gen. 9 Dec. 1881. C.B. 1866; K.C.B. 21 June, 1887; G.C.B. 22 June, 1897. Created a baronet of the U.K. 22 Jan. 1858, in recognition of his father's services. Assumed the additional surname of Allan 17 Mar. 1880. M.P. for Sunderland 1874-81, for S.E. Durham 1885-92, and again in 1895. Shot by tribesmen near Ali Masjid 30 Dec. 1897, while visiting British troops on the Afghan frontier.

KEMMIS, Arthur William Marsh (1881-). Bt. Lieut.-Colonel.

Born at Buenos Aires 7 July, 1881; only son of William Kemmis, 84th (York and Lancaster) Regt. and later of Las Rosas, Argentine Republic. 2nd Lieut. The Royal Irish Regt. 20 Jan. 1900; appointed to 10th Bengal Lancers July, 1901. Burma Mil. Police 1904-9. D.A.A.G. 4th (Quetta) Div. 1916. Accompanied 10th Lancers to Mesopotamia Oct. 1916, and served with the regiment throughout its four years of active operations in that country, commanding it from Aug. to Dec. 1917, Aug. 1918 to Feby. 1919 and Apr. 1919 till its return to India in Nov. 1920 including the Arab war. (3 Medals, Despatches 3 times, D.S.O. and Brev. Lt.-Col.) Retired 20 Jan. 1922.

LOCKHART, Sir William Stephen Alexander (1841-1900). General.

Born 2 Sept. 1841; 3rd son of Rev. Lawrence Lockhart, D.D., of Milton Lockhart, co. Lanark. Entered Bengal Army 4 Oct. 1858. Served with 5th Fusiliers in Oudh Dec. 1858 and Jan. 1859. Ensign 44th Bengal N.I. 19 June, 1859; appointed to 10th Bengal Cav. 23 Mar. 1863. Transferred to 14th Bengal Lancers, 1864. Bhutan campaign 1864-6 (Medal with clasp). Abyssinia campaign 1867-8 (Medal). Hazara expedition 1868-9, as D.A.Q.M.G. (Clasp). Afghan war 1878-80 (C.B., Medal with clasp). D.Q.M.G., Intelligence Branch, 1880-5. Commd. Brigade in Burma war 1886-7 (K.C.B., Clasp). Asst. Military Sec. for Indian affairs at the Horse Guards 1889-90. Commanded Punjab Frontier Force 1890-5; commanded Isazai F.F. 1892; Waziristan expedition 1894-5 (K.C.S.I.) Commanded forces in Tirah expedition 1897 (G.C.B.). Appointed C.-in-C. in India 1898. C.B. 1880; C.S.I. 1887; K.C.B. 1 July, 1887; K.C.S.I. 25 May, 1895; G.C.B. 20 May, 1898. General 1896. Died at Calcutta 18 Mar. 1900.

LOW, Robert Balmain (1864-1927). Lieut.-Colonel.

Born 7 Oct. 1864; son of Gen. Sir Robert Cunliffe Low, G.C.B. Lieut. Roy. Irish Rifles, 7 Feb. 1885. Appointed to the 9th Bengal Lancers 10 Dec. 1888. N.E. Frontier of India (Lushai), 1889 (Medal and clasp.) N.W. Frontier (Hazara), 1891 (Clasp). Chitral relief force, 1895, A.D.C. to G.O. Comdg. (Despatches; clasp; D.S.O.). N.W. Frontier, 1897-8 (Tirah; operations in the Bara Valley) (2 clasps). China, 1900; Prov. Marshal and H.Q. Camp Comdt. (Despatches, medal with clasp; brev. major). Comdt. 9th Hodson's Horse, 3 Oct. 1911 to 3 Oct. 1916;

Great War, comdg. the regiment in France and Flanders, Nov. 1914-Apr. 1915 (Givenchy, 1914), invalided. (2 Medals and star). Retired 31 May 1919.
Lieut.-Colonel 3 Oct. 1911. D.S.O. 21 Jan. 1896.
Died at Camberley, 20 Apr. 1927.

McDowell, Charles Theophilus Metcalfe (1829-1858). Lieutenant.

Born at Calcutta 29 Oct. 1829; son of Dr James McDowell, Senior Member of Bengal Medical Board. Entered Bengal Army 24 Feb. 1846; Ensign 2nd Bengal European Regt. 28 Dec. 1846. Second Sikh War 1848-9; action of Ramnagar; passage of the Chenab; battles of Chilianwala and Gujrat (Medal with 2 clasps). Mutiny Campaign; action of Badli-ki-Serai. Appointed to do duty with Hodson's Horse early in July 1857; Adjt. 16 July and afterwards second in command. Served with the regt. throughout the campaign until his death, repeatedly distinguishing himself by his gallantry. Mortally wounded at Shamshabad 27 Jany. 1858 and died at Fatehgarh the next day.

MacGregor, Sir Charles Metcalfe (1840-1887). Major-General.

Born at Agra 12 Aug. 1840; 2nd son of Major Robert Guthrie MacGregor, Bengal Artillery, a lineal descendant of Rob Roy. Entered Bengal Army 20 Oct. 1856; Ensign 57th Bengal N.I. 5 Jan. 1857. Served throughout Mutiny campaign 1857-9; siege and capture of Lucknow; operations in Oudh; three times wounded (Medal with clasp). Commanded a Sqdn. of Hodson's Horse in Aug. 1858; doing duty with 1st Regt. H.H. in 1859. Second China war 1860; with Fane's Horse; twice severely wounded (Medal with two clasps). Second-in-Comd. 10th Bengal Cav. 27 Apr. 1861 till Jan. 1864. Q.M.G.'s dept. 1864. Bhutan campaign 1864-5, as Bde. Major and D.A.Q.M.G.; twice severely wounded (Medal with clasp). Abyssinia campaign 1867-8 (Medal). Compiled the Gazetteer of Central Asia for the Indian Govt. 1868-73. Second Afghan war 1878-9; commanded a Bde. in Kabul to Kandahar march (K.C.B., medal and star). Commanded the Marri expedition. Q.M.G. of India 1880; G.O.C. Punjab Frontier Force 1885. C.S.I. 31 Dec. 1875; C.I.E. 1 Jan. 1878; K.C.B. 22 Feb. 1881. Maj.-Gen. 23 Jan. 1887. Compiled the History of the Second Afghan war, and wrote 'The Defence of India,' 'Our Native Cavalry' and 'Mountain Warfare.' Died at Cairo 5 Feb. 1887.

MAXWELL, William Lockhart (1862-1914). Lieut.-Colonel.

Born 20 Nov. 1862; eldest son of Surg. Major Thomas Maxwell, I.M.S. Lieut. R. Munster Fusiliers 25 Aug. 1883. Burma War 1886-7; operations of 1st Bde.; Wuntho expedition (Medal with 2 clasps). Squadron Officer 7th Bengal Cav. 20 Aug. 1889. Appointed to 10th Bengal Lancers as Squadron Comdr. 17 Dec. 1894 and continued to serve with the regt. till his death. Operations N.W.F. 1897 (Clasp). Commandant 24 Dec. 1911. Died at Loralai, Baluchistan, 8 Mar. 1914.

MECHAM, Clifford Henry (1831-1865). Captain.

Born 24 Nov. 1831; son of Capt. George Mecham, 3rd D.G. Entered Madras Army 20 Jan. 1849; Ensign 27th Madras N.I. Appointed Adjt. 7th Oudh Irregular Inf. Feb. 1856. Served throughout defence of Residency at Lucknow June-Nov. 1857; with 1st Madras Fus. throughout the occupation and defence of the Alambagh 1857-8; with Hodson's Horse at capture of Lucknow; and in campaign in Oudh 1858 (severely wounded, and horse shot, at Nawabganj on 13 June, 1858). (Medal with two clasps.) Offg. Adjt. 2nd Regt. H.H. Aug. 1858; offg. 2nd-in-Comd. 3rd Regt. 12 Oct. 1858; 2nd-in-Comd. 2nd Regt. Mar. 1859; Comdt. 3rd Regt. May-June, 1859. Comdt. 9th Bengal Cav. Apr. 1863 till Sept. 1864; appointed to do duty with 10th Bengal Cav. July, 1865. Author of 'Sketches and Incidents of the Siege of Lucknow.' Died at Kalka 12 Sept. 1865.

MONEY, Ernle Edmund (1849-1894). Lieut.-Colonel.

Born at Ode, Norfolk, 23 Jan. 1849; son of Rev. William Money. Ensign 12th Foot 18 Dec. 1867; transferred to Bengal Staff Corps 5 Apr. 1871, and appointed to 11th Bengal Lancers; Second Afghan war 1878-9; engagements at Maidanak and Pesh Bolak (Medal). Appointed D.A.Q.M.G. May, 1884; Offg. A.Q.M.G. at Army H.Q. Mar. 1888. Served in Black Mountain, Hazara, 1888, as D.A.Q.M.G. 1st Bde. (Medal with clasp.) Appointed 2nd-in-Comd. 9th Bengal Lancers Jan. 1889; Comdt. 1 Feb. 1893. Lt.-Col. 18 Dec. 1893. Shot dead by a Dafadar of the regiment at the Cavalry Camp of Instruction at Muridki 20 Dec. 1894.

PALLISER, Sir Charles Henry (1830-1895). Major-General.

Born at Devonport 20 Aug. 1830; son of Maj.-Gen. Henry Palliser, R.A. Ensign 63rd Bengal N.I. 11 June, 1847.

Severely wounded in action against the Sheoranis 14 Mar. 1853, whilst serving as adj. Sind Camel Corps on the Derajat frontier (Medal with clasp). 2nd in command 13th Irregular Cav. Indian mutiny 1857-9; 2nd in command of Benares Horse; advance from Allahabad to Cawnpore; first relief of Lucknow; defence of Residency; Alambagh under Outram Nov. 1857-Mar. 1858; capture of Lucknow; with Hodson's Horse during campaign in Oudh 1858. (Wounded three times; medal with two clasps; brev. majority on promotion to captain.) Appointed 2nd in command 1st H.H. 23 July 1858. Appointed to command 2nd H.H. 13 May, 1859, and held command till 28 Dec. 1880. Abyssinia campaign 1868 (Medal). Afghan War 1878-80; commanded cavalry brigade, southern Afghanistan F.F.; march on Khandahar; battle of Ahmad Khel, action of Urzu (Medal with clasps, K.C.B.). Commanded Sialkot brigade 1881. Retired 1882. C.B. 20 May, 1871; K.C.B. 22 Feb. 1881; G.C.B. 26 May, 1894. Died 22 Nov. 1895.

PALMER, Sir Arthur Power (1840-1904). General.

Born at Karnal 25 June, 1840; eldest son of Capt. Nicholas Power Palmer, 54th Bengal N.I. Entered Bengal Army 20 Feb. 1857; Ensign 5th European Regt. Indian Mutiny 1857-9; joined Hodson's Horse June, 1858; actions of Nawabganj and Barabanki and minor affairs in the Oudh campaign with 2nd Regt. Hodson's Horse (Medal). Adjt. 10th Bengal Cav. Aug. 1862 to Mar. 1869. Abyssinia 1868 (Medal). Transferred to 9th Bengal Cav. 1869 and continued to serve with that regt. till 1888. Commandant 1885-8 including expedition to Suakin 1885 (Medal with clasp and Khedive's star; despatches; C.B.) Extra-regtl. services: Dafla expedition 1874-5. Dutch war in Achin 1876-7 (Dutch cross with two clasps). Afghan War 1878-80; Peiwar Kotal; Khost Valley; Q.M.G. with Kuram F.F. (Medal with clasp). A.A.G. Bengal, 1880-5. Commanded force in N. Chin Hills, Burma, 1892-3. Tirah Expedition 1897-8 (Medal with 2 clasps). Commanded Punjab Frontier Force 1898-1900. C. in C. in India Mar. 1900 to Dec. 1902. K.C.B. 8 May, 1894; G.C.I.E. 9 Nov. 1901; G.C.B. 26 June, 1903. General 27 June, 1899. Died in London 28 Feb. 1904.

PENNINGTON, Arthur Watson (1867-1927). Brig.-General.

Born 20 Mar. 1867; son of Lt.-Gen. Sir Charles Richard Pennington, K.C.B., Col. 14th Murray's Jat Lancers. 2nd Lieut. The Border Regt. 11 Feb. 1888. Transferred to 9th Bengal Cav. Relief of Chitral 1895 (Medal with clasp). Tirah campaign 1897-8; operations against the Khani Khel Chamkannies (Medal with 2 clasps). D.A.A.G. of Indian Contingent at the coronation of King Edward; and in 1907 in charge of the King's Indian Orderly Officers. M.V.O. Served in the Great War with Indian Troops, first in Egypt, then in command of the 9th H.H. in France and Belgium from April 1915 to June 1916. Subsequently with Patiala Lancers in Egypt and in Mesopotamia from July 1917 till the end of the war. (Despatches five times; two medals and star). Retired in 1922 with hony. rank of Brig.-General. Died at Camberley, 8 May, 1927.

PEYTON, Algernon George (1859-). Colonel.

Born 3rd Oct. 1859; 2nd son of Captain L. W. Peyton, R.N. First commission, 6th Foot, 11 Aug. 1880; 2nd Lt. 70th Foot (afterwards 2nd batt. E. Surrey Regt.) 11 Sept. 1880. Transferred to the Indian Army and appointed to the 9th Bengal Cav. 25 Apr. 1884. Suakin 1885 (Medal and two clasps; bronze star). Chitral campaign 1895 (Medal and clasp). N.W. Frontier 1897 (Clasp). D.A.A.G. Peshawar, 1900-04; Brig. Major, Ambala Cav. Brigade, 1905. Commandant 9th Hodson's Horse, 1 Dec. 1907-3 Oct. 1911. Retired 19 Aug. 1912. Colonel 6 April, 1910.
Great War. Commanded and trained 2/7th Worcestershire Regt. in England and France. Commanded reinforcement depôts at Étaples, Marseilles and Rouen, 1915-18. (Despatches twice; 2 medals).

RICKETTS, Robert Lumsden (1872-). Brig.-General.

Born at Allahabad 9 July, 1872; 4th son of George Henry Mildmay Ricketts, C.B., I.C.S. (who raised a large contingent of Ludhiana Sikhs for Hodson). 2nd Lieut., Unattached List, 3 Sept. 1892; attached 7th D.G. and 2nd D.G.; joined 10th Bengal Lancers 24 Dec. 1893. Relief of Chitral 1895 (Medal with clasp); N.W.F. 1897 (Clasp). Guardian to H.H. Maharajah of Alwar 25 June, 1900. Bde. Major 1st Cav. Bde., Risalpur, 1 Sept. 1909; Comdt. Imperial Cadet Corps 26 Feb. 1912; Bde. Major 6th Cav. Bde., Mesopotamia, 27 Feb. 1915; battle of Barjisiyeh, march to Awaz (Despatches); D.A.A.G. Base, Basra, 20 July, 1915. G.S.O.2 Cadet Coll., Quetta, 17 Feb. 1916; G.S.O.1, Asst. Comdt., 17 Apr. 1916; G.S.O.1, 2nd (Rawal Pindi) Div., 17 Aug. 1916. Comdt. 10th Lancers in Mesopotamia, March to August, 1918; action of Khan Baghdadi (2 medals and star); Comdt. L. of C., Dunsterforce, Hamadan,

Persia, 14 Aug. 1918 (Despatches). Bt. Col. 1 Jan. 1919. Special duty at War Office Apr. 1919. Brig.-Gen. Gen. Staff, Army of Black Sea, Constantinople, 24 Mar. 1920; Comdt. 84th Inf. Bde. 14 Aug. till 3 Dec. 1921. Colonel 1922. Retired with rank of Brig.-Gen. 22 June, 1922.

ROBERTSON, Divie Henry (1842-1913). Colonel.

Born at Madras 5 Nov. 1842; son of Andrew Robertson, Madras C.S. Ensign, General List, 4 Nov. 1860. Bhutan expedition 1865-6 (Medal with clasp). Served with 44th B.N.I.; with 7th B.C. 22 Dec. 1868 till 1870. Sqdn. Officer 9th B.C. 8 Aug. 1873. Suakin 1885; action at Hashin—severely wounded (Medal with clasp, and Khedive's Star; Despatches; Brevet of Lt.-Col.) Comdt. 9th B.L. 10 Dec. 1888 till 1 Feb. 1893. Retired 1 Feb. 1893. Died 28 May, 1913.

ROWCROFT, Claude Harold (1872-). Colonel.

Born at Delhi 2 Mar. 1872; only son of Capt. Harry Crommelin Rowcroft, R.E. Commissioned 73rd Field Batty. R.A. 19 Feb. 1892; transferred to 31st Field Batty. Sept. 1892; to No. 3 Batty. Hyderabad Contingent Apr. 1896; to Poona Horse Nov. 1896. Appointed to the 9th Bengal Lancers Dec. 1898. Served with the regiment in the Great War from Nov. 1914 to the end; present at the battles of Festubert Dec. 1914; in reserve at Neuve Chapelle 1915; present at 1st, 2nd and 3rd battles of the Somme 1916; Hindenburg Retreat 1917; 1st and 2nd battles of Cambrai; Jordan Valley 1918; battle of Megiddo 1918; capture of Nazareth; capture of Damascus; march to and capture of Aleppo 1918. (Despatches several times; D.S.O.; 2 medals and star). Commanded the 9th H.H. from Dec. 1917 onwards. O.C. in Marash Feb.-Nov. 1919. Transferred as Comdt. to 26th (K.G.O.) Light Cav. Feb. 1920; retransferred as Comdt. to 9th H.H. 21 June, 1920, and held the command till 2 Mar. 1922. Offg. Comdt. Ambala Brigade Area Apr.-Oct. 1921. Retired 8 Oct. 1922. Colonel 19 Feb. 1922. D.S.O. 1919.

STRONG, Dawsonne Melancthon (1840-1903). Major-General.

Born at Brampton Abbots, 9 Nov. 1840; eldest son of the Rev. Clement Dawsonne Strong. Ensign 10 Dec. 1859. Did duty with 2nd Bn. Rifle Brigade 1860-2; in civil employ 1862-5. Appointed to 10th Bengal Cav. as 2nd Doing Duty Officer 14 Oct. 1865. Abyssinia campaign

1868 (Medal). Second Afghan war 1879-80; action of Kam Dakka (recommended for the V.C.); several actions with the Ghilzais in Dec. 1879 (Medal—mentioned in Despatches, Brevet Major 22 Sept. 1879; Brevet Lt.-Col. 2 Mar. 1881). A.Q.M.G. Gwalior District, and at A.H.Q. Simla, 1884-7. Commanded 10th Lancers 29 Dec. 1887 till 14 Oct. 1894. Commanded Multan Brigade Feb.-Sept. 1888. Maj.-Gen. 26 Aug. 1894. Retired 25 Apr. 1895. C.B. June 1893; Good Service Pension Dec. 1894. Died at Haslemere 11 May, 1903.

WATSON, Thomas James (1832-1905). Major-General.

Born at Cherrapunji 6 Apr. 1832; son of Lt.-Col. Thomas Colclough Watson, 53rd Bengal N.I. Entered Bengal Army 20 Feb. 1851; Ensign 57th B.N.I.; Lieut. 46th B.N.I. Indian Mutiny 1857-9; Gangiri, Patiali, Mainpuri, and all engagements of Bdr. Seaton's column between Delhi and Fatehgarh; capture of Lucknow; with Oudh F.F. under Sir Hope Grant; temporarily commanded a Sqdn. of Hodson's Horse (Medal with clasp). Appointed to 17th Bengal Cav. Bhutan expedition 1864 (Medal). N.W.F. 1877-8; Jowaki expedition (Clasp). Afghan war 1879-80 (Medal). Commanded 9th Bengal Cav. 18 Oct. 1882 till 24 Apr. 1885. Maj.-Gen. 23 Sept. 1892. Died at Oulton, Tasmania, Oct. 1905.

WOOD, Edward James Fandon (1850-). Lieut.-Colonel.

Born in Assam 4 Feb. 1850; 3rd son of Browne Wood, B.C.S. Sub-Lieut. 4th Hussars 1 Jan. 1873; transferred to Bengal Staff Corps Oct. 1875. Jowaki campaign 1877-8, with 17th Bengal Cav. (Medal with clasp). Afghan war 1878-80, first with 19th Bengal Lancers, including advance on and occupation of Kandahar and Kalat-i-Ghilzai; afterwards with 10th Bengal Lancers, Gandamak, Jagdalak (Medal, Despatches). Appointed to 10th B.L. as Sqdn. Officer 8 Apr. 1879; Adjt. 24 Oct. 1879 till 31 Dec. 1884. Zhob Valley 1884-5 (Despatches). Comdt. 10th B.L. 14 Oct. 1894 till 13 Oct. 1901. Lt.-Col. 1 Jan. 1899. Retired 1 Jan. 1905.

YOUNG, Henry George (1870-). Brig.-General.

Born 7 Mar. 1870; 3rd son of Rt. Hon. John Young. Lieut. Royal Fusiliers 8 Oct. 1890; transferred to Ind. Army, and appointed to 10th Bengal Lancers 25 Nov. 1892. Chitral Relief force 1895, as transport officer (Medal with clasp). N.W.F. operations 1897-8, as Adj. of the regt. (Clasp). Served in S. Africa with Imperial Yeomanry, 1902 (Medal with 2 clasps). With 10th Lancers in Mesopotamia 1916-17; commanded composite Cav. Regt. with III. Corps; operations round Kut, crossing of Tigris, action of Imam Mahdi; tempy. comdt. 10th Lancers Apr. to Aug. 1917. Transferred to command 22nd (Sam Browne's) Cav. Aug. 1917; actions of Ramadi, Tekrit and Tuskurmatli. Brig.-Gen. comdg. 7th Cav. Bde. Mesopotamia 1919-21 (2 Medals, despatches 5 times, D.S.O., Croix-de-Guerre, C.I.E.). Retired, 1921, with hony. rank of Brig.-General.

YOUNG, Wilfred Edward (1868-). Lieut.-Colonel.

Born at Canterbury 12 July, 1868; youngest son of Major George Augustus Young. 2nd Lieut. 7th D.G. 21 Sept. 1889; joined 10th Bengal Lancers as Offg. Sqdn. Officer 24 May, 1892. N.W.F. 1897-8; Malakand F.F.; Utman Khel expedition; Buner expedition (Medal with clasp). Comdt. 10th Lancers 8 Mar. 1914 till 7 Mar. 1918; Mesopotamia; Tigris Defences Oct. 1916 (Despatches, 2 medals). Invalided July, 1917. On duty with Remount Dept. Apr. 1918 till Feb. 1919. Lt.-Col. 8 Mar. 1914. Retired Jan. 1921.

MAN SINGH, Sardar Bahadur (....-1892).

Son of Sardar Dava Singh of Ruriala in the Gujranwala district. Served with distinction as an officer of cavalry in the Sikh army, including a campaign against the Afghans and the principal battles of the 1st Sikh war against the English. Entered the Punjab mounted police, 1852. At the request of Mr Montgomery raised the first risala of horse for Lieut. Hodson in June, 1857, and served with great distinction in command of it throughout the Mutiny campaign, including the siege of Delhi, the capture of the king and princes of Delhi, the action of Shamshabad, capture of Lucknow, action of Nawabganj, where he displayed conspicuous gallantry in charging and capturing three enemy guns (severely wounded) and numerous other actions. (Medal with clasps; Despatches; Order of Merit 1st Class and two jaghirs). Appointed first Risaldar Major, 9th Bengal Cav. 9 Mar. 1866. Order of British India, 1st Cl.
Retired, 1877. Hony. Magistrate at Amritsar. Appointed manager of the Darbar Sahib at Amritsar (a post which had formerly been held for thirteen years by his elder brother Sardar Jodh Singh), and held that office till his death on Mar. 16, 1892. Was a C.I.E. and a Viceregal Darbari.

MIRZA ATA-ULLAH KHAN, Sardar Bahadur (1831-1902). Hony. Lieutenant-Colonel and Risaldar Major.

Born 1 June, 1831; eldest son of Rajah Faquir Ullah Khan, the last ruler of Rajauri state. Raised a risala for Lieut. Hodson, June, 1857, and served throughout the Mutiny campaign with Hodson's Horse (Medal with clasps; Order of Merit). Appointed first Risaldar Major, 10th Bengal Lancers, 9 Mar. 1866. Abyssinia 1868 (Medal). Order of British India, 2nd Cl., 1869; 1st Class, 1873. Afghan war, 1878-80 (Medal). Appointed British Agent in Afghanistan with honorary rank of Lieut.-Colonel 1885 and held that post till 1892. Chief Justice of Bahawalpur state 1895- . Died at Waziribad, 7 Mar. 1902.

MANOWAR KHAN, Sardar Bahadur (........). Risaldar Major.

Born . . . son of . . . Served throughout the 1st Afghan war, 1838-42, including the taking of Ghazni, 1839, and Sir Wm. Nott's advance in 1841-2, the recapture of Kandahar, Ghazni and Kabul. (Medal with clasps, and star.) Operations in Sindh under Sir Charles Napier, 1844-5; Sutlej campaign 1846 with 12th Irregular Cavalry; battles of Firozshahr, Aliwal, and Mudki (Medal and clasps). 2nd Sikh war, 1848-9; battle of Gujerat and pursuit of Afghans across the Indus (Medal with clasps.) On the outbreak of the Indian Mutiny Manowar Khan was serving as a risaldar in the 12th Irregular Cavalry and was with a detachment of that regiment at Gorakhpur, where he displayed marked loyalty when the greater part of the regiment mutinied. He received a certificate from the District Judge to the effect that the safety of the treasury and the lives of the Europeans at the station were due to his efforts. (Order of British India, 1st Class, and Order of Merit, 1st Class.) He later served with Havelock's force at the first relief of Lucknow, and (having been appointed to Hodson's Horse) took part in the final relief of Lucknow, the defence of the Alambagh by Sir J. Outram and the recapture of Lucknow in March, 1858, as well as in the subsequent operations in Oudh. (Medal and 3 clasps.) Appointed Risaldar Major, 9th Bengal Cav. 1877. Retired 1880.

MIRZA MUHAMMAD ABDULLA KHAN, Sardar Bahadur (1834-1893). Risaldar Major.

Born 1834, younger son of Rajah Faquir Ullah Khan of Rajauri. Joined as a dafadar the risala raised for Lieut. Hodson by his brother Sardar Ata-ullah Khan (*q.v.*) and served with it throughout the Mutiny campaign. (Medal with clasps.) Promoted Jemadar, 10th Bengal Lcrs., 17 May, 1865; Ressaidar, 8 Dec. 1878; Risaldar, 30 Jan. 1880; appointed Risaldar Major, 1 June, 1885. Served with the regt. in Abyssinia, 1860 (Medal), Afghan war, 1878-80 (Medal). Order of British India 2nd Cl. 1880, 1st Cl. 1883. Retired, Oct. 1890. Died at Waziribad, 26 Dec. 1893.

KHAN BAHADUR KHAN, Sardar Bahadur (1839-1903). Risaldar Major.

Born 25 March, 1839, son of Ali Mardan Khan. Joined the 4th Sikh Infantry, Aug. 1859; Kani Kurram expedition, N.W.F. Transferred to 10th Bengal Lancers, 25 Mar. 1864. Promoted Jemadar, 19 Apr. 1878; Wordi Major, 20 Aug. 1879; Risaldar, 1 June, 1885. Appointed Risaldar Major, 18 Oct. 1890. Abyssinia, 1868 (Medal). Afghan war, 1878-80 (Medal). Zhob Valley, 1884. A.D.C. to Lieut.-General, Punjab Command, 1895. Was one of the personal escort of H.M. Queen Victoria at her Diamond Jubilee, 1897 (Medal). Tirah Expedition 1897-8 (Despatches, medal with clasp). A.D.C. to the C. in C. in India, 1899. Order of British India 2nd Cl. 1894; 1st Cl. 1898. Retired, 1 Jany. 1901. Died at Jhelum, 26 July, 1903.

RAM SINGH, Sardar Bahadur (....-1917). Hony. Captain and Risaldar Major.

Son of Chet Singh. Joined the 9th Bengal Cav. 1 Nov. 1882. Promoted Jemadar, 8 Mar. 1891, Ressaidar, 1 Dec. 1892, Risaldar, 21 Dec. 1894. Appointed Risaldar Major 16 Oct. 1907. Suakin, 1885 (Medal and bronze star). Chitral Relief Force, 1895 (Medal and clasp). Tirah Expedy. Force, 1897-8 (Clasp). One of a representative detachment to Australia on declaration of Commonwealth, 1900. Proceeded to England with coronation contingent, 1911 (Coronation medal). Great War: France and Flanders, 1914-5. (Two medals and star.) Order of British India, 1st Cl. Retired with hony. rank of Captain, 22 Feb. 1916. Died 12 June 1917.

KASHI NAND, Rai Bahadur (1858-1923). Risaldar.

Born 1 Aug. 1858, son of Sheo Mal of Peshawar. Joined 10th Bengal Lcrs. 1 May, 1880. Promoted Jemadar 2 Apr. 1886, Ressaidar 24 Sept. 1890, Wordi Major 18 Oct. 1890, Risaldar 2 May, 1893. Inspector of Dir Levies, 17 July, 1898. Extra Assist. Commr. N.W.F. Province 7 Nov.

1902. For several years Personal Assist. to the Ch. Commr. (Sir Harold Deane). Accompanied Brit. Mission to Kabul 1910. Retired 1 Aug. 1913. Subsequently Hony. Magistrate and Municipal Commissr., Peshawar. Received the title of Rai Bahadur (1900), and was a provincial Darbari (1920). Died 11 May, 1923.

LABH SINGH, Sardar Bahadur (1879-). Hony. Captain and Risaldar Major.

Born 14 Jan. 1879; son of Ressaidar Buddhu, late 10th Bengal Lcrs. Joined 10th B. L. as Lce. Dafr. 3rd Dec. 1896. Promoted Jemadar, 16 Nov. 1902. Wordi Major 10 Mar. 1910, Ressaidar 1 July 1915, Risaldar 11 Apr. 1916. Appointed Risaldar Major 2 Oct. 1921. Hony. Lieut. 8 Nov. 1924; Hony. Capt. 15 Oct. 1927. N.W.F. operations 1897-8 (Medal and clasp). Great War: Mesopotamia, 1916-21. A.D.C. to Army Commander, Aug. 1919-May, 1921. (Despatches 3 times; 3 medals, I.D.S.M. for gallantry and devotion to duty.) Order of Brit. India, 2nd Cl. 4 June, 1924, 1st Cl. 23 Oct. 1926. Retired 16 Nov. 1927. A.D.C. to the Gov. of the Punjab, 19 Jan. 1928.

Note.—This officer on his retirement presented a challenge cup, a sword and three money prizes for annual competition in the regiment.

MIR JAFAR KHAN, Khan Sahib, Sardar Bahadur (....). Hony. Capt. and Risaldar Major.

Born . . .; son of Jhang Baz Khan, Yusufzai, of Peshawar. Joined the 9th Bengal Cav. 12 Feb. 1884. Promoted Jemadar 18 Oct. 1897, Ressaidar 20 June 1907, Risaldar 1 Mar. 1910, Risaldar Major 8 Oct. 1915. Hony. Capt. 1 July 1920. Suakin 1885 (wounded; medal, bronze star). Chitral relief force, 1895 (Medal and clasp). Tirah exped. force, 1897-8 (Clasp, Order of Merit for gallant conduct, 18 Oct. 1897, *v.* p. 148). Proceeded to England with Coronation contingent, 1902 (Coronation medal). Great War, France, 1915 (Bronze star and two medals; despatches twice). Retired, Jan. 1917. Contributed 30 rupees p.m. throughout the war to the funds for comforts for the troops. Hony. Recruiting officer during the war and received a sanad for his work, as well as letter of thanks from the C. in C. in India. Order of Brit India, 1st Cl., Member of Order of Brit. Empire, 1919, Hony. rank of Captain; received jaghir grant of Rs. 600 per ann. Title of Khan Sahib 1919. A.D.C. to C. in C. in India for 7 years and later appointed

Hony. A.D.C. for life. Acting President of Prov. Advisory Committee, Territorial Force. Honorary Magistrate, Peshawar.

LAURASIB KHAN (1883-). Risaldar.

Born 26 July 1883; son of Risaldar Nawab Khan, late of the 10th Bengal Lcrs. and Guides Cav., and grandson of Ris. Maj. Khan Bahadur Khan, 10th B. L. (see p. 369). Joined the 10th B. L. 28 Dec. 1900. Promoted Jemadar 29 Oct. 1914. Appointed Wordi Major 14 Jan. 1916. Ressaidar 9 Jul. 1918, Risaldar 1 Apr. 1921. Mesopotamia 1916-1920 (Despatches, Mil. Cross for conspicuous gallantry and devotion to duty on 1 Mar. 1920, see p. 306. 3 medals). Roy. Victorian medal. Orderly officer to H.M. the King, 1921.

NUR AHMAD KHAN (1881-). Hony. Lieut. and Risaldar Major.

Born 20 Dec. 1881; son of Khan Atta Muhammad Khan, Ghori Khel, of Peshawar. Joined the 9th Bengal Lcrs. 11 Dec. 1901. Promoted Jemadar 16 Nov. 1906, Wordi Major 1 Apr. 1914, Ressaidar 1 Feb. 1915, Risaldar 18 Dec. 1917, Risaldar Major 15 Nov. 1921; 2nd Lieut. 14 Jan. 1922, Lieut. 29 July 1924. Great War: France and Palestine 1914-20. Orderly Officer to Genl. Lord Allenby. (Various actions. Raid on St Helène trenches 1917. Ind. Ord. Merit 2nd Cl. p. 173. Cambrai 30 Nov. '17, I.O.M. 1st Cl. p. 182. Adv. in Palestine, 1918. Mil. Cross, p. 217. 3 medals and 1914 star). Appointed A.D.C. to C. in C. in India, 1 Apr. 1921. Quartermaster of 4th Hodson's Horse, 1924-27. Asst. Recruiting Officer, Rawal Pindi, 16 Nov. 1927.

APPENDIX IV.

HONOURS AND REWARDS FOR FIELD SERVICE.

THE INDIAN MUTINY.

Captain W. S. R. Hodson	Brevet Majority.
Captain C. J. S. Gough .	Victoria Cross and Brevet Majority.
Lieut. H. H. Gough . .	Victoria Cross.
Lieut. C. H. Palliser . .	Brevet Majority on promotion to Captain.

ORDER OF MERIT.

1st Class.

Risaldar Man Singh.
Risaldar Muhammad Raza Khan (and pension of 200 rs. per mensem in perpetuity).
Jemadar Changan Singh.*
Sowar Sardul Singh.

2nd Class.

Risaldar Harditt Singh.
Risaldar Fateh Ali Shah.
Jemadar Nihal Singh.
Sowar Warriam Singh.

3rd Class.

Risaldar Mirza Ata-ullah Khan.
Ressaidar Mirza Jiwan Beg.
Naib Risaldar Jahangir Khan.
Jemadar Deva Singh.
Jemadar Husain Ali Khan.
Jemadar Mirza Ahmed Beg.
Jemadar Jawala Singh.
Jemadar Madat Ali.

3rd Class (continued).

Jemadar Parkha Singh.
Kotfadar Ibrahim Khan.
Kot Dafadar Karak Singh.
Kot Dafadar Jawala Singh.
Kot Dafadar Safdar Khan.
Dafadar Mokam Singh.
Dafadar Gurmukh Singh.
Dafadar Sardar Singh.
Sowar Hazara Singh.
" Shemdeh Shah.
" Fateh Singh.
" Hur Buj.
" Partab Singh.
" Dhip Singh.
" Bhan Singh.
" Narayan Singh.
" Bhag Khan.
" Jawala Singh.
" Kazan Singh.
" Karimula.
" Nattab Singh.
" Bahadur Singh.
" Man Singh.
" Rattan Singh.

* This officer (before receiving a commission) was severely wounded at Chibberamau on 29th December, 1857, when he was one of the escort accompanying Hodson and McDowell on their ride to the C. in C.'s camp. He was afterwards promoted and was again wounded at Jewa in March, 1859.

ABYSSINIA, 1868. (10th Bengal Lancers).

Major C. H. Palliser	Brevet Lieut.-Colonel.
Risaldar Major Mirza Ata-ullah Khan	Order of British India.

AFGHANISTAN, 1878-80. (10th Bengal Lancers.)

	Colonel C. H. Palliser, C.B.	K.C.B. (for service in command of a brigade).
	Captain D. M. Strong	Brevet Major and Brevet Lieut.-Colonel.
	Captain S. D. Barrow	Brevet Major.
884	Sowar Bhagwan Singh	Order of Merit, 3rd Class.
	Risaldar Man Singh	Order of British India, 2nd Class.

SUAKIN, 1885. (9th Bengal Cavalry.)

	Colonel A. P. Palmer	C.B.
	Major D. H. Robertson	Brevet Lieut.-Colonel.
	Risaldar Hukm Singh	Order of Merit, 3rd Class.
1007	Lance Dafadar Indar Singh	do.
1453	Lance Dafadar Samandar Khan	do.
1199	Lance Dafadar Ajab Khan	do.
1217	Lance Dafadar Sayyid Ghulam	do.
1133	Lance Dafadar Puran Singh	do.
1137	Trumpeter Kesar Singh	do.
1357	Sowar Amin Khan	do.
1545	Sowar Abdulla Khan	do.
1531	Farrier Inayat Khan	do.

CHITRAL EXPEDITION, 1895.

Captain R. B. Low, 9th Bengal Lancers	D.S.O. (for service in staff employ).

FRONTIER OPERATIONS, 1897. (9th Bengal Lancers.)

	Kot Dafadar Mir Jafar Khan	Order of Merit, 3rd Class.
2076	Sowar Rahmat Khan	do.

CHINA EXPEDITION, 1900-01.

Major F. W. P. Angelo, 9th Bengal Lancers	Brevet Lieut.-Colonel (for service on staff).
Captain R. B. Low, D.S.O. do.	Brevet Major (for service on staff).

THE GREAT WAR.

9th Hodson's Horse.

Lieut.-Colonel G. A. H. Beatty . .	C.M.G., D.S.O. and bar, Legion of Honour.
Lieut.-Colonel C. H. Rowcroft . .	D.S.O.
Major H. L. Dyce	M.C.
Major J. C. Russell	D.S.O.
Major A. I. Fraser	D.S.O.
Major F. St J. Atkinson	D.S.O.
Captain E. de Burgh	Brevets of Major and Lieut.-Colonel, D.S.O.
Captain M. D. Vigors	D.S.O., M.C. and Croix de Guerre.
Captain G. de la P. Beresford (10th Lcrs.)	M.C., Order of the Nile.
Captain T. W. Corbett	Brevet of Major, M.C. and bar.
Captain L. C. T. Graham	M.C.
Captain C. F. L. Stevens (10th Lcrs.)	M.C.
Captain F. W. Messervy	Order of the Nile.
Captain S. Dutt, I.M.S.	M.C.
Captain F. K. Moody (13th Lcrs.) .	M.C.
Captain R. A. Carr-White (31st Lcrs.)	M.C.
Captain J. A. Ewart (I.A.R.O.) . .	M.C.
Captain M. Dudding (I.A.R.O.) . .	Order of the Nile.
Risaldar Major Ram Singh, Sardar Bahadur	Hony. rank of Captain on retirement.
Risaldar Mir Dad Khan	O.B.I. 2nd class.
Risaldar Major Malik Khan Muhammad	O.B.I. 2nd class and hony. rank of Lieutenant.
Risaldar Jai Ram	O.B.I. 2nd class.
Risaldar Muhammad Akram Khan .	O.B.I. 2nd class and Ind. Dist. Serv. Medal.
Risaldar Ram Singh	O.B.I. 2nd class.
Risaldar Major Dost Muhammad Khan	Ind. Order of Merit, 2nd class.
Risaldar Nur Ahmad Khan . . .	M.C., I.O.M. 2nd class and 1st class.
Ressaidar Harditt Singh (killed) .	I.D.S.M.
Risaldar Tek Singh	I.D.S.M.
Risaldar Sardar Khan	I.O.M. 2nd class, I.D.S.M., and grant of land.
Risaldar Hassan Shah	M.C. and Nadha order of the Hedjaz.
Ressaidar Bur Singh (16th Cav.) .	M.C. and grant of land.
Jemadar Habib Gul	I.D.S.M.
Jemadar Bhagwan Singh	I.O.M. 2nd class
Jemadar Nawab Ali Khan	I.O.M. 2nd class, I.D.S.M. and Croix de Guerre of Belgium.

Other Ranks.

Indian Order of Merit (2nd Class).

2889 Lc. Daf. Jit Singh.
2514 Lc. Daf. Ganda Ram.
Lc. Daf. Saidan Shah.
Sr. Sarfaraz Khan.
2793 Sr. Hayat Muhammad (10th Lcrs.)
3456 Sr. Abdulla Khan.
2967 Daf. Sarfaraz Khan (10th Lcrs.)
2764 Daf. Hakim Singh.
2782 Daf. Amin Muhammad (killed).
1388 S.A.S. Atta Muhammad Khan.
2804 Lc. Daf. Muhammad Azam.
3095 Daf. Fateh Khan (11th Lcrs.)

Indian Distinguished Service Medal.

3341 Sr. Shamsuddin Khan.
3150 Lc. Daf. Pritam Singh.
3090 Sr. Sajjan Singh.
2588 K. Daf. Abdul Sattar Khan.
2726 Lc. Daf. Surain Singh.*
2986 Sr. Mahain Singh (11th Lcrs.)
3189 Sr. Baz Singh.
3568 Sr. Ramzan Khan.
3086 Sr. Firoz Khan.
2965 Sr. Kapur Singh.
2839 Daf. Ram Singh.
1390 K. Daf. Gujar Singh.
154 Daf. Dalip Singh.
3331 Sr. Mir Badshah.
3253 Sr. Kabul Khan.
2792 Daf. Mehtab Singh.
2870 Sr. (ward orderly) Mir Hussain.
335 Sr. Wali Muhammad.
2939 Sr. Hashim Khan.
2785 Sq. Q.M. Daf. Sher Shah (for service with the Shereefian forces).
3259 Actg. L.D. Mula Singh.
2722 Sr. Kalu Ram.
3172 Sr. Sarwan Singh.
3327 Act. L. D. Durab Khan.
2148 Act. L. D. Karm Singh.
2775 Daf. Maj. Yakub Khan.
2511 Daf. Chur Singh.
3445 Sr. Nur Muhammad.
3035 Daf. Namda Khan.
2787 Daf. Karm Dad.
2482 Daf. Maluk Singh.
3009 Daf. Abdul Razaq.
2625 Far. Kapur Singh.
3486 Sr. Afzal Khan.
3077 Act. L.D. Sundar Singh.
2581 Far. Maj. Rabbal Ram.
2715 Lc. Daf. Harnam Singh.
3496 Act. L.D. Kirpa Singh.
3219 Sr. Sundar Singh.
3064 Sal. Shib Singh.
2868 Far. Suraj Din.
3133 Lc. Daf. Karm Dad.
2933 Tr. Maj. Sant Ram.
2533 Daf. Maj. Ghulam Muhammad.
2372 Daf. Harnam Singh.
2881 Far. Niamat Ullah.
3024 Lc. Daf. Indar Singh.
3167 Sr. Ram Singh.
2949 Daf. Surain Singh.

* *Was also awarded the Belgian Croix de Guerre.*

Indian Meritorious Service Medal.

2596 K. Daf. Imam Din.
2688 Daf. Malang Khan.
2701 Daf. Ghulam Rasul Khan.
2794 K. Daf. Chanan Singh.
2430 K. Daf. Mangal Singh.
600 Daf. (Head Sal.) Jetha Singh (22nd Cav.)
3095 Daf. Fateh Khan.
2489 Daf. Maj. Firoz Din.
2626 Daf. Santa Singh.
2569 Daf. Amar Singh.
3423 Daf. Ajab Khan.

There was also thirty-four grants of pecuniary rewards for war service (*jangi inam*).

10th D.C.O. Lancers (Hodson's Horse).

Lieut.-Colonel H. G. Young . . .	D.S.O.
Lieut.-Colonel R. L. Ricketts . .	Brevet Colonel.
Major A. D. Strong	D.S.O. and Brevet Lieut.-Colonel.*
Major P. R. Chambers	D.S.O. Brevets of Major and Lieut.-Colonel.*
Major A. W. M. Kemmis	D.S.O. and Brevet Lieut.-Colonel.
Major K. O. Goldie	O.B.E.*
Major D. O. W. Lamb	O.B.E.
Captain M. G. P. Willoughby . .	M.C. and Brevet Major.*
Captain S. D. N. Cahusac . . .	M.C.
Captain R. T. Lawrence	M.C.
Captain O. B. P. Russell	M.C.
Risaldar Major Uttam Singh . .	O.B.I. 2nd class, I.D.S.M and grant of land (2 squares).
Risaldar Major Mahan Singh . .	Hony. rank of Lieutenant.
Risaldar Major Sant Singh . . .	Grant of land (2 squares).
Risaldar Labh Singh	I.D.S.M. and grant of land (2 squares).
Risaldar Mal Singh	Grant of land (2 squares).
Risaldar Nur Khan	O.B.I. 2nd class and grant of land (2 squares).
Risaldar Bishn Singh	Grant of land (2 squares).
Risaldar Muhammad Ali Khan . .	Grant of land (2 squares).
Risaldar Hayat Khan	I.O.M. 2nd class.
Risaldar Laurasib Khan	M.C. and grant of land (2 squares).
Jemadar Saidan Shah	Grant of land (2 squares).
1243 Daf. (Hony. Jemadar) Sharf Din	Grant of land (1 square).

Other Ranks.

3242 Lc. Daf. Sundar Singh, I.D.S.M and I.O.M. 2nd class.
3109 Act. L.D. Mansa Ram, I.O.M. 2nd class.

* *Vid.* pp, 331, 332.

I.D.S.M.

2620 Daf. Sohbat Khan.
2681 Daf. Ghulam Baquir Khan, I.D.S.M. and bar and grant of land (1 square).
2779 Lc. Daf. Amin Chand.
3750 Sr. Mir Akbar Khan.
3994 Sr. Mazammal Din.
3056 Daf. Badan Singh.
3782 Act. L.D. Bashambar Ram.
2938 Act. L.D. Lochan Singh.
2754 Tr. Kaim Din.
3089 Sr. Raghunath Singh.
3287 Sr. Hazura Singh.
3049 Daf. Gulistan Khan, I.D.S.M., bar.
2642 Daf. Maj. Kirpal Singh.
4261 Sr. Yasin Khan.
4805 Sr. Sher Dil.
3648 Sr. Sarwan Singh.
3180 Sr. Bahal Singh.

2403 Daf. Sayyid Aftab Shah . . . . I.M.S. Medal.
2464 Daf. Ganda Singh . . . . I.M.S. Medal.

Grants of Land (1 square each).

3069 Lc. Daf. Kartar Singh.
2592 Daf. Saif Ali.
2911 Sr. Basant Singh.
2096 S. Qr.-Mr. D. Siri Chand.
2651 Sr. Chanan Singh.
2780 Daf. Mehar Muhammad.
2967 Daf. Sarfaraz Khan, I.O.M.
2724 Daf. Harnam Singh.
2880 Sr. Narayan Singh.
2715 Daf. Fazal Karim.
1741 Daf. Teja Singh.

There were also twenty grants of pecuniary rewards (*jangi inam*).

APPENDIX V.

CASUALTIES AMONGST OFFICERS IN THE FIELD.

Indian Mutiny—			
Delhi	1857	Naib Risaldar Jodh Singh	Wounded
,,	,,	Naib Risaldar Khadar Khan	,,
Rohtak	15 Aug. 1857	Lieut. H. H. Gough	,,
,,	,,	Jemadar Mirza Ahmad Beg	,,
Lucknow	Nov. 1857	Lieut. R. D. Craigie-Halkett	Killed
Gangiri	14 Dec. 1857	Risaldar Muhammad Taki Khan	,,
Shamshabad	27 Jan. 1858	Lieut. C. T. M. McDowell	,,
,,	,,	Captain W. S. R. Hodson	Wounded
,,	,,	Captain C. J. S. Gough	,,
,,	,,	Jemadar Jawala Singh	,,
Mianganj	23 Feb. 1858	Naib Risaldar Hukm Singh	Killed
,,	,,	Risaldar Muhammad Raza Khan	Wounded
Jalalabad	25 Feb. 1858	Lieut. H. H. Gough	,,
,,	,,	Risaldar Harditt Singh	,,
Lucknow	11 Mar. 1858	Major W. S. R. Hodson	Killed
,,	Mar. 1858	Naib Risaldar Deva Singh	Wounded
Nawabganj	13 June 1858	Lieut. C. H. Mecham	,,
,,	,,	Lieut. the Hon. J. H. Fraser	,,
,,	,,	Risaldar Man Singh	,,
,,	,,	Jemadar Husain Ali Khan	,,
Dariabad	18 Sept. 1858	Lieut. C. M. MacGregor	,,
Jabrauli	25 Oct. 1858	Lieut. R. C. W. Mitford	,,
Sultanpur	27 Oct. 1858	Lieut. C. H. Palliser	,,
,,	,,	Jemadar Man Singh (Gurkha)	,,
Basantpur	23 Dec. 1858	Ressaidar Ghulam Muhammad Khan	Killed
,,	,,	Jemadar Changan Singh	Wounded
Afghanistan (10th Bengal Lancers)	1879-80	Brev. Maj. H. C. Greenaway	All these officers died of disease in Afghanistan except Lieut. Pollock, who was invalided and died the next year.
	,,	Lieut. A. Burlton-Bennet	
	,,	Risaldar Randar Singh	
	,,	Risaldar Man Singh, Bahadur	
	,,	Jemadar Sher Narayan Singh	
	,,	Jemadar Henry Ling	
	,,	Risaldar Isar Singh	
	,,	Lieut. C. E. Pollock	
	,,	Capt. S. D. Barrow	Wounded (while employed on staff).
Suakin (9th Bengal Cavalry)	20 Mar. 1885	Ressaidar Shibdeo Singh	Killed
	,,	Major D. H. Robertson	Wounded
Mamani(TirahFrontier) (9th Bengal Lancers)	18 Oct. 1897	Jemadar Sarwar Khan	Killed
S. Africa (9th Bengal Lancers)	27 Apr. 1900	Capt. G. P. Brasier-Creagh	Killed (while on special service)

France (9th Hodson's Horse)	22 Nov. 1914	Lieut. T. W. Corbett	Wounded
	"	Ressaidar Sultan Muhammad Beg	Very severely wounded
	June 1915	Capt. F. St J. Atkinson	Slightly wounded
	"	Lieut. M. N. Morris, I.A.R.O.	"
	"	Jemadar Tek Singh	"
(10th Lancers)	5 Sept. 1915	Lieut. E. C. Braddyll	Killed (while serving with Flying Corps)
(9th Hodson's Horse)	12 Aug. 1916	Jemadar Samand Singh	Killed
	"	Lieut. P. V. Douetil	Severely wounded
	1916	Lieut. G. B. Reeves	Accidentally killed
	18 June 1917	Lieut. G. Wilson, I.A.R.O.	Wounded
	1917	Major J. C. Russell	Killed
	30 Nov. 1917 2 Dec.	Major A. I. Fraser	"
	"	Major F. St J. Atkinson	"
	"	Ressaidar Harditt Singh	Died of wounds
	"	Captain M. Dudding, I.A.R.O.	Wounded
	"	Lieut. J. R. K. Murphy	"
	"	Risaldar Harbant Singh	"
	"	Jemadar Mir Alam Khan	"
	"	Jemadar Sardar Khan	"
Palestine (9th Hodson's Horse)	23 May 1918	Capt T. W. Corbett	"
	"	Ressaidar Bur Singh	"
	"	Jemadar Indar Singh	"
	"	Jemadar Bhagwan Singh	"
	20 Sept. 1918	Lieut. W. S. Shepherd, I.A.R.O.	"
Mesopotamia (10th Lancers)	26 Jan. 1917	Major J. E. Moir	Died
	1918	Jemadar Mal Singh	Wounded
	1 Mar. 1920	Lieut. A. T. Oates	Killed
	5 Mar. 1920	Lieut. C. E. Waters	"
	7 Mar. 1920	Capt. J. C. Platts (17th Cav.)	"
	"	Risaldar Bishn Singh	Wounded
	"	Risaldar Rai Singh	"
	May 1920	Lieut. T. C. Crichton (5th Cav.)	"
	3 Sept. 1920	Capt. O. B. P. Russell, M.C.	Killed
	"	Risaldar Rai Singh	"
	"	Jemadar Bhagwan Singh	Died

APPENDIX VI.

A LIST OF BRITISH OFFICERS WHO HAVE SERVED WITH HODSON'S HORSE, OMITTING (EXCEPT IN SPECIAL CASES) THE NAMES OF THOSE TEMPORARILY ATTACHED FOR BRIEF PERIODS.

	Date.	Rank.	Unit.
Abercrombie, D. E. . .	1917-22	2nd Lt.—Lieut.	X.
Allan, Sir H. M. Havelock-	*vid.* H	avelock.	
Anderson, A. C. . . .	1865-66	Lieut.	X.
Anderson, R. B. . . .	1859	Lieut.	H. H.
Anderson, T.	1857-58	Asst.-Surg.	H. H.
Angelo, F. W. P. . . .	1880-1907	Lieut.—Col.	IX.
Armstrong, A. T. . .	1864-76	Capt.—Lt. Col.	X.
Atkins, R.	1860-65	Lieut.	H. H. & IX.
Atkinson, F. St J. . .	1904-17	Lieut.—Major	IX.
Babington, C. W. . .	1864-73	Lieut.—Capt.	IX.
Baker, G. A. A. . . .	1857-58	Lieut.	H. H.
Barker, W. G. . . .	1919-23	2nd Lt.—Lieut.	X.
Barnes, O.	1865-87	Lieut.—Col.	X.
Barrow, S. D. . . .	1870-86	Cornet—Lt. Col.	X.
Beatson, C. H. . . .	1878, 1887-1905	Surgn.—Lt. Col.	X.
Beatty, G. A. H. . . .	1892-1920	Lieut.—Lt. Col.	IX.
Bellers, E. V.	1882-83	Lieut.	IX.
Bendle, M. S.	1918- —	Lieut. ——	IX.
Beresford, G. de la P. .	1907- —	2nd Lt. ——	X. & H. H.
Bidie, A. G. C. . . .	1914- —	2nd Lt. ——	X. & H. H.
Blyth, F. A.	1881-1907	Lieut.—Col.	X.
Braddyll, E. C. . . .	1908-16	2nd Lt.—Lieut.	X.
Brasier-Creagh, G. P. .	1886-1900	Lieut.—Capt.	IX.
Brown, M. A.	1917-22	Lieut.—Capt.	IX.
Burlton-Bennet, A. . .	1875-79	Lieut.	X.
Cahusac, S. D. N. . .	1912- —	Lieut. ——	X. & H. H.
Campbell, H. L. . . .	1864-81	Major—Col.	IX.
Cardew, F. G. . . .	1887-1907	Lieut.—Major	X.
Carleton, H. A. . . .	1882-83	Lieut.	X.
Caulfield, J. P. . . .	1859-63	Capt.—Major	H. H. & IX.
Chambers, P. R. . . .	1901-24	Lieut.—Col.	X. & H. H.
Clarke, T. V.	1917-20	2nd Lt.—Lieut.	IX.
Clifford, R. C. . . .	1860-64	Lieut.	IX.
Clifford, R. M. . . .	1882-86	Major—Lt.-Col.	IX.
Coape-Smith [Coape-Ludlow] . .	1887-92	Lieut.	IX.
Coghlan, H.	1863-64	Lieut.	X.

	Date.	Rank.	Unit.
Coldstream, J. C. . .	1895-1904	2nd Lt.—Capt.	X.
Colvin, J. R. C. . . .	1882-90	Lieut.	IX.
Cook, L. A. C. . . .	1881-90	Capt.—Major	X.
Corbett, T. W. . . .	1908- —	Lieut. ——	IX. & H. H.
Cowper, M.	1885-1911	Lieut.—Col.	X.
Craigie-Halkett, R. D. .	1857	Lieut.	H. H.
Crocker, S. F. . . .	1888-1910	Lieut.—Lt. Col.	IX.
Crofts, A. M.	1880-86	Surgn.	X.
Cumming, R. H. R. . .	1918- —	2nd Lieut. ——	IX. & H. H.
Currie, G. V.	1869-84	Surg.—Brig. Surg.	X.
Daly, G. H.	1859-60	Assist. Surg.	H. H.
Daly, H. D.	1858-59	Major—Lt. Col.	H. H.
Dawson, H. L. . . .	1880-1901	Lieut.—Lt. Col.	IX.
Dayrell, T.	1859-82	Lieut.—Major	H. H. & IX.
de Burgh, E.	1904-25	2nd Lt.—Lt. Col.	IX. & H. H.
De Lisle, F. G. . . .	1880-82	Lieut.	X.
Drake, J. A.	1859,63-64	Lieut.	H. H. & X.
Drummond, F. H. R. .	1878-80	Lieut.	X.
Drummond, W. L. P. .	1862-64	Lieut.	IX.
Durrant-Steuart, J. N. .	1902-03	2nd Lieut.	X.
Dutt, S.	1917-18	Lieut. (I.M.S.)	IX.
Dyce, H. L.	1899-1920	2nd Lt.—Major	IX.
Ekin, W. J.	1918-20	2nd Lieut.—Lieut.	X.
Elliot, G. H.	1870-71	Ensign—Lieut.	X.
Emerson, G. A. . . .	1885-86	Surgn.	IX.
England, A.	1860-80	Lieut.—Major	H. H. & X.
Evans, W. N. . . .	1890-96	Lieut.	X.
Fagan, C. G. F. . . .	1880-85	Lieut.	X.
Fagan, H. H. F. . . .	1885-98	Lieut.—Major	X.
Fasken, W. H. . . .	1885-1909	Lieut.—Lt. Col.	X.
Finch, C.	1893-94	Lieut.	X.
Foster, W. M. A. . .	1906-21	Lieut.—Major	X.
Fraser, A. I.	1901-17	2nd Lt.—Major	IX.
Fraser, The Hon. J. H. .	1858-60	Lieut.	H. H.
Furse, G. A.	1864-65	Lieut.	IX.
Garstin, G. L. . . .	1877-1901	Lieut.—Col.	IX.
Gastrell, E. H. . . .	1917-27	Lieut.—Capt.	IX.
Gee, A. J.	1858	Asst. Surg.	H. H.
Godby, R. F.	1859-60	Lieut.	H. H.
Goldie, K. O.	1903-23	2nd Lt.—Lt. Col.	X. & H. H.
Gough, C. J. S. . . .	1858	Lieut.—Capt.	H. H.
Gough, H. H. . . .	1857-58	Lieut.	H. H.
Graham, L. C. T. . .	1911- —	2nd Lt. ——	IX. & H. H.
Grant, W. C.	1859	Capt.	H. H.
Greenaway, H. C. . .	1865-79	Lieut.—Capt.	X.
Halliday, G. T. . . .	1863-64	Lieut.	X.
Hallowes. J. H. . . .	1900-01	Lieut.	IX.
Hankin, G. C. . . .	1859-60	Capt.	H. H.
Havelock, Sir H. M., Bt.	1858-59	Major	H. H.
Hobson, A. W. F. . .	1920-23	2nd Lt.—Lt.	X. & H. H.
Hodson, V. C. P. . .	1904-20	Lieut.—Major	X.
Hodson, W. S. R. . .	1857-58	Lieut.—Major	H. H.
Hogg, G. E. M. . . .	1900-08	2nd Lt.—Capt.	X.
Hughes, W. T. . . .	1859	Major	H. H.
Hurst, G. S.	1918- —	Lieut. ——	IX.
Hutcheson, G. . . .	1884-85	Surg. Major	IX.
Hutchings, C. M. . . .	1918-22	Lieut.	IX.
Hutchison, H. S. . . .	1915-16	Capt. (I.M.S.)	X.

	Date.	Rank.	Unit.
Johnston, J. W.	1882-83	Surg. Major	IX.
Judd, E. L.	1918-22	Lieut.	IX.
Keighley, H. D. S. . .	1901-02	2nd Lieut.	IX.
Kemmis, A. W. M. . .	1901-21	2nd Lt.—Lt. Col.	X.
Lamb, D. O. W.	1905- —	2nd Lt. ——	X. & H. H.
Lane-Ryan, S. E. . .	1918-22	2nd Lieut.—Lieut.	IX.
Lawford, F. A. . . .	1858-9	Lieut.	H. H.
Lawrence, R. T. . . .	1911- —	2nd Lieut. ——	X. & H. H.
Lockhart, W. S. A. . .	1863-64	Lieut.	X.
Long, D. T.	1917-22	Capt.	IX.
Low, R. B.	1886-1916	Lieut.—Lt. Col.	IX.
Lukin, R. C. W.	1892-1918	Lieut.—Lt. Col.	IX.
Luttman-Johnson, H. W.	1913-23	2nd Lt.—Capt.	IX. & H. H.
Macartney, H. F. T. . .	1883-85	Lieut.	IX.
MacGregor, C. M. . .	1858-64	Lieut.	H. H. & X.
Mackenzie, H. M. . .	1873-89	Lieut.—Capt.	IX.
Macleod, J. N. . . .	1895-1901	Surg. Lt.—Capt.	X.
Mantel, R.	1866-80	Asst. Surg.—Surg. Maj.	IX.
Marsh, J. T.	1897-1905	Lieut.	IX.
Mathews, E. A. C. . .	1901-18	Lt.—Major (I.M.S.)	X.
Maxwell, W. L. . . .	1894-1914	Capt.—Lt. Col.	X.
McDowell, C. T. M. . .	1857-58	Lieut.	H. H.
McKellar, E.	1862-68	Asst. Surg.—Surg.	X.
Mecham, C. H. . . .	1858-64	Lieut.—Capt.	H. H. & IX.
Melville, C. W. F. . .	1904-13	Capt. (I.M.S.)	IX.
Messervy, F. W. . . .	1914- —	2nd Lt. ——	IX. & H. H.
Middleton, Lord . . .	*vid.* Willoughby		
Miller, A. T.	1911-16	2nd Lt.—Lieut.	X.
Mitford, R. C. W. . .	1858-62	Lieut.	H. H. & X.
Moir, J. E.	1898-1917	2nd Lt.—Major	X.
Money, E. E.	1889-94	Major—Lt. Col.	IX.
Moore, M. J.	1866-69	Lieut.	IX.
Morgan, H. S. . . .	1917-22	Lieut.	IX.
Muhd. Mubaraz Khan, Malik	1902- —	Lieut.—Capt.	IX.
Murphy, J. R. K. . .	1917-22	Lieut.—Capt.	IX.
Nicholetts, R. C. . . .	1865-66	Lieut.	IX.
Ninis, G. W.	1917-22	Lieut.—Capt.	IX.
Oates, A. T.	1917-20	2nd Lieut.—Lieut.	X.
Oldfield, H. T. . . .	1864-76	Capt.—Lt. Col.	IX.
Onslow, R. C. . . .	1879-99	Lieut.—Major	X.
Oswald, R. A. . . .	1918- —	Lieut. ——	X. & H. H.
Palliser, C. H. . . .	1858-80	Lieut.—Col.	H. H. & X.
Palmer, A. P. . . .	1858-88	Ensign—Col.	H. H. & IX. & X.
Palmer, E.	1886-1901	Surg. Maj.—Surg. Lt. Col.	IX.
Parsons, J. H. . . .	1882-84	Lieut.	X.
Pennington, A. W. . .	1890-1917	Lieut.—Lt. Col.	IX.
Peters, J.	1901-22	2nd Lt.—Major	X.
Peyton, A. G. . . .	1884-1911	Lieut.—Col.	IX.
Pierce, I. G. F. . . .	1918-22	Lieut.—Capt.	IX.
Pollock, C. E. . . .	1878-81	Lieut.	X.
Poole, C. A.	1859-66	Assist. Surg.	H. H. & IX.
Porter, R. R. M. . . .	1914-17	Lieut. (I.M.S.)	IX.
Probyn, F. H. . . .	1882-84	Lieut.	X.
Rampur, H.H. the Nawab of	1896- —	Major—Col.	IX. & H. H.
Ramsay, H. L. . . .	1876-77	Lieut.	IX.

	Date.	Rank.	Unit.
Reeves, G. B. . . .	1912-15	2nd Lt.—Lieut.	IX.
Ricketts, R. L. . . .	1894-1922	Lieut.—Col.	X.
Robertson, D. H. . .	1872-93	Capt.—Col.	IX.
Roche, C. S. de F. . .	1860-69	Lieut.—Capt.	IX.
Ross, G. C.	1884-85	Lieut. Col.	X.
Rowcroft, C. H. . . .	1898-1922	Lieut.—Col.	IX. & H. H.
Russell, J. C.	1902-17	Lieut.—Major.	IX.
Russell, O. B. P. . . .	1915-20	2nd Lt.—Capt.	X.
Sampson, D. T. H. . .	1860-80	Lieut.—Major	H. H. & IX.
Sarel, H. A.	1858	Major	H. H.
Scott-Moncrieff, W. E. .	1895-1904	Surg. Lt.—Capt.	IX.
Seymour, E. V. E. . .	1909-23	2nd Lt.—Capt.	IX. & H. H.
Shakespear, G. R. J. .	1888-94	Lt.-Col.—Col.	X.
Sheldon, H. G. . . .	1917-22	2nd Lt.—Capt.	IX.
Sheppard, G. S. . . .	1891-1900	Lieut.—Capt.	IX.
Slade-Baker, J. B. . .	1919-23	Capt.	IX.
Smith, C. A.	1896-1901	2nd Lt.—Lieut.	IX.
Smith, O. F.	1902-21	2nd Lt.—Major	IX.
Soltau, G. A.	1906-07	Lieut. (I.M.S.)	IX.
Stainforth, P. T. . . .	1897-1901	2nd Lt.—Lieut.	X.
Stephen, J. R. M. . .	1918-22	Lieut.	IX.
Stevens, C. F. L. . . .	1912- —	2nd Lt. ——	X. & H. H.
Stevenson, F.	1912-13	Capt. (I.M.S.)	X.
Stewart, W.	1883-99	Lieut.—Major	X.
Strong, A. D.	1896-1921	2nd Lt.—Lt. Col.	X.
Strong, D. M. . . .	1865-94	Lieut.—Col.	X.
Tha, R. R. H. O. . .	1916-18	Lieut.—Capt. (I.M.S.)	X.
Thomas, T. I. G. . . .	1908-14	Lieut.	IX.
Thornton, C. E. . .	1890-92	Lieut.	IX.
Travers, A. de la C. . .	1886-88	Lieut.	IX.
Trench, F.	1858	Lieut.	H. H.
Vigors, M. D.	1906- —	2nd Lt. ——	IX. & H. H.
Wake, E. St A. . . .	1892-1906	Lieut.—Major	X.
Walker, J. E. . . .	1917- —	Lieut. ——	IX. & H. H.
Ward, G.	1857	Lieut.	H. H.
Ward, W. J.	1859	Capt.	H. H.
Warde, S. G.	1858-59	Lieut.	H. H.
Waters, E. J. F. . . .	1919-20	2nd Lt.—Lieut.	X.
Watson, N.	1918-19	2nd Lt.—Lieut.	IX.
Watson, T. J. . . .	1882-85	Colonel	IX.
Webb, W. B. S. . . .	1920-22	2nd Lt.—Lieut.	IX.
Webber, E. J. . . .	1863-65	Lieut.	X.
Welchman, A. J. T. . .	1873-75	Capt.	X.
Wells, L. F.	1858	Lieut.	H. H.
Wethered, T. A. . . .	1859-61	Surg.—Surg. Major	H. H. & IX.
Whistler, A. R. . . .	1915-23	2nd Lt.—Capt.	X. & H. H.
Wikeley, J. M. . . .	1890-91	Lieut.	X.
Willis, J. L. N. . . .	1869-85	Lieut.—Major	IX.
Willoughby, M. G. P. .	1909-24	2nd Lt.—Major	X. & H. H.
Wise, D. W.	1857-58	Lieut.	H. H.
Wood, A.	1902-07	2nd Lt.—Lieut.	IX.
Wood, E. J. F. . . .	1878-1901	Lieut.—Lt. Col.	X.
Woodwright, W. H. E. .	1893-95	Surg. Capt.	X.
Wright, T.	1860-61	Asst. Surg.	IX.
Wylie, H.	1864-65	Lieut.	X.
Young, H. G.	1892-1917	Lieut.—Lt. Col.	X.
Young, W. G. P. . . .	1904-11	Lieut.	X.
Young, W. E. . . .	1892-1918	Lieut.—Lt. Col.	X.
Younghusband, L. N. .	1887-88	Lieut.	X.

GLOSSARY OF INDIAN TERMS.

Alkhálak; a long, close-fitting coat, buttoned down the front.

Assámi; an appointment, berth, post. In the silladar cavalry each recruit paid a sum of money which was supposed to represent the cost of his horse, and by this payment he secured his "assami" in the regiment. When he left the regiment the price of his "assami" was returned to him.

Bahádur; a champion, hero.

Bhisti; a water-carrier.

Bunnia (banya); a tradesman and money-lender. In the silladar cavalry there were bunnias permanently attached to every regiment, who arranged all supplies of food for the men and grain for the horses. They also very commonly made loans of money and supplied goods on credit to the men.

Chowdry (chaudhri); the headman of a craft; in a silladar cavalry regiment the chief bunnia was so termed.

Dafadár; a non-commissioned officer, the equivalent of sergeant.

Khálsa; lit. pure, genuine. The term was specially applied by the Sikhs to their government and army.

Kot dafadar; the senior dafadar of a troop; equivalent to troop sergeant-major.

Kotal; a mountain pass, col.

Kulla; a small close-fitting cap, generally pointed in shape, round which the lunghi or pugri is wound.

Kurta; a loose frock or blouse, reaching to the knees, with an opening in front down to the waist.

Jemadár; a junior rank of Indian officer, a subaltern.

Kucherry; a court-house.

Lascár; a camp-follower, tent-pitcher.

Lumbai; from lumba = long; a term to describe the long strides or bounds of a plunging horse.

Lumbardár; the responsible headman of a village.

Lunghi; originally a waistcloth, but commonly used as synonymous with pugri. A lunghi is tied more loosely and is bigger than a turban, which is a tightly tied, close-fitting head-dress.

Mulla; a teacher and doctor of the Korán.

Multáni mutti ; lit. "Multan earth" ; a term to describe an ochre dye.

Munshi ; a secretary, reader, writer.

Mutsaddi ; an accountant.

Nagarchi ; a kettle-drummer.

Naib ; deputy. Naib Risaldar = deputy risaldar.

Nakeeb (naquîb) ; a herald, a person who announces orders.

Nishánburdár ; a standard-bearer.

Nullah ; a ravine, water-course, a "wadi."

Pandy ; Pandé is a common clan name among Brahmans. Two prominent mutineers at the beginning of the troubles of 1857 were men of this name, and from this fact all the mutineers came to be called generically "Pandies."

Posteen ; a coat of sheepskin, the tanned surface outside and the hair inside.

Pug ; a small kerchief wound round the head by Sikhs (who never wear a cap). The lunghi is tied over the pug.

Ressaidár ; a troop commander of lower grade than a risaldar. The term has taken the place of "naib-risaldar."

Risála ; any formed body of horse. The word is used either of a single troop or of a whole regiment.

Risaldár ; the commander of a risala or troop ; in the Indian cavalry, the senior rank of Indian officer. The Risaldar-Major is the senior risaldar of the regiment.

Saloo ; thin muslin cloth.

Sanad ; properly a treaty, but the word is used to describe an official testimonial.

Sardár ; a leader, commander, chief.

Silladár ; a man who bears arms, "armiger." Silladar cavalry is that in which each man brings (or pays for) his own arms, equipment and horse.

Sowár ; a horseman, trooper.

Wordi Major ; the Indian assistant to the adjutant.

Zamindár ; a land-holder.

INDEX.

A

B

C

D

E

F

I

J

K

L

M

N

O

P

R

S

T

U, V

W

Y

Z

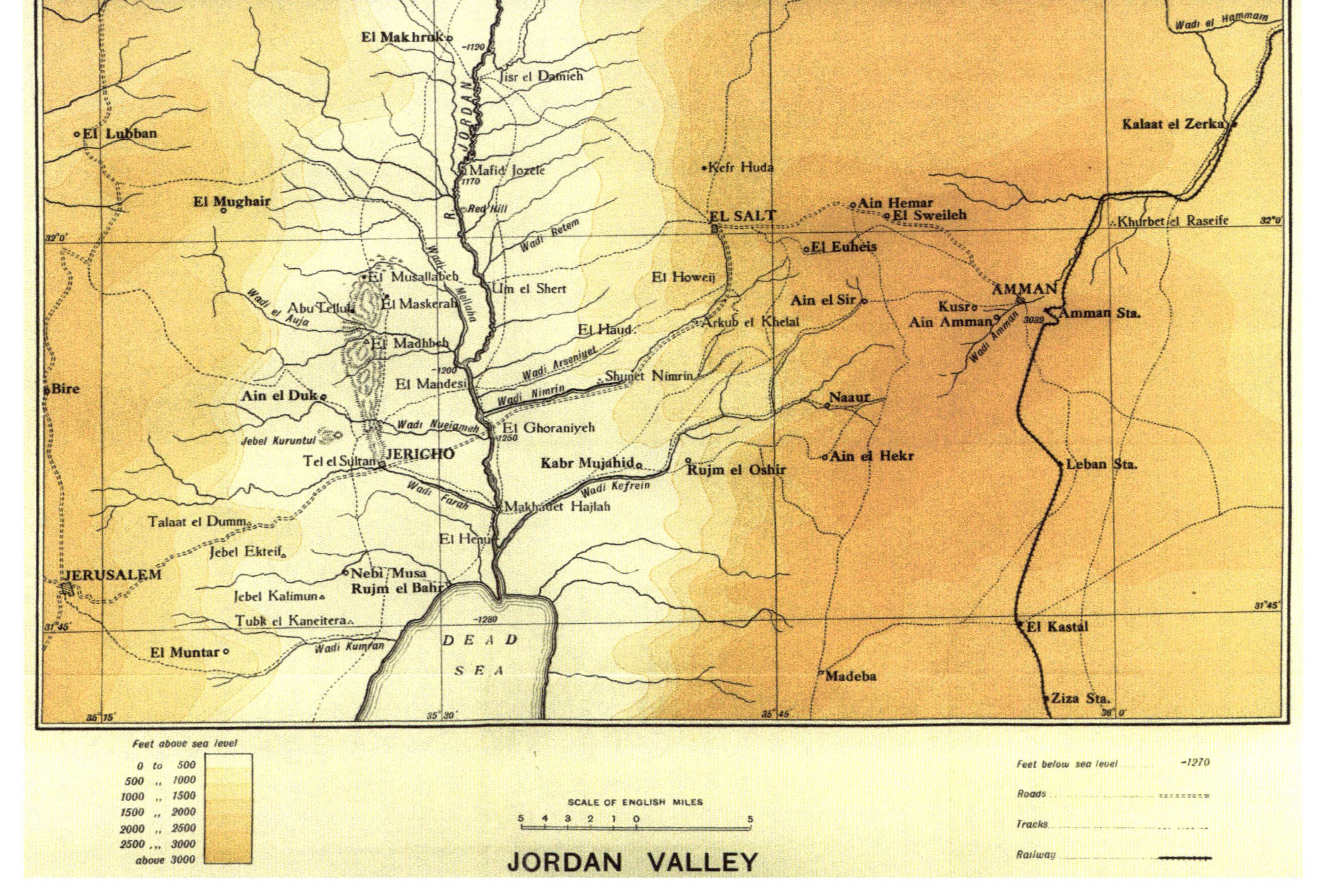

JORDAN VALLEY

NORTHERN PALESTINE & SOUTHERN SYRIA

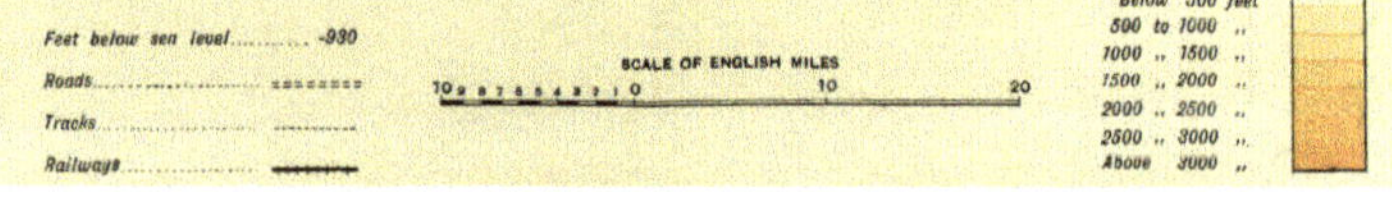

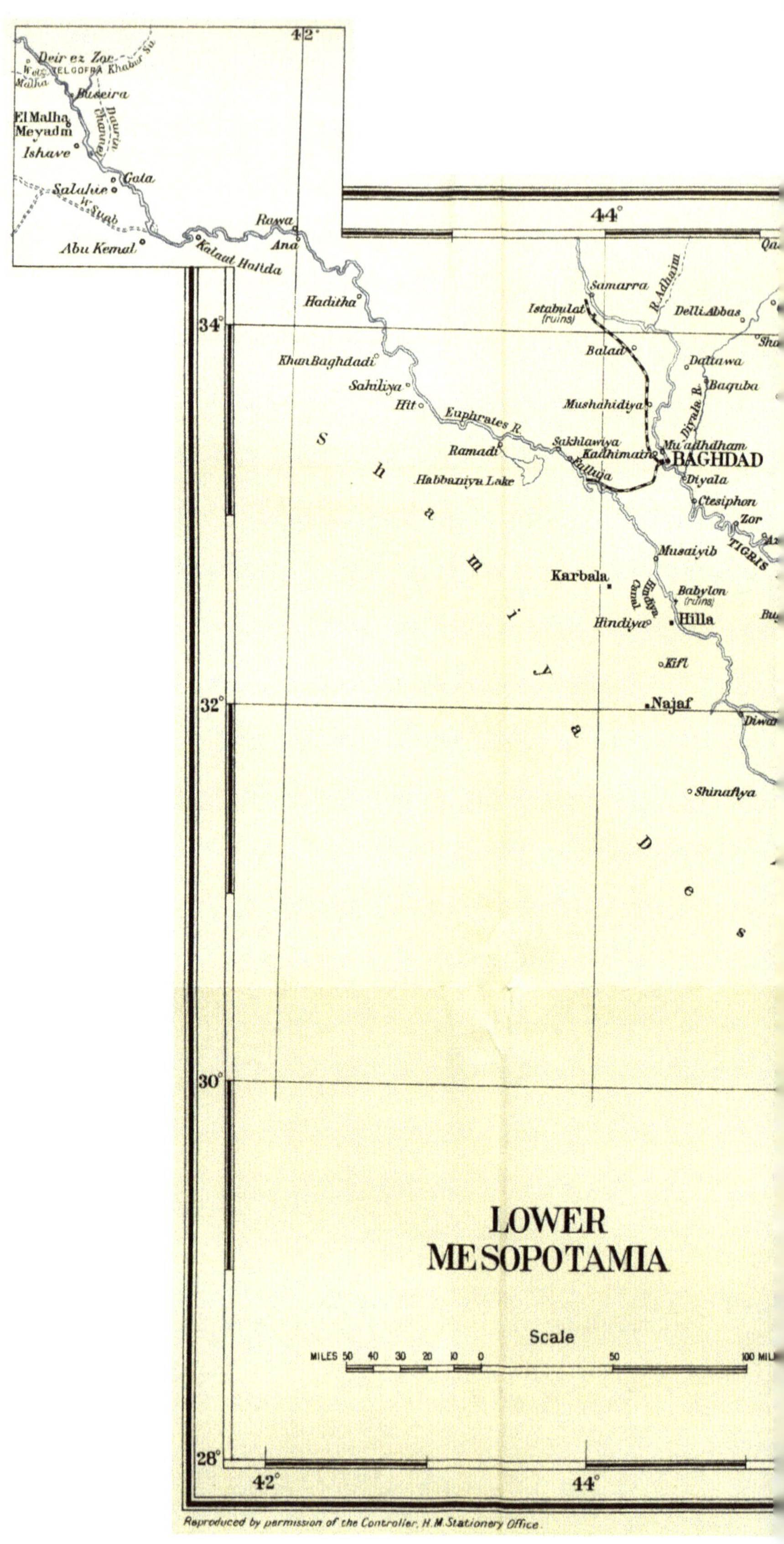

Reproduced by permission of the Controller, H.M. Stationery Office.

Printed at the War Office, 1928.

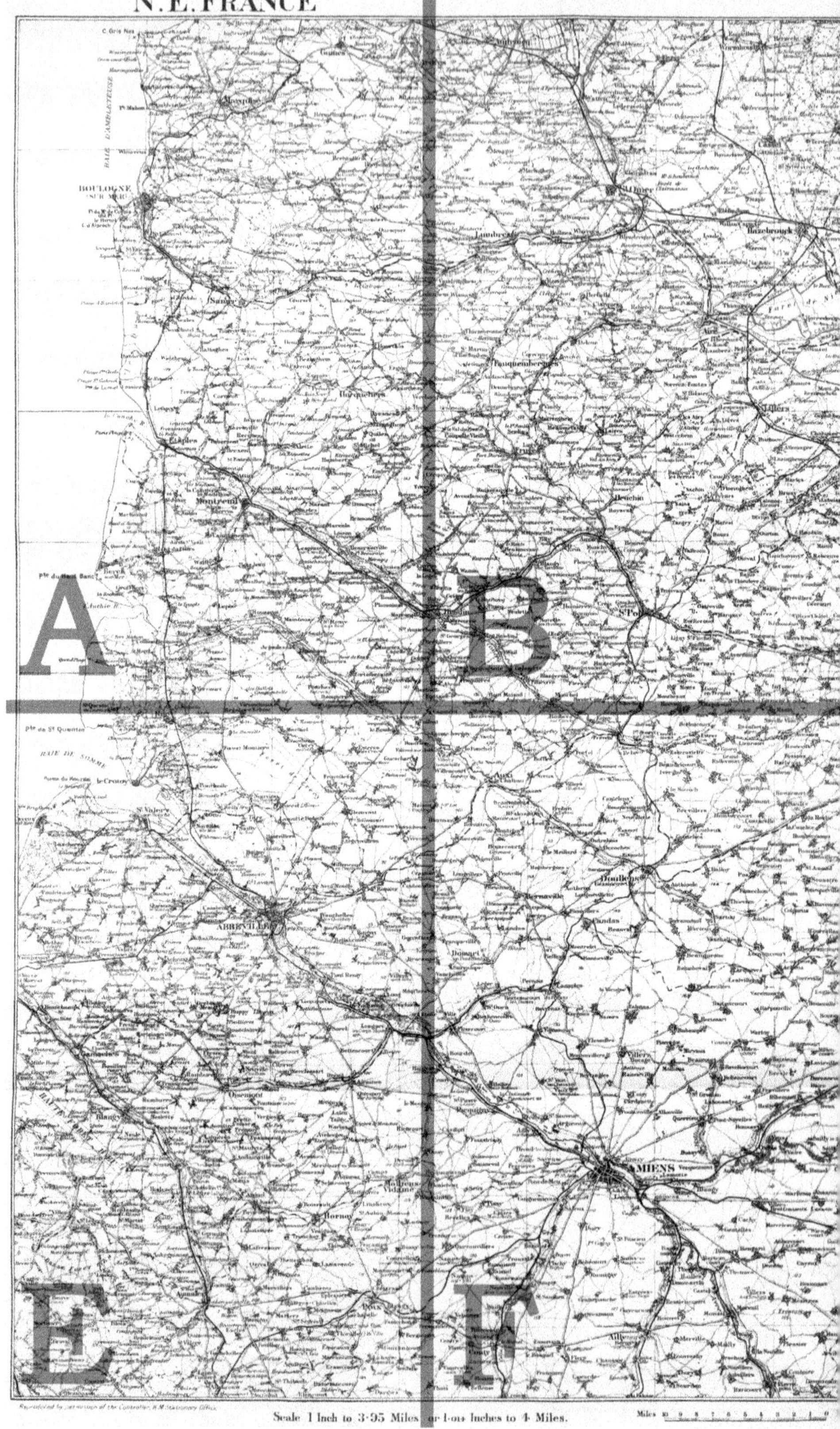

Scale 1 Inch to 3·95 Miles or 1·01 Inches to 4 Miles.

C
D
G
H
COURTRAI
Audenarde
Ypres
Menin
Halluin
Renaix
TOURCOING
ROUBAIX
Armentières
LILLE
TOURNAI
Leuze
St Amand
Condé
DOUAI
VALENCIENNES
ARRAS
le Quesnoy
CAMBRAI
Guise
OISE R.
St QUENTIN
Mont d'Origny
Ribemont
Ham
Kilometres 10 9 8 7 6 5 4 3 2 1 0 Kilometres
1 Centimetre to 2·5 Kilometres

N. E. FRANCE

C. Gris Nez
Guines
Marquise
Pt Mahon
BAIE D'AMBLETEUSE
Wimereux
BOULOGNE
(SUR MER)
Pt du Mt de Couple
le Portel
C.d'Alprech
Desvres
Samer
Hucqueliers
Etaples
Montreuil
Paris Plage
la Canche R.
Pte de Lornel
Camiers
Berck
sur Mer
Pte du Haut Banc
l'Authie R.
St Quentin

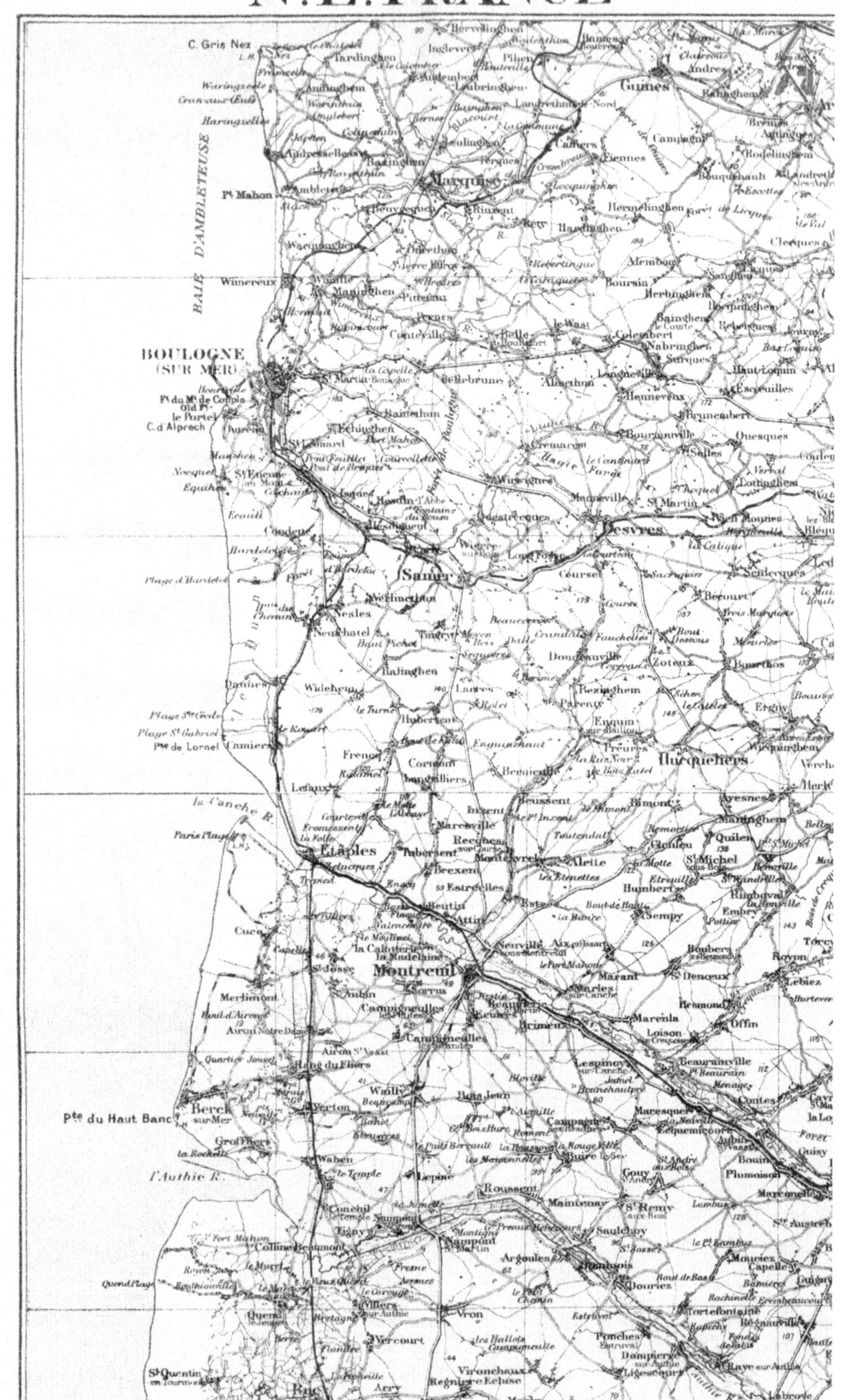

A

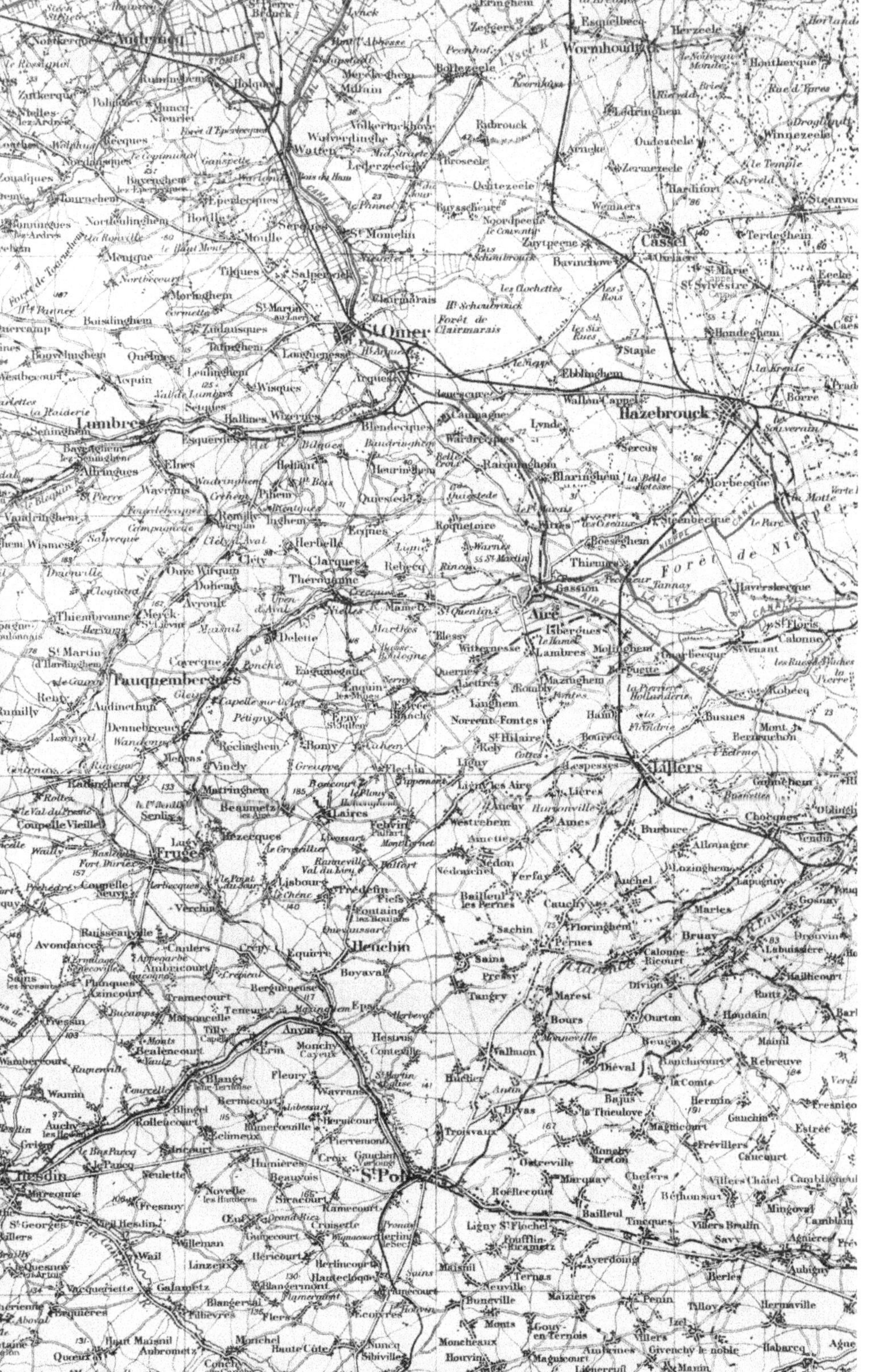

Ardres
Audruicq
St Omer
Forêt de Clairmarais
Clairmarais
Arques
Wormhoudt
Esquelbecq
Herzeele
Houtkerque
Winnezeele
Bollezeele
Zeggers
Merckeghem
Millain
Watten
Wulverdinghe
Lederzeele
Cassel
Oxelaere
Terdeghem
Steenvoorde
Zermezeele
Hardifort
Wemaers
Noordpeene
Zuytpeene
Bavinchove
Ste Marie Cappel
St Sylvestre Cappel
Eecke
Hondeghem
Staple
Ebblinghem
Wallon Cappel
Hazebrouck
Renescure
Campagne
Wardrecques
Lynde
Sercus
Blaringhem
Morbecque
Steenbecque
Forêt de Nieppe
Thiennes
Haverskerque
St Venant
Calonne
Robecq
Busnes
Lillers
Aire
Isbergues
Lambres
Molinghem
Berguette
Mazinghem
Blendecques
Heuringhem
Quiestède
Racquinghem
Roquetoire
Ecques
Herbelle
Clarques
Rebecq
Thérouanne
Delettes
Mametz
St Quentin
Blessy
Witternesse
Quernes
Lumbres
Esquerdes
Elnes
Wavrans
Remilly
Setques
Wisques
Leulinghem
Acquin
Quelmes
Tatinghem
Longuenesse
Zudausques
Moringhem
Tilques
Salperwick
Serques
Moulle
Houlle
Eperlecques
Tournehem
Nordausques
Recques
Zutkerque
Polincove
Ruminghem
Holque
Bonningues
Nortleulinghem
Mentque
Boisdinghem
Quercamp
Bouvelinghem
Westbecourt
Affringues
Vaudringhem
Wismes
Ouve Wirquin
Dohem
Avroult
Merck St Liévin
Thiembronne
Fauquembergues
St Martin d'Hardinghem
Renty
Rumilly
Audincthun
Dennebroeucq
Reclinghem
Vincly
Bomy
Erny St Julien
Enquin les Mines
Enguinegatte
Estrée Blanche
Linghem
Norrent Fontes
Ham
St Hilaire
Ligny
Lespesses
Lières
Auchy
Ames
Amettes
Burbure
Allouagne
Lozinghem
Chocques
Gonnehem
Auchel
Marles
Lapugnoy
Bruay
Labuissière
Calonne Ricouart
Divion
Haillicourt
Ruitz
Houdain
Maisnil
Rebreuve
Ranchicourt
Bours
Marest
Pernes
Floringhem
Sachin
Cauchy
Ferfay
Nédon
Nédonchel
Bailleul les Pernes
Fiefs
Fontaine les Boulans
Heuchin
Boyaval
Equirre
Lisbourg
Prédefin
Febvin Palfart
Laires
Beaumetz les Aire
Matringhem
Senlis
Radinghem
Coupelle Vieille
Fruges
Lugy
Hezecques
Verchin
Coupelle Neuve
Ruisseauville
Avondance
Canlers
Crépy
Ambricourt
Planques
Azincourt
Tramecourt
Bergueneuse
Teneur
Maisoncelle
Fressin
Béalencourt
Tilly Capelle
Erin
Anvin
Monchy Cayeux
Hestrus
Conteville
Fleury
Wavrans
Valhuon
Sains
Pressy
Tangry
Diéval
Ourton
Bajus
La Thieuloye
Hermin
Brias
Troisvaux
Huclier
Wambercourt
Wamin
Blangy
Blingel
Bermicourt
Rollencourt
Éclimeux
Incourt
Auchy les Hesdin
Grigny
Hesdin
Marconne
Le Parcq
Neulette
Huméroeuil
Hernicourt
Pierremont
Croix
Gauchin Verloingt
St Pol
Troisvaux
Ostreville
Monchy Breton
Magnicourt
Frévillers
Caucourt
Chelers
Marquay
Roëllecourt
Bailleul
Ligny St Flochel
Tincques
Villers Brulin
Bethonsart
Villers Châtel
Mingoval
Camblain
Savy
Aubigny
Berles
Averdoing
Ternas
Ramecourt
Siracourt
Beauvois
Fresnoy
Vieil Hesdin
St Georges
Wail
Willeman
Linzeux
Galametz
Hericourt
Herlincourt
Hautecloque
Blangermont
Framecourt
Blangerval
Flers
Croisette
Guinecourt
Ecoivres
Buneville
Monts
Gouy en Ternois
Penin
Tilloy
Hermaville
Izel
Villers
Givenchy le noble
Habarcq
Manin
Noyelle Vion
Lignereuil
Averdoingt
Maizières
Moncheaux
Houvin
Sars
Magnicourt
Nuncq
Sibiville
Séricourt
Ligny
Conchy sur Canche
Boubers sur Canche
Monchel
Haute Côte
Aubrometz
Haut Maisnil
Quœux
Le Quesnoy en Artois
Vacqueriette
Fillièvres
Canal de Neuffossé
Canal d'Aire
L'Aa R.
La Lys

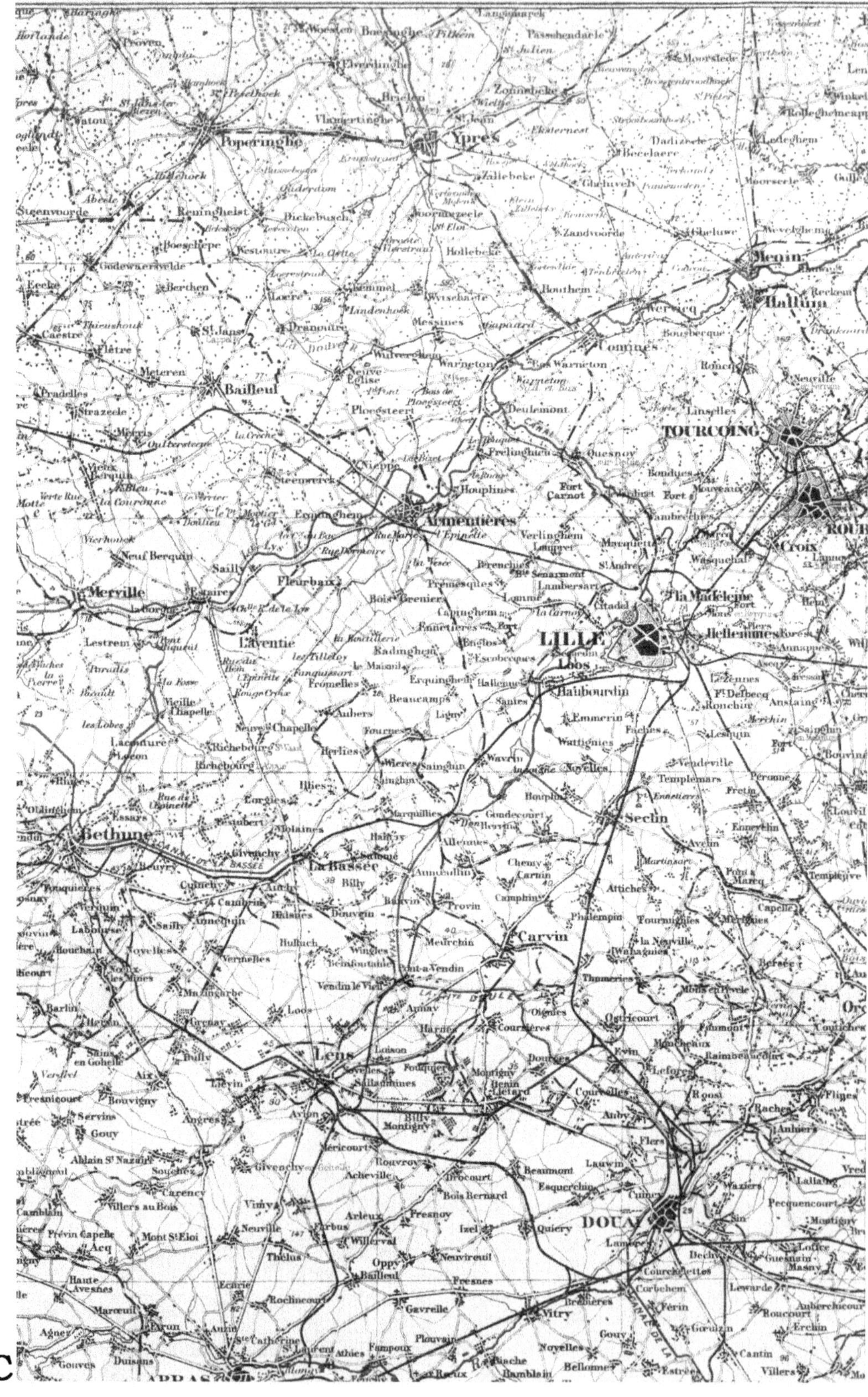

Langemarck
Passchendaele
Boesinghe
Elverdinghe
Proven
Poperinghe
Vlamertinghe
Ypres
St. Julien
Zonnebeke
Moorslede
Dadizeele
Becelaere
Ledeghem
Zillebeke
Gheluvelt
Reninghelst
Dickebusch
Voormezeele
Hollebeke
Zandvoorde
Gheluwe
Wevelghem
Steenvoorde
Boeschepe
Westoutre
Godewaersvelde
Berthen
Kemmel
Wytschaete
Menin
Halluin
Eecke
Dranoutre
Lindenhoek
Messines
Houthem
Wervicq
Caestre
Fletre
St. Jans Cappel
Wulverghem
Warneton
Bas Warneton
Comines
Bousbecque
Roncq
Meteren
Bailleul
Neuve Eglise
Ploegsteert
Deulemont
Linselles
Neuville
Strazeele
Merris
Outtersteene
La Crèche
Nieppe
Le Bizet
Frelinghien
Quesnoy
TOURCOING
Vieux Berquin
Steenwerck
Houplines
Fort Carnot
Bondues
Mouveaux
Armentières
Erquinghem
Verlinghem
Lompret
Marquette
Wambrechies
Wasquehal
Croix
Neuf Berquin
Sailly
Fleurbaix
St. André
Merville
Estaires
Bois Grenier
Prémesques
Lambersart
Lomme
La Madeleine
Lestrem
Laventie
Capinghem
Ennetières
LILLE
Hellemmes
Radinghem
Escobecques
Loos
Fromelles
Erquinghem
Hallennes
Haubourdin
Beaucamps
Ronchin
Annappes
Vieille Chapelle
Aubers
Ligny
Santes
Emmerin
Lesquin
Fournes
Lacouture
Richebourg
Herlies
Wavrin
Wattignies
Noyelles
Vendeville
Sainghin
Templemars
Fretin
Illies
Béthune
Essars
Festubert
Lorgies
Marquillies
Gondecourt
Seclin
Givenchy
Violaines
La Bassée
Hantay
Salomé
Allennes
Annœullin
Chemy
Carnin
Avelin
Beuvry
Billy
Cuinchy
Auchy
Cambrin
Annequin
Sailly
Bauvin
Provin
Camphin
Attiches
Phalempin
Labourse
Hulluch
Douvrin
Meurchin
Carvin
Tourmignies
Wingles
Vermelles
Noyelles
Bénifontaine
Pont-à-Vendin
Wahagnies
Mazingarbe
Vendin le Vieil
Thumeries
Loos
Annay
Ostricourt
Barlin
Grenay
Harnes
Courrières
Oignies
Bully
Lens
Loison
Fouquières
Montigny
Dourges
Evin
Moncheaux
Raimbeaucourt
Aix
Liévin
Sallaumines
Hénin Liétard
Courcelles
Leforest
Roost
Flines
Fresnicourt
Bouvigny
Servins
Gouy
Angres
Avion
Billy Montigny
Auby
Raches
Anhiers
Ablain St. Nazaire
Souchez
Givenchy
Méricourt
Rouvroy
Acheville
Drocourt
Beaumont
Flers
Lauwin
Esquerchin
Carency
Villers au Bois
Vimy
Bois Bernard
Arleux
Fresnoy
Cuincy
Waziers
Camblain
Neuville
Farbus
Willerval
Izel
Quiéry
DOUAI
Sin
Pecquencourt
Montigny
Mont St. Eloi
Acq
Thélus
Oppy
Neuvireuil
Lambres
Dechy
Guesnain
Lallaing
Courchelettes
Bailleul
Fresnes
Haute Avesnes
Ecurie
Roclincourt
Gavrelle
Brebières
Corbehem
Lewarde
Marœuil
Vitry
Férin
Roucourt
Erchin
Anzin
St. Catherine
Plouvain
Gouy
Goeulzin
Agnez
Duisans
Gouves
St. Laurent
Athies
Fampoux
Noyelles
Bellonne
Cantin
Villers
Estrées
Biache
Hamblain
ARRAS

C

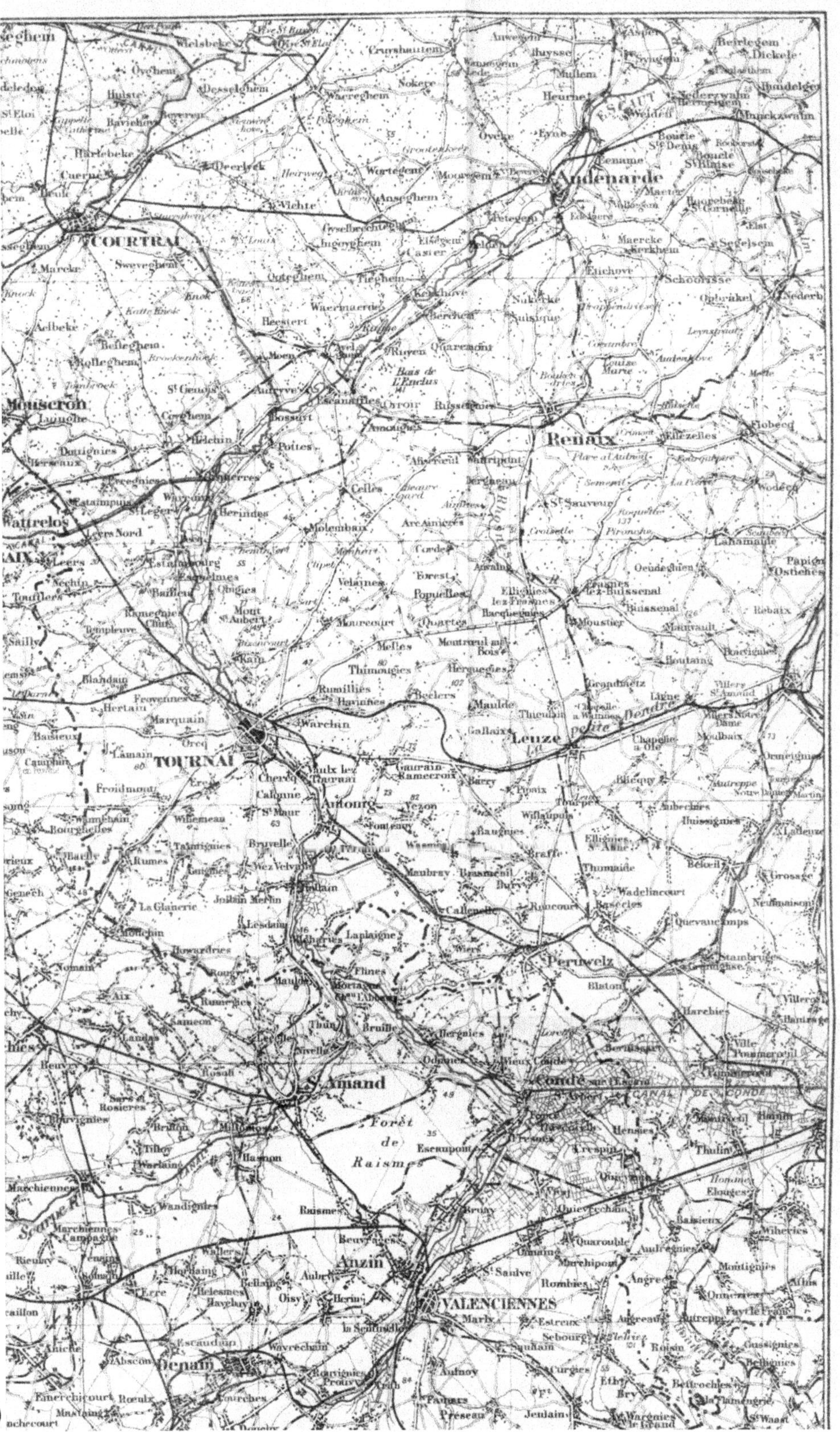

Wielsbeke
Desselghem
Waereghem
Nokere
Cruyshautem
Huysse
Mullem
Heurne
Beirlegem
Dickele
Harlebeke
Cuerne
Deerlyck
Vichte
Wortegem
Anseghem
Audenarde
Ename
Mater
COURTRAI
Sweveghem
Outrighem
Ingoyghem
Tieghem
Kerkhove
Berchem
Quaremont
Waermaerde
Heestert
Bellegem
Rolleghem
Aelbeke
Moen
Bois de l'Enclus
Escanaffles
Orroir
Russeignies
Mouscron
Luingne
Dottignies
Bossuyt
Helchin
Pottes
Amougies
Renaix
Ellezelles
Flobecq
Schoorisse
Nederbr
Nukerke
Elst
Segelsem
Celles
Herseaux
Estaimpuis
St Léger
Wattrelos
Molembaix
Velaines
Popuelles
Forest
Arc Ainières
Anvaing
St Sauveur
Frasnes lez-Buissenal
Buissenal
Moustier
Oeudeghien
Lahamaide
Rebaix
Blicquy
Leers
Nechin
Toufflers
Estaimbourg
Bailleul
Templeuve
Blandain
Kain
Mourcourt
Melles
Thimougies
Hérinnes
Ramegnies
Hertain
Froyennes
Marquain
Orcq
Warchin
TOURNAI
Lamain
Camphin
Froidmont
Chercq
Calonne
Antoing
Vaulx lez Tournai
Gaurain Ramecroix
Barry
Leuze
Ligne
Gallaix
Maulde
Thieulain
Tourpes
Willaupuis
Ellignies St Anne
Thumaide
Bruyelle
Willemeau
Bourghelles
Rumes
Taintignies
Wez Velvain
Jollain Merlin
La Glanerie
Hollain
Baugnies
Braffe
Maubray
Wasmes
Bury
Callenelle
Wadelincourt
Basècles
Beloeil
Grosage
Quevaucamps
Stambruges
Grandglise
Lesdain
Laplaigne
Wiers
Peruwelz
Blaton
Harchies
Pommeroeul
Flines
Mortagne
Maulde
Rumegies
Howardries
Mouchin
Nomain
Aix
Sameon
Landas
Lecelles
Thun
Bruille
Hergnies
Odomez
Vieux Condé
Condé
Bernissart
Rosult
S. Amand
Forêt de Raismes
Escaupont
Fresnes
Hensies
Crespin
Thulin
Quiévrain
Quievrechain
Vicq
Bruay
Millonfosse
Hasnon
Wandignies
Marchiennes
Raismes
Beuvrages
Anzin
Onnaing
Quarouble
Marchipont
Rombies
Angre
Élouges
Wiheries
Baisieux
Audregnies
St Saulve
VALENCIENNES
Marly
Estreux
Sebourg
Aulnoy
Wallers
Hérin
Oisy
Haveluy
Helesmes
Denain
Escaudain
Abscon
Wavrechain
Rouvignies
Prouvy
Trith
Famars
Preseau
Jenlain
Curgies
Saultain
Rombies
Onnezies
Fayt-le-Franc
Gussignies
Bellignies
Eth
Bry
Roeulx
Erre
Fenain
Emerchicourt
Mastaing
CANAL DE CONDÉ
ESCAUT
Dendre
Scarpe R.

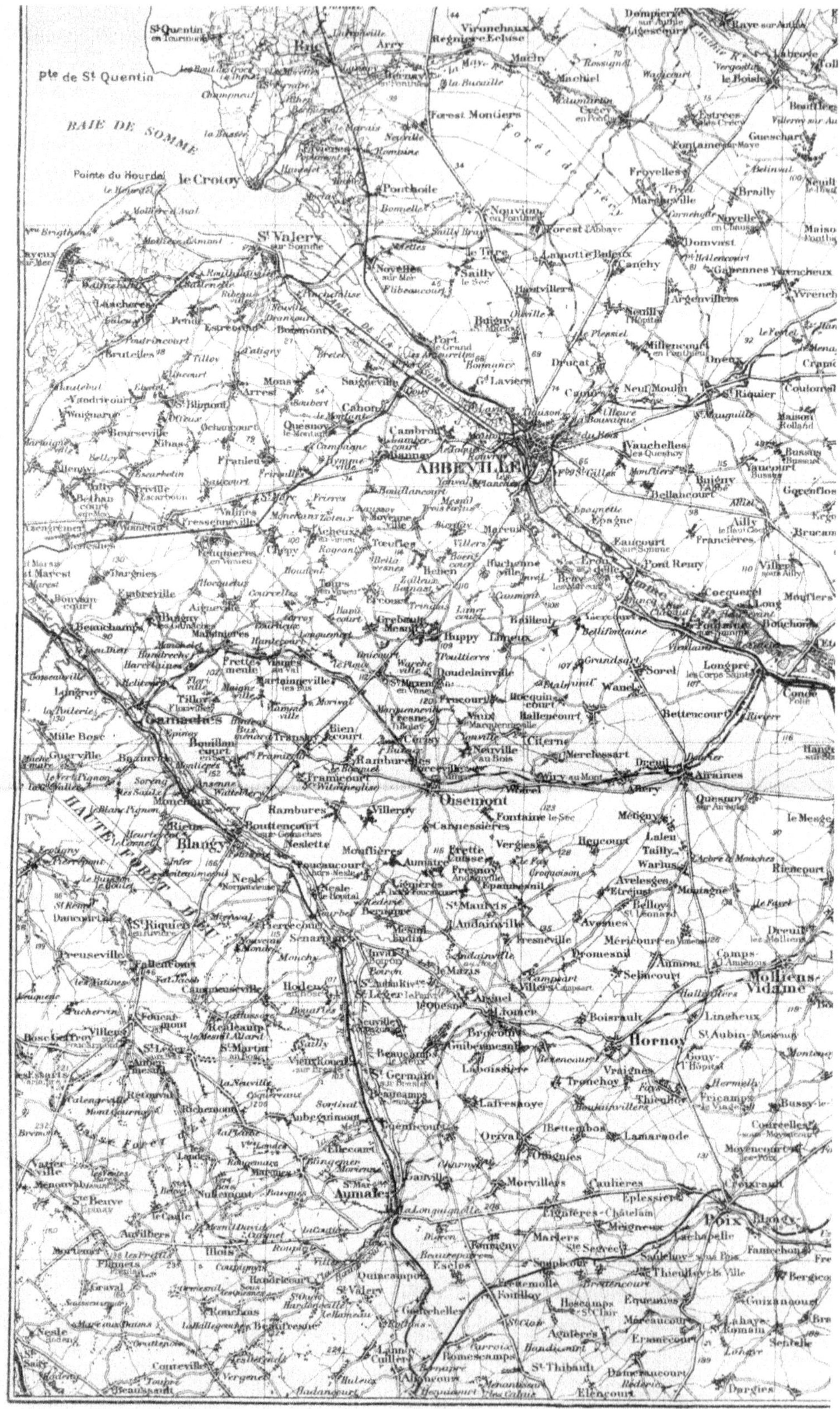

E

Scale 1 Inch to 3·95 Miles

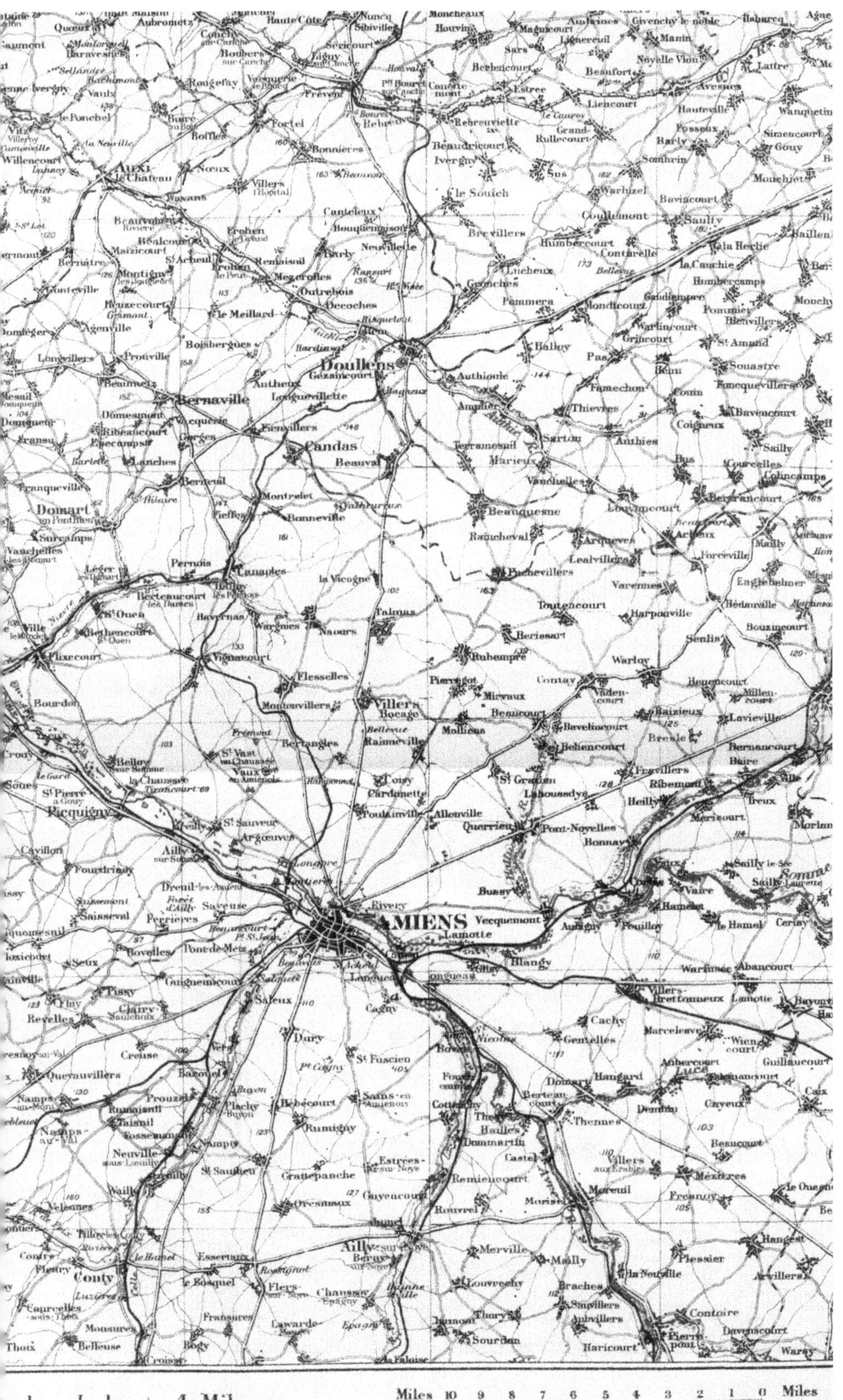

…r 1·014 Inches to 4 Miles.

ARRAS
Scarpe
Bapaume
Albert
Peronne
Somme R.
Ham
Nesle
Roisel
Combles
Bray
Le Catelet
Miles
Kilometres 10 9 8 7 6 5 4 3 2 1 0 Kilometres
1 Centimetre to

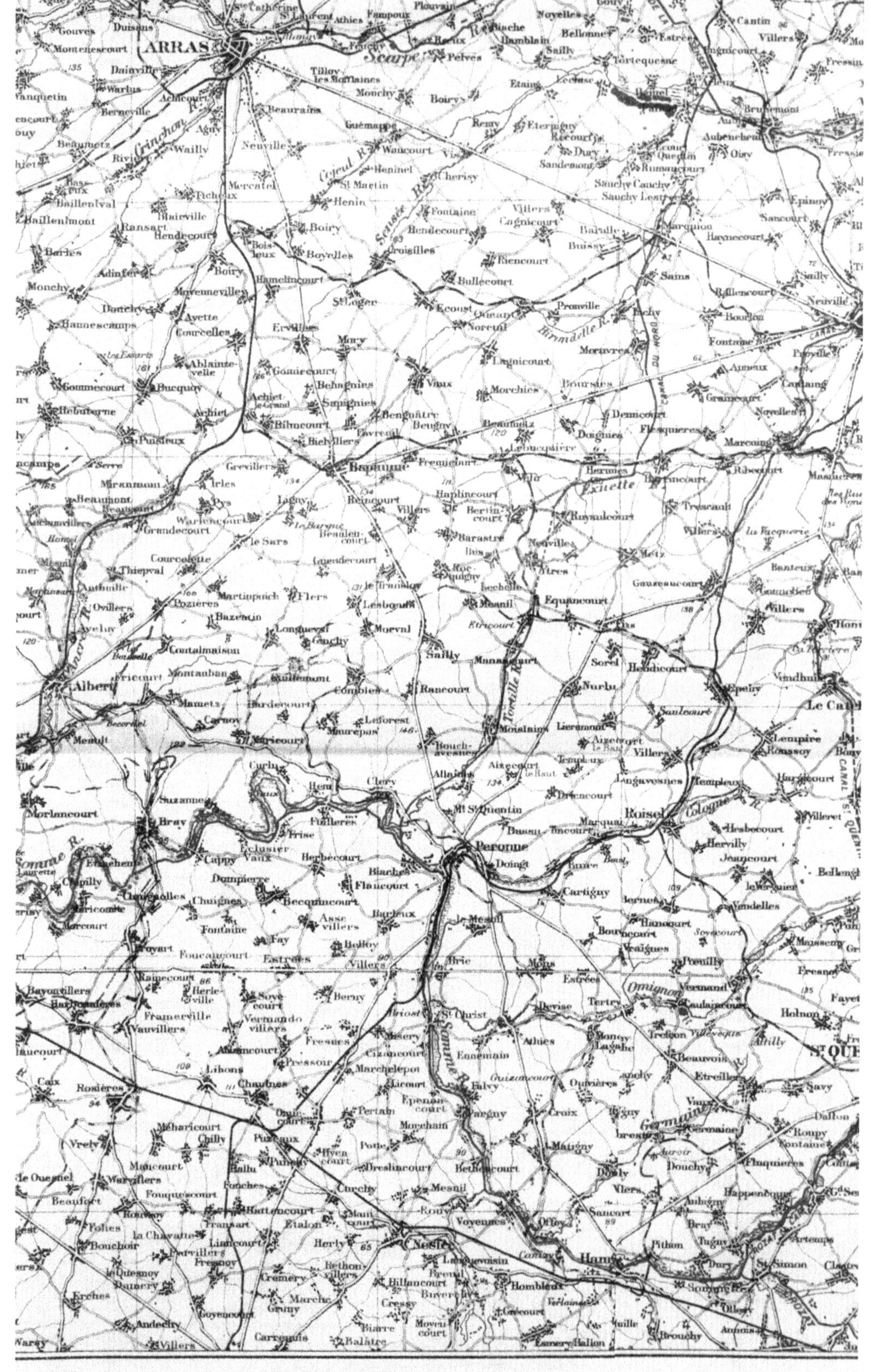

G

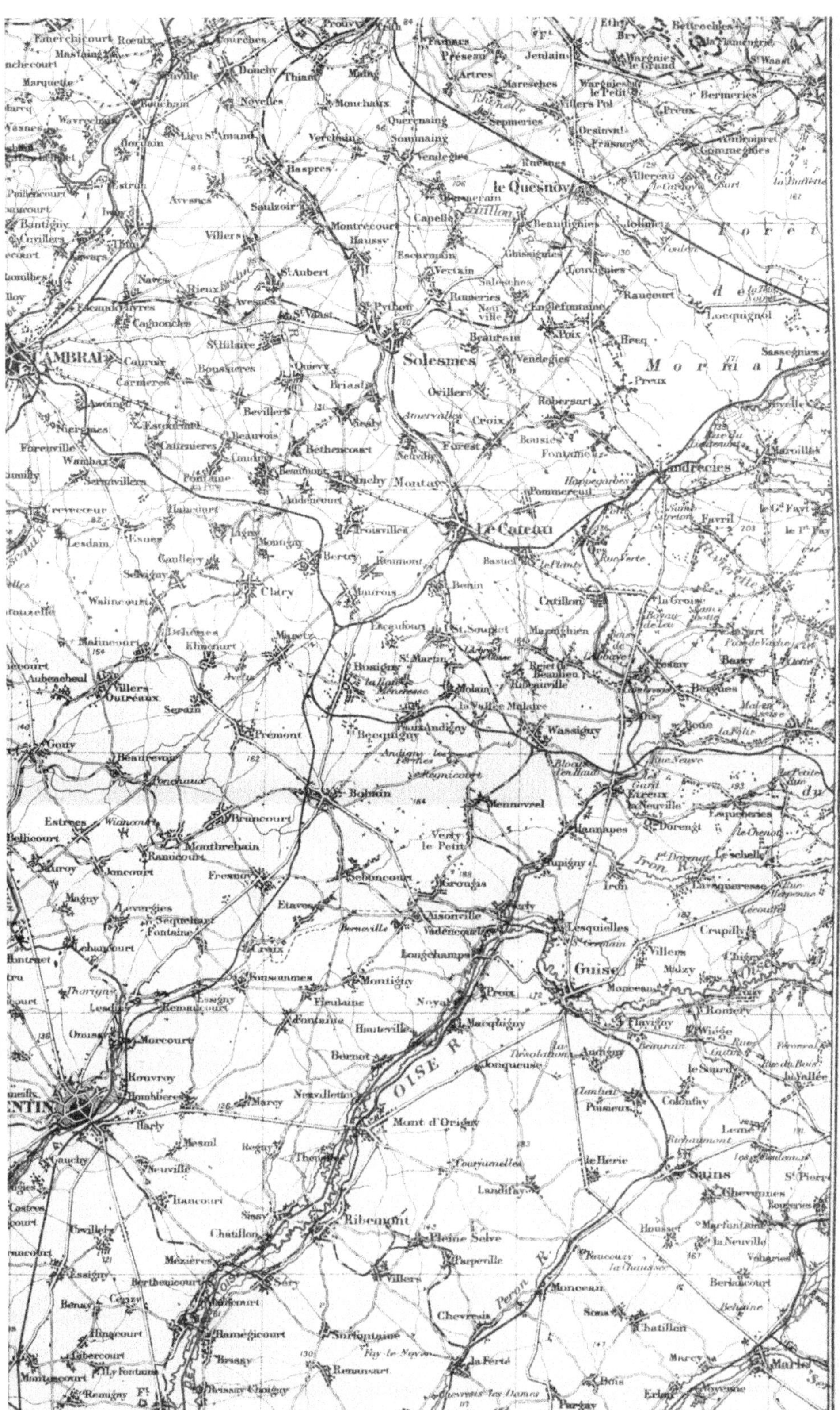

2·5 Kilometres

www.ingramcontent.com/pod-product-compliance
Ingram Content Group UK Ltd.
Pitfield, Milton Keynes, MK11 3LW, UK
UKHW021839270726
14058UKWH00002B/247